EARTH

This is an outstanding overview of the history of Earth from a unique planetary perspective for introductory courses in the earth and space sciences. *Earth: Evolution of a Habitable World* tells how Earth has come to its present state, why it differs from its neighboring planets, what life's place is in Earth's history, and how humanity affects the processes that make our planet livable. Today's human influences are contemplated in the context of natural changes on Earth.

The text considers the burning issues in our quest to understand Earth. It covers Earth's origin as part of the cosmic phenomenon of star formation, the early history of our planet including the formation of the Moon, the development of continents and oceans, the origin of life and its fundamental alteration of Earth's atmosphere and climate, the rise of humankind, and the challenges we face in the future as a technological species. The book shows how the myriad disciplines of science are interwoven to understand our world and its sister planets. It points the way to how science and engineering must be applied to today's challenges if humankind is to have a promising future on Earth.

This book brings a fresh perspective to the study of Earth for students who wish to learn how our planet evolved to its present form.

Jonathan I. Lunine is Professor of Planetary Sciences at the University of Arizona. He undertakes research on a range of topics in the solar system and planetary systems beyond our own. In addition, he participates extensively in advisory activities for NASA. In 1994, *Time* magazine named him to its "50 for the Future" list of emerging American leaders.

EARTH

EVOLUTION OF A HABITABLE WORLD

JONATHAN I. LUNINE

Original illustrations by Cynthia J. Lunine

PUBLISHED BY THE PRESS SYNDICATE OF THE UNIVERSITY OF CAMBRIDGE
The Pitt Building, Trumpington Street, Cambridge, United Kingdom

CAMBRIDGE UNIVERSITY PRESS
The Edinburgh Building, Cambridge CB2 2RU, UK
40 West 20th Street, New York, NY 10011-4211, USA
10 Stamford Road, Oakleigh, VIC 3166, Australia
Ruiz de Alarcón 13, 28014 Madrid, Spain
Dock House, The Waterfront, Cape Town 8001, South Africa

http://www.cambridge.org

© Cambridge University Press 1999

This book is in copyright. Subject to statutory exception and
to the provisions of relevant collective licensing agreements,
no reproduction of any part may take place without
the written permission of Cambridge University Press.

First published 1999
Reprinted 2000

Printed in the United States of America

Typeset in Sabon 9.75/13 pt. and Eurostile in LaTeX 2_ε [TB]

A catalog record for this book is available from the British Library

Library of Congress Cataloging in Publication Data
Lunine, Jonathan Irving.
  Earth : evolution of a habitable world / Jonathan I. Lunine ;
original illustrations by Cynthia J. Lunine.
    p.   cm.
  Includes index.
  ISBN 0-521-47287-3 (hardbound). – ISBN 0-521-64423-2 (pbk.)
  1. Earth. 2. Earth sciences. 3. Environmental sciences.
I. Lunine, Cynthia J. II. Title.
QB631.L836 1999
525 – dc21   98-28902
                                                                   CIP

ISBN   0 521 47287 3   hardback
ISBN   0 521 64423 2   paperback

To Professor Carl Sagan, who saw beyond the problems of today to articulate an extraordinary future for humankind, and to Dr. Eugene Shoemaker, who instructed us in the hazards and opportunities of growing up in a crowded solar system.

CONTENTS

Preface		*page* xvii
Acknowledgments		xix

PART ONE. THE ASTRONOMICAL PLANET: EARTH'S PLACE IN THE COSMOS — 1

1 An Introductory Tour of Earth's Cosmic Neighborhood — 3
   1.1 Ancient Attempts to Determine the Scale of the Cosmos — 3
   1.2 Brief Introduction to the Solar System — 4
   1.3 Questions — 7
   1.4 Readings — 7
       1.4.1 General Reading — 7
       1.4.2 References — 7

2 Largest and Smallest Scales — 8
   2.1 Introduction — 8
   2.2 Scientific Notation — 8
   2.3 Motions of Earth in the Cosmos — 8
   2.4 Cosmic Distances — 13
       2.4.1 The Planets — 13
       2.4.2 Nearby Stars — 14
       2.4.3 Nearby Galaxies — 14
       2.4.4 Beyond the Galactic Neighborhood — 15
       2.4.5 To the Farthest Edge of the Universe — 15
   2.5 Microscopic Constitution of Matter — 17
   2.6 Questions — 21
   2.7 Readings — 22
       2.7.1 General Reading — 22
       2.7.2 References — 22

3 Forces and Energy — 23
   3.1 Introduction — 23
   3.2 Forces of Nature — 23
   3.3 Radioactivity — 28

		3.4	Conservation of Energy, and Thermodynamics	29
		3.5	Electromagnetic Spectrum	30
		3.6	Abundances in the Sun	33
		3.7	Questions	34
		3.8	Readings	34

4 Fusion, Fission, Sunlight, and Element Formation — 35
- 4.1 Introduction — 35
- 4.2 Stars and Nuclear Fusion — 35
- 4.3 Element Production in the Big Bang — 38
- 4.4 Element Production During Nuclear Fusion in Stars — 39
- 4.5 Production of Other Elements in Stars: s, r, and p Processes — 40
- 4.6 Nonstellar Element Production — 42
- 4.7 Element Production and Life — 42
- 4.8 Questions — 43
- 4.9 Readings — 43
  - 4.9.1 General Reading — 43
  - 4.9.2 References — 44

PART TWO. THE MEASURABLE PLANET: TOOLS TO DISCERN THE HISTORY OF EARTH AND THE PLANETS — 45

5 Determination of Cosmic and Terrestrial Ages — 47
- 5.1 Overview of Age Dating — 47
- 5.2 The Concept of Half-Life — 47
- 5.3 Carbon-14 Dating — 49
- 5.4 Measurement of Parents and Daughters: Rubidium-Strontium — 51
- 5.5 Caveat Emptor — 52
- 5.6 Questions — 53
- 5.7 Readings — 53
  - 5.7.1 General Reading — 53
  - 5.7.2 References — 53

6 Other Uses of Isotopes for Earth History — 54
- 6.1 Introduction — 54
- 6.2 Stable Isotopes, Seafloor Sediments, and Climate — 54
  - 6.2.1 Carbon — 54
  - 6.2.2 Oxygen — 54
  - 6.2.3 Hydrogen — 55
- 6.3 A Possible Temperature History of Earth from Cherts — 56
- 6.4 Questions — 58
- 6.5 Readings — 58

7 Relative Age Dating of Cosmic and Terrestrial Events: The Cratering Record — 60
- 7.1 Introduction — 60
- 7.2 Process of Impact Cratering — 60
- 7.3 Using Craters to Date Planetary Surfaces — 63
  - 7.3.1 Relative Ages of Events on a Planetary Surface — 66
  - 7.3.2 Absolute Chronology of Solar System Events — 69

7.4	Cratering on Planetary Bodies with Atmospheres		73
7.5	Impactors Through Time		73
7.6	Questions		73
7.7	Readings		74

8 Relative Age Dating of Terrestrial Events: Geologic Layering and Geologic Time — 75

8.1	Introduction		75
8.2	Catastrophism versus Uniformitarianism		75
8.3	Estimating the Age of Earth, without Radioisotopes		75
8.4	Geologic Processes and Their Cyclical Nature		76
8.5	Principles of Geologic Succession		78
8.6	Fossils		79
8.7	Radioisotopic Dating of Earth Rocks		81
8.8	Geologic Timescale		82
8.9	A Grand Sequence		82
8.10	The Geologic Timescale as a Map		82
8.11	Questions		84
8.12	Readings		84
	8.12.1	General Reading	84
	8.12.2	References	84

9 Plate Tectonics: An Introduction to the Process — 85

9.1	Introduction		85
9.2	Early Evidence for and Historical Development of Plate Tectonics		85
9.3	Genesis of Plate Tectonics After World War II		86
	9.3.1	Seafloor Topography	86
	9.3.2	Magnetic Imprints on Rocks	86
	9.3.3	Geologic Record on Land	91
	9.3.4	Earthquakes and Subduction	91
9.4	The Basic Model of Plate Tectonics		93
9.5	Past Motions of the Plates and Supercontinents		95
9.6	Driving Forces of Plate Motions		96
9.7	An End to Techniques and the Start of History		98
9.8	Questions		98
9.9	Readings		98
	9.9.1	General Reading	98
	9.9.2	References	98

PART THREE. THE HISTORICAL PLANET: EARTH AND SOLAR SYSTEM THROUGH TIME — 99

10 Formation of the Solar System — 101

10.1	Introduction		101
10.2	Timescale of Cosmological Events Leading Up to Solar System Formation		101
10.3	Formation of Stars and Planets		102
	10.3.1	Molecular Clouds and Star Formation	102

		10.3.2	The Start of Star Formation	102
		10.3.3	A Star Is Born	103
		10.3.4	Figure Skaters and Astrophysicists: The Formation of Planets	104
		10.3.5	Disks Around Protostars – the Source of Planets?	105
		10.3.6	The End of Planet Formation	107
	10.4	Primitive Material Present in the Solar System Today		107
		10.4.1	Remnants of the Beginning: Meteorites	108
		10.4.2	Comets and Kuiper Belt Objects	109
		10.4.3	Interplanetary Dust Particles	111
	10.5	The Search for Other Planetary Systems		111
		10.5.1	Indirect Techniques	112
		10.5.2	Direct Techniques	113
	10.6	Summary of Planet Formation		114
	10.7	Questions		114
	10.8	Readings		114
		10.8.1	General Reading	114
		10.8.2	References	114
11	The Hadean Earth			115
	11.1	Introduction		115
	11.2	Bulk Composition of the Planets		115
		11.2.1	Solid Planets	116
		11.2.2	The Giant Planets	118
	11.3	Internal Structure of Earth		120
	11.4	Accretion: The Building Up of Planets		123
	11.5	Early Differentiation After Accretion		124
	11.6	Radioactive Heating		125
	11.7	Formation of an Iron Core		127
	11.8	Formation of the Moon		127
	11.9	Origin of Earth's Atmosphere, Ocean, and Organic Reservoir		130
	11.10	From the Hadean into the Archean: Formation of the First Stable Continental Rocks		132
	11.11	Questions		132
	11.12	Readings		132
		11.12.1	General Reading	132
		11.12.2	References	132
12	The Archean Eon and the Origin of Life: I. Properties of and Sites for Life			134
	12.1	Introduction		134
	12.2	Definition of Life and Essential Workings		134
		12.2.1	What Is Life?	134
		12.2.2	Basic Structure of Life	136
		12.2.3	Information Exchange and Replication	136
		12.2.4	Formation of Proteins	137
		12.2.5	Mutation and Genetic Variation	138
	12.3	The Basic Unit of Living Organisms: The Cell		138

	12.4	Energetic Processes That Sustain Life	139
		12.4.1 Common Metabolic Mechanisms	139
		12.4.2 Photosynthesis	139
	12.5	Other Means of Utilizing Energy	141
	12.6	Elemental Necessities of Life: A Brief Examination	142
		12.6.1 Why Carbon?	142
		12.6.2 Why Water?	143
		12.6.3 Is Free Oxygen Essential?	143
	12.7	Solar System Sites for Life	144
		12.7.1 Atmospheres of the Giant Planets	144
		12.7.2 Interior of Europa	144
		12.7.3 Titan	145
		12.7.4 The Mars of Today and Yesterday	145
		12.7.5 Earth	151
	12.8	Questions	152
	12.9	Readings	152
		12.9.1 General Reading	152
		12.9.2 References	152
13		**The Archean Eon and the Origin of Life: II. Mechanisms**	**153**
	13.1	Introduction	153
	13.2	Thermodynamics and Life	153
	13.3	The Raw Materials of Life: Synthesis and the Importance of Handedness	155
	13.4	Two Approaches to Life's Origin	156
	13.5	The Vesicle Approach and Autocatalysis	156
	13.6	The RNA World: A Second Option	158
		13.6.1 The Promise: RNA as Replicator and Catalyst	158
		13.6.2 The Problem: Invention of RNA	159
	13.7	The Essentials of a Cell and the Unification of the Two Approaches	161
	13.8	The Archean Situation	163
	13.9	Questions	164
	13.10	Readings	164
14		**The First Greenhouse Crisis: The Faint Early Sun**	**165**
	14.1	The Case for an Equable Climate in the Archean	165
	14.2	The Faint Early Sun	165
	14.3	The Greenhouse Effect	166
	14.4	Primary Greenhouse Gases	169
	14.5	Implications for Earth During the Faint-Early-Sun Era	169
	14.6	Paleosols and the Carbon Dioxide Abundance	171
	14.7	Carbon Dioxide Cycling and Early Crustal Tectonics	172
		14.7.1 Basic Carbon-Silicate Weathering Cycle	172
		14.7.2 Negative Feedbacks in the Carbon-Silicate Cycle	173
		14.7.3 The Carbon-Silicate Cycle During the Archean	174
	14.8	A Balance Unique to Earth, and a Lingering Conundrum	174
	14.9	Questions	176

14.10	Readings		176
	14.10.1	General Reading	176
	14.10.2	References	176

15 Climate Histories of Mars and Venus, and the Habitability of Planets — 177

15.1	Introduction		177
15.2	Venus		177
	15.2.1	Origin of Venus' Thick Atmosphere	177
	15.2.2	Overview of the Surface of Venus	181
15.3	Mars		183
	15.3.1	Mars Today	183
	15.3.2	Geological Hints of a Warmer Early Mars	183
15.4	Was Mars Really Warm in the Past?		188
	15.4.1	Limits to a Carbon Dioxide Greenhouse	188
	15.4.2	Abodes for Life on Early Mars	189
	15.4.3	Searching for Evidence of Life, and the Early Climate	190
15.5	Putting a Martian History Together		190
15.6	Implications of Venusian and Martian History for Life Elsewhere		191
15.7	The Finite Life of Our Biosphere		193
15.8	Questions		193
15.9	Readings		195

16 Earth in Transition: From the Archean to the Proterozoic — 196

16.1	Introduction		196
16.2	Abundances of the Elements in Terrestrial Rocks		196
16.3	Mineral Structure		197
16.4	Partial Melting and the Formation of Basalts		198
16.5	Formation of Andesites and Granites		200
	16.5.1	Rock Relationships	200
	16.5.2	Seismic Waves and Composition	200
	16.5.3	Role of Water in Partial Melting	200
	16.5.4	The Puzzle of Granite Formation	201
16.6	Formation of Protocontinents in the Archean		203
16.7	The Archean-Proterozoic Transition		204
16.8	After the Proterozoic: Modern Plate Tectonics		206
16.9	Venus: An Earth-Sized Planet without Plate Tectonics		206
16.10	Water and Plate Tectonics		208
16.11	Continents, the Moon, and the Length of Earth's Day		209
16.12	Entree to the Modern World		210
16.13	Questions		210
16.14	Readings		210
	16.14.1	General Reading	210
	16.14.2	References	210

17	The Oxygen Revolution		211
	17.1	Introduction	211
	17.2	Modern Oxygen Cycle	211
	17.3	Oxygen Balance with and without Life	213
	17.4	Limits on the Oxygen Levels on Early Earth	213
		17.4.1 Minerals Unstable in the Presence of Oxygen	213
		17.4.2 Banded Iron Formation	214
		17.4.3 Redbeds	215
		17.4.4 Fossils of Aerobic Organisms	215
	17.5	History of the Rise of Oxygen	215
	17.6	Balance Between Oxygen Loss and Gain	215
	17.7	Reservoirs of Oxygen and Reduced Gases	217
	17.8	History of Oxygen on Earth	218
		17.8.1 Stage 1	218
		17.8.2 Stage 2	218
		17.8.3 Stage 3	219
		17.8.4 Stage 4	219
	17.9	Shield Against Ultraviolet Radiation	219
	17.10	Onset of Eukaryotic Life	220
	17.11	Questions	222
	17.12	Readings	222
18	The Phanerozoic: Flowering and Extinction of Complex Life		223
	18.1	Introduction to the Phanerozoic	223
	18.2	Evolution	225
		18.2.1 Traditional, Darwinian, Model of Evolution	225
		18.2.2 Punctuated Equilibrium Approach to Evolution	226
	18.3	Vendian-Cambrian Revolution	229
		18.3.1 Taxonomy for the Restless	229
		18.3.2 Establishment of the Basic Plans	229
		18.3.3 Clues from the Vendian	230
		18.3.4 Causes of the Vendian-Cambrian Revolution	230
		18.3.5 Why Has It Not Happened Again?	232
	18.4	Mass Extinction Events in the Phanerozoic	232
	18.5	Cretaceous-Tertiary Extinction	232
		18.5.1 Boundary Sediments	233
		18.5.2 Interpretation of the K/T Boundary as an Impact Event	234
		18.5.3 Biological Effects of the Impact	235
		18.5.4 Where Is the Crater?	237
		18.5.5 Impacts and Other Extinction Events	238
	18.6	Questions	239
	18.7	Readings	239
		18.7.1 General Reading	239
		18.7.2 References	239

19	Climate Change Across the Phanerozoic		240
	19.1	Introduction	240
	19.2	The Supercontinent Cycle	240
	19.3	Effects of Continental Breakups and Collisions	242
	19.4	Evidence of Ice Ages on Earth	243
	19.5	Causes of the Ice Ages	243
		19.5.1 Positive Feedbacks in the Basic Climate System	243
		19.5.2 Negative Feedbacks in the Climate System	244
		19.5.3 Additional Influences on Global Glaciation	244
		19.5.4 Snowball Earth?	245
	19.6	Cretaceous Climate	245
		19.6.1 Evidence for the Cretaceous Climate Pattern	245
		19.6.2 Plate Tectonic Effects on Cretaceous Climate Change	246
		19.6.3 Additional Important Effects on Cretaceous Climate	246
		19.6.4 Causes for Climate Change That Probably Are Not Important in the Cretaceous	246
		19.6.5 Model for the Warm Cretaceous	247
	19.7	The Great Tertiary Cooldown	248
	19.8	Causes of the Pleistocene Ice Age and Its Oscillations	249
	19.9	Saved From Instability: Earth's versus Mars' Orbital Cycle	252
	19.10	Effects of the Pleistocene Ice Age: A Preview	253
	19.11	Questions	253
	19.12	Readings	253
20	Toward the Age of Humankind		255
	20.1	Introduction	255
	20.2	Pleistocene Setting	255
	20.3	The Vagaries of Understanding Human Origins	255
	20.4	Humanity's Taxonomy	256
	20.5	The First Steps: Australopithecines	256
	20.6	The Genus *Homo*: Out of Africa I	257
	20.7	Out of Africa II	258
	20.8	Final Act: Neanderthals and an Encounter with Our Humanity	260
		20.8.1 Climate Setting	260
		20.8.2 Physical Features of Neanderthals	262
		20.8.3 Neanderthal Lifestyle	263
		20.8.4 Interaction of Neanderthals with Moderns	264
		20.8.5 Who Were the Neanderthals?	265
	20.9	This Modern World	266
	20.10	Questions	266
	20.11	Readings	266
		20.11.1 General Reading	266
		20.11.2 References	266

PART FOUR. THE ONCE AND FUTURE PLANET — 267

21	Climate Change Over the Past 100,000 Years		269
	21.1	Introduction	269

	21.2	The Record in Ice Cores	269
	21.3	Climate from Plant Pollen and Packrat Midden Studies	272
	21.4	Tree Rings	273
	21.5	Climate Variability in the Late Holocene	277
	21.6	The Younger Dryas: A Signpost for the Oceanic Role in Climate	278
	21.7	Into the Present	279
	21.8	Questions	280
	21.9	Readings	280
		21.9.1 General Reading	280
		21.9.2 References	280
22	Human-Induced Global Warming		281
	22.1	The Records of CO_2 Abundance and Global Temperatures in Modern Times	281
	22.2	Modeling the Response of Earth to Increasing Amounts of Greenhouse Gases	283
		22.2.1 Review of Basic Greenhouse Physics	283
		22.2.2 Some Complications	284
		22.2.3 General Circulation Models	285
	22.3	Predicted Effects of Global Warming	287
		22.3.1 Large Stratospheric Cooling	287
		22.3.2 Global Mean Surface Temperature Warming	289
		22.3.3 Global Mean Increase in Precipitation	289
		22.3.4 Northern Polar Winter Surface Warming	289
		22.3.5 Rise in Global Mean Sea Level	289
		22.3.6 Summer Continental Warming and Increased Dryness	289
		22.3.7 Regional Vegetation Changes	290
		22.3.8 More Severe Precipitation Events	290
		22.3.9 Changes in Climate Variability	290
		22.3.10 Regional-Scale Changes Will Look Very Different from the Global Average, but Their Nature Is Uncertain	290
		22.3.11 Biosphere-Climate Feedbacks	290
		22.3.12 Details of Life in the Next Quarter Century	290
	22.4	The Difficulty of Proof: Weather versus Climate	291
	22.5	Role of the Oceans in Earth's Climate	292
		22.5.1 Basics of Ocean Circulation	293
		22.5.2 El Niño Phenomenon	294
		22.5.3 Prolonged Global Warming and Ocean Circulation Shutdown	295
	22.6	Global Warming: A Long-Term View	295
	22.7	Postscript: Human Effects on the Upper Atmosphere – Ozone Depletion	295
	22.8	Questions	296
	22.9	Readings	296
23	Limited Resources: The Human Dilemma		298
	23.1	The Expanding Human Population	298

	23.2	Prospects for Agriculture		299
	23.3	Energy Resources		300
		23.3.1	Fossil Fuels	300
		23.3.2	The Challenges of Fossil Fuels	302
		23.3.3	Alternative Energy Sources	303
		23.3.4	Energy Use in the Future	303
	23.4	Economically Important Minerals		305
	23.5	Pollution		306
	23.6	Can We Go Back?		307
	23.7	Questions		308
	23.8	Readings		308
		23.8.1	General Reading	308
		23.8.2	References	308
24	Coda: The Once and Future Earth			309
Index				311

Note: An eight-page color plate section appears between pages 84 and 85.

PREFACE

At the close of the second millennium A.D., we live in extraordinary times. A generation ago, human beings first ventured beyond Earth's atmosphere into the vast emptiness of space, began to unlock the remarkable secrets beneath the oceans of how Earth's geology works, and to crack in earnest the genetic code that determines the fundamental nature of all life. Today, these and other frontiers remain open to us, yet we also are consumed with a multitude of problems seemingly of our own making. Increasingly, too many people compete for too few resources and make fundamental changes to the life-giving air and oceans of our planet. Further, we find ourselves confused about science and technology: Are they the cause of, or the solution to, these daunting problems?

Regardless of which way one chooses to answer this question, of deepest concern is that science and technology are understood by few, even in the technologically advanced industrial nations. We use computers and cellular phones with ease, yet how many of us understand the basic principles by which they work? We look forward to the change of seasons and scan the weather reports for tomorrow's outlook on rain, yet remarkably few of us can explain the motions of Earth in the cosmos, and the underlying causes of the atmospheric changes we call weather and climate. We talk glibly about the promise and problems of genetic engineering, yet most such conversations are conducted in the absence of any familiarity with what the genetic code actually is and how it functions. Increasingly, it seems, the decisions to proceed or not with the most advanced technologies lie by default in the hands of a scientifically literate few.

It is our responsibility as citizens of this planet Earth to comprehend the underpinnings of science and technology today, sufficiently so that at least one can hold a thoughtful conversation on important science- and technology-based issues. This does not mean that each of us must store an encyclopedic quantity of material on science in our heads, available for instant recall. It does mean that we should be capable of finding sources of information on science and technology, as required, from our local libraries; and we should be capable of reading articles in major newspapers on the subject of new scientific developments in such a way as to understand the information and issues being presented. We expect no less of an informed citizenry in other areas of common knowledge such as politics.

How can any one of us do that across the bewildering spectrum of science and engineering, from biology to physics, astronomy to geology, hydrology to chemistry to electronics? The key is to have the right framework within which to study a broad core of material from each of the major disciplines of science and engineering. One excellent framework, the focus of this book, is the history of Earth. Broadly taken, the origin and evolution of Earth from the beginning to the present day is a story that encompasses all major disciplines of science and, placing humankind's technological tools in the picture, of engineering as well.

In what follows, we hit some of the high points of what might be called the "whole" history of Earth, show how the myriad disciplines of science are interwoven to understand this world and its sister planets, and how science and engineering must be applied to today's challenges if humankind is to have a promising future on (and off) this planet. We explore Earth's history chronologically through the bulk of the book, but to undertake this journey requires that we equip ourselves with the necessary gear. This includes introductions to the basic concepts of physics, chemistry, and geology that underpin our understanding of the world around us, and a little "orienteering" to the cosmos in which we find ourselves. These introductions precede our chronological journey, and readers are urged to at least dip into the chapters corresponding to material most far afield from their own experience.

In writing this book, I have tried to steer a course between the very basic and cursory examination of science that is increasingly popular in college courses for non-scientists, and a more specialist review of the literature. I recognize that this may make parts of the book difficult for some students encountering science for the first time, but I urge upon them the patience to work through the text – with the promise that they will be exposed to the most interesting issues associated with Earth's history that other students at their level may never encounter. Conversely, I anticipate the book to be useful to working scientists who may wish an introduction to some of the current issues associated with Earth's evolution as a planet. I have tried to provide enough references to guide such readers into the literature of the various subfields, without being exhaustive. I hope that all readers will appreciate that, within the broad framework of Earth's history, I have but delved into a selected set of problems that stand at the forefront of our understanding of this planet's extraordinary evolution.

ACKNOWLEDGMENTS

My wife, Cynthia Lunine, graciously and expertly drafted the original illustrations appearing in the text; her contribution has been so large that it is acknowledged on the title page. We both thank Karen Swarthout, who drafted figures in the earliest phases but then was called to new adventures in Peru. Individuals who read the entire text and provided valuable comments were Philip Eklund, Professor E.H. Levy (Dean of the College of Science of the University of Arizona), and Professor Michael Drake (Director of the Lunar and Planetary Laboratory and Head of the Department of Planetary Sciences, University of Arizona). An additional note of gratitude is in order to Mike, for providing teaching release time and much other support enabling completion of this work. Many other scientists read and criticized portions of the manuscript: Vic Baker, Julio Betancourt (U.S. Geological Survey), Christopher Chyba (SETI Institute), Michael Cusanovich, Robert Dickinson, Annita Harlan, James Kasting (Pennsylvania State University), Cathy Allen Manduca (Carleton College), Paul Martin, Joann Miller, Roger Phillips (Washington University), Robert Strom, Thomas Swetnam, Timothy Swindle, and Steven Zegura. (Those whose affiliations are not specifically identified are from the University of Arizona.) Professors Chris Chyba and Ed McCullough used drafts of this book in their nonscience introductory classes in Spring and Fall 1997; permission of Cambridge University Press to do so is gratefully acknowledged.

Separate acknowledgments for use of figures is given in each of the figure captions, but I especially want to thank Firouz Naderi, Jet Propulsion Laboratory Origins Program Manager, for the cover art. Both the Jet Propulsion Laboratory and Mary Guerreri of the University of Arizona's Space Imagery Center helped in the procurement and production of other images, for which I am grateful. Further thanks are due Maria Schuchart, who ably converted color images into attractive black and white figures.

Professor Alex Dessler, University of Arizona and member of the Editorial Board for the Cambridge Space Science Series, approached me to write this book, for which (in the end) I am grateful. My editor at Cambridge, Catherine Flack, was always cheerful, expert, and tolerant of delays heaped on top of delays. Thomas Harlan and Thomas Swetnam, University of Arizona Laboratory of Tree-Ring Research, encouraged my interest in climate studies, tolerating my sometimes naive questions while allowing me on some fascinating field excursions.

I want to reserve a final expression of gratitude to the faculty and students of the Lunar and Planetary Laboratory, and beyond them to the larger College of Science of the University of Arizona of which they are a part. Without the extraordinary collegiality across discipline and departmental boundaries that so exemplifies this institution, I would not have dared to write such a book. For endless hours of discussion, sharing of data, welcome invitations to field trips and laboratories, and the like – many many thanks.

Jonathan I. Lunine
Lunar and Planetary Laboratory
The University of Arizona
Tucson, Arizona USA

I THE ASTRONOMICAL PLANET: EARTH'S PLACE IN THE COSMOS

1 | AN INTRODUCTORY TOUR OF EARTH'S COSMIC NEIGHBORHOOD

1.1 ANCIENT ATTEMPTS TO DETERMINE THE SCALE OF THE COSMOS

The science of astronomy developed in many different cultures and from many different motivations. Because even in cities of the preindustrial world, the stars could be seen readily at night, the pageant of the sky was an inspiration for, and embodiment of, the myths and legends of almost all cultures. Some people tracked the fixed stars and moving planets with great precision, some for agricultural purposes (the ancient Egyptians needed to prepare for the annual flooding of the Nile River valley) and more universally to attempt to predict the future. The regularity of the motions of the heavens was powerfully suggestive of the notion that history itself was cyclical, and hence predictable. The idea of human history linked to celestial events remains with us today as the practice of astrology. In spite of a lack of careful experimental tests, or demonstrated physical mechanisms, this powerfully attractive belief system is pursued widely with varying amounts of seriousness, extending in the early 1980s to the level of the presidency of the United States.

Although ancient understanding of the nature of the cosmos varied widely and was usually a reflection of particular mythologies of a given culture, the classical Greeks distinguished themselves by their (often successful) attempts to use experiment and deduction to learn about the universe. Some Greek philosophers understood the spherical nature of Earth and something of the scale of nearby space. Aristotle, in the fourth century B.C., correctly interpreted lunar eclipses as being due to the shadow of Earth projected on the surface of the Moon. By noting that the shadow was rounded, he deduced that Earth must be spherical; in fact, another acceptable shape based on that one observation is a disk (figure 1.1). Others, such as Plato, had much earlier endorsed a spherical shape on aesthetic grounds.

Eratosthenes, who lived in the third century B.C., made a remarkably accurate determination of the size of our planet without having to travel too far. He used the observation that at high noon on summer solstice (June 21 in our calendar, when the Sun reaches its northernmost point in the sky of Earth), the Sun was directly overhead at a site in Syene (now Aswan), Egypt, because no shadow could be seen in the vertical well shaft. Eratosthenes lived in Alexandria, due north of Syene, and there he could observe that the Sun cast a shadow at noon on that same date of June 21 (figure 1.1).

What did this mean? If Earth were a sphere, then different people standing at different locations on Earth at the same time would see the Sun in different parts of the sky. By measuring as an angular distance in the sky, the change in the position of the Sun from one place to another and knowing the distance between the two stations, one could then by a simple calculation work out the circumference of the whole globe. In his home city, Eratosthenes carefully measured the size of a shadow cast by an obelisk of known height, at the same time on the same day that no noontime shadow occurred at Syene. The angular position of the Sun, from the size of the shadow at Alexandria, gave an angle of 7.2 degrees between the position of the Sun at the two stations, or one-fiftieth of the entire angular extent of the sky (360 degrees). Therefore, Earth's circumference, he knew, must be 50 times the distance between Syene and Alexandria.

The distance was, however, known only approximately from the number of days it took a camel to travel between the two towns and the distance a typical camel walks in a typical day. Furthermore, to compare the result with the value we know today, the units of measurement used

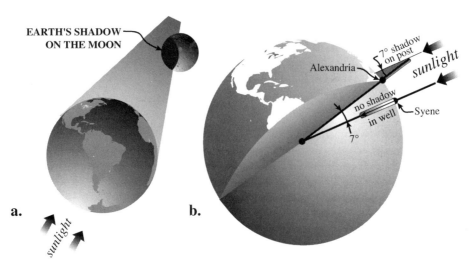

Figure 1.1. Two ancient Greek observations of the cosmos: (a) Aristotle's determination of Earth's sphericity via a lunar eclipse; (b) Eratosthenes' measurement of the size of Earth. Adapted from Snow (1991).

by the Greeks must be converted to modern ones, which is also an uncertain exercise. In modern units, the intercity distance is 570 miles, or 918 kilometers (km), and hence Eratosthenes' experiment yields an Earth circumference of 46,000 km, just 12% too large. This represents an extraordinary achievement, 2,300 years before human beings could view the round globe of Earth from space.

Not everything about the cosmos that the Greek philosophers deduced or inferred came out right. The most celebrated mistake was that of Ptolemy, who lived 400 years after Eratosthenes and is associated most closely with the cosmological system in which the Sun and the planets (in fact, the whole cosmos) were thought to orbit Earth. However, this was just the penultimate round in a long debate on the topic: Aristarchus of Samos, a generation before Eratosthenes, put the Sun at the center with Earth and the other planets orbiting it. This correct model of the solar system was discredited at the time because the Greeks could not see the stars shift in position as Earth moved from one point in its orbit to the opposite side. In fact, the stars do appear to shift position, as we describe later, but they are so far away that the shift cannot be detected with the unaided eye. This the Greeks did not know, and the failed experiment led them down the wrong path of an Earth-centered cosmos that would not finally be discarded until the times of Copernicus and Galileo, over 1,500 years later.

We should not fault the Greeks for their wrong interpretations, but should admire their startling successes, which were based on observations unaided by the technologies available at present, coupled with the disciplined logic of inductive and deductive reasoning which was the foundation of the scientific method. Few of us today could repeat the insights of these extraordinary philosophers. In point of fact, we in the industrialized world still have a mindset in essence of an Earth-centered universe: We think little of the sky, now obscured by the lights of cities and hence unfamiliar to us, unless it is to wonder when the Sun will set today, or what the local newspaper horoscope claims our immediate future will hold.

1.2 BRIEF INTRODUCTION TO THE SOLAR SYSTEM

The solar system consists of 9 planets, some 60 natural satellites (or *moons*), and innumerable small bodies, all orbiting the Sun. Robotic spacecraft have traversed the distance to the farthest planet in the solar system, some 6 billion km. The distance to the nearest star, Proxima Centauri, is 6,000 times greater; hence, we have no hope of seeing spacecraft reach such targets in the foreseeable future. In view of this, the solar system is our cosmic neighborhood, accessible for study by spacecraft and constituting the setting within which Earth has evolved through time.

Here the solar system is summarized in tutorial form to provide a context for what follows. The information presented is the result of at least three millennia of observations and insights, capped by three decades of intense scientific study from the ground and space. Some of this effort is described in the book, but to present a complete history of the exploration of the solar system would require a separate volume.

Figure 1.2 is a map of the solar system. The nine planets fall roughly into three classes according to their size

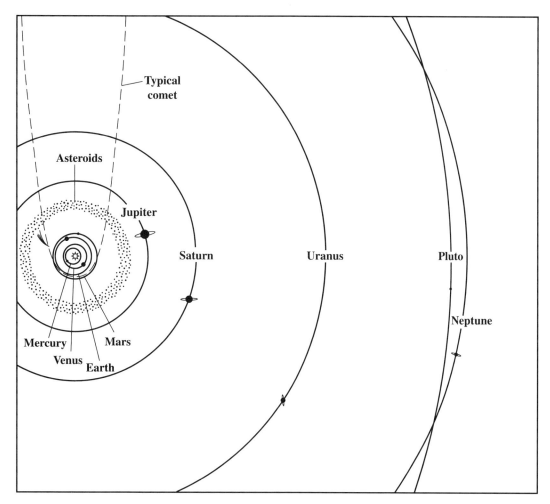

Figure 1.2. Schematic map of our solar system, showing the correct relative sizes of orbits but not of the bodies themselves. Note the small scale of the orbits of terrestrial planet compared to the vast realm of the outer planets. The Kuiper Belt objects just beyond Neptune's orbit and the Oort cloud of comets are described and depicted in chapter 10. Reproduced courtesy of Nancy and Larry Lebofsky, from *Our Solar System*, by permission of Arizona Board of Regents.

and composition. The four *terrestrial* planets – Mercury, Venus, Earth, and Mars – range in diameter from 4,800 km (Mercury) to 12,700 km (Earth). They occupy a small, inner region of the solar system, and are composed of a mixture of rocky and metallic materials.

The four *giant* or *Jovian* planets – Jupiter, Saturn, Uranus, and Neptune – are substantially bigger than Earth, ranging in diameter from 49,000 km (Neptune) to 142,000 km (Jupiter). They are much farther from the Sun than are the inner planets: Jupiter's distance from the Sun is 5 times that of Earth's and hence is abbreviated as 5 *astronomical units* (AU); Neptune is 30 AU from the Sun. In terms of common units of distance, Earth lies 150 million km from the Sun, and thus Neptune is more than 4 billion km from the solar system's center.

The giant planets are composed of a mixture of rocky and icy material and varying amounts of gases; Jupiter and Saturn are mostly hydrogen and helium gas whereas Uranus and Neptune are predominantly icy and rocky material with lesser amounts of hydrogen and helium gas. (Rocky and icy material is used here to mean atoms of silicon, magnesium, iron, oxygen, carbon, nitrogen, sulfur, and others that tend to form rocky and icy materials under conditions of normal pressure. Because of the intense pressures deep within these giant planets, much of the icy and rocky material is in atomic form, rather than the molecular form with which we are familiar.)

Beyond Neptune is the "oddball" planet Pluto, about 2,400 km in diameter (smaller than Earth's Moon), which may be the largest of a class of debris left over from the formation of the solar system. In size and density (amount of mass per volume in the object), Pluto is remarkably similar to Triton, the largest moon of Neptune.

Seven of the planets have at least one moon, with almost 60 such objects known to be orbiting the planets. Some are small, irregular fragments kilometers across; others – two moons of Jupiter, one of Saturn – are larger than the planet Mercury. The giant planets have multiple satellite systems, some in very regular, circular orbits, which can be considered as miniature solar systems. Saturn's largest moon, Titan, possesses an atmosphere thicker than ours on Earth; several other moons have tenuous atmospheres, including our own Moon which exhibits an extremely rarefied atmosphere of sodium and potassium. All of the planets have atmospheres, though that of Mercury is like our Moon's in being very tenuous.

The four giant planets have ring systems composed of debris from house-sized to dust, which orbits in the equatorial plane of the planet. Saturn's famous ring system is considerably more massive than those of the other major planets. None of the terrestrial planets possesses an organized ring system.

The solar system exhibits several regularities in its structure, which are important in understanding its origin, as we discuss later. All planets orbit the sun in nearly circular orbits, close to the plane of the Sun's equator, with the exception of Pluto, the orbit of which is both *inclined* (tilted relative to the sun's equator) and *eccentric* (significantly noncircular). All orbits are in the same direction; by convention, they are counterclockwise around the Sun when viewed from above the Sun's *northern hemisphere*. With two exceptions, Venus and Uranus, all planetary spins are in the same, counterclockwise, direction. However, the planetary rotational axes are all tilted relative to their orbital planes by varying degrees.

There is a strong correlation between the properties of the planets and their location in the solar system. The four terrestrial planets, which contain proportionately little water and gases, are closest to the Sun and not very massive compared to the giant planets. From Jupiter outward, solid objects (moons and Pluto) contain significant amounts of water ice and more volatile species. (Here, volatile refers to the tendency for a material to transform from a condensed state to a vapor.) The four giant planets seem to be of two classes, with the more gaseous planets, Jupiter and Saturn, closer to the Sun.

Viewed from a neighboring star, the most notable characteristic of the solar system would not be the planets, but the debris of small solid bodies outside Pluto's orbit. Only in recent years has the structure of these debris regions become evident. The orbit of Pluto extends from just inside Neptune's (29 AU) outward to nearly 50 AU. Within this region over 60 objects with diameters of a hundred kilometers or so have been detected orbiting the Sun. They are thought to be representative of a class of material, referred to as *Kuiper belt objects*, that are the leftover debris from the formation of the outer planets. The inner edge of this thick belt of material is defined by the giant planets, whose strong gravitational fields have swept the region from 5 to 30 AU clear of debris.

Well beyond the Kuiper objects lies more icy and rocky material in distant orbits ranging out to perhaps 100,000 AU from the Sun. The presence of such material is inferred from the existence of comets, rock-ice bodies perhaps 1–10 kilometers in diameter that come into the inner solar system on highly noncircular, that is, elliptical, orbits. Careful plotting of the paths of comets indicates that most of the orbits originate in an ill-defined shell of material termed the *Oort cloud*. The comets are the small fraction of Oort cloud objects that fall inward to the Sun after having been perturbed by close-passing stars. The total number of comet-sized Oort cloud objects may approach one trillion.

Remote observation of comets as they pass through the inner solar system suggests that they are accumulations of dust, organic material, water ice, and frozen gases. The Oort cloud material is thought to have been ejected from the 5- to 30-AU region by the giant planets after their formation and, in addition to comet-sized bodies, both larger and smaller objects may reside in this cloud.

Between the orbits of Mars and Jupiter lie belts of rocky objects known as asteroids. The largest asteroids are several hundred kilometers across; in number and total mass they are minuscule compared to the Oort cloud and the Kuiper belt. They are thought to be debris that never formed into a planet because of the proximity of Jupiter, whose gravitational field prevented efficient growth of a large body from smaller ones. Another collection of asteroids crosses the orbit of Earth–the so-called *near-Earth asteroids*, some of which may be old comets that have lost their mantles of ice after many passes by the Sun. Finally, lanes and regions of dust released from comets or asteroids lace the solar system; the precise distribution of this material, some of which can be seen faintly after sunset as the *zodiacal light*, remains somewhat uncertain.

The history of collisions between the numerous bits of small debris and the planets is recorded by the ubiquitous existence of craters throughout the solar system. Even Earth shows the scars, Meteor Crater in Arizona being a famous recent example. As we shall see, impacts may have played key roles in the origin and evolution of life on this planet Earth.

1.3 QUESTIONS

a. Consider how you have responded to a controversial scientific or technological issue. Did you try to weigh rationally the pros and cons, or did you respond on the basis of your instincts or emotions?
b. Imagine that the knowledge leading to atomic energy had never been achieved. What are some of the things that might have been different about the period from World War II to today? Can you say whether the world would have been better or worse off?

1.4 READINGS

1.4.1 General Reading

Boorstein, D.J. 1983. *The Discoverers*. Vintage Books, New York.

Sagan, C. 1996. *The Demon-Haunted World: Science as a Candle in the Dark*. Ballantine Books, New York.

1.4.2 References

Snow, T.P. 1991. *The Dynamic Universe: An Introduction to Astronomy*, 4th ed. West Publishing, St. Paul, MN.

2 LARGEST AND SMALLEST SCALES

2.1 INTRODUCTION

In chapter 1, we became acquainted with the scale of the solar system – the stage upon which planetary evolution is set. However, the formation of elements out of which planets and life came into being involved the universe of stars and galaxies – a scale much larger than the solar system – and the microscopic world of atoms, which involves size scales much smaller than that of our ordinary experiences. In this chapter we explore how cosmic distances are gauged, and then begin to acquaint ourselves with the basic building blocks of matter.

2.2 SCIENTIFIC NOTATION

Although the book is written with the nonmathematically inclined reader in mind, the discussion of numbers, both large and small, cannot be avoided if we are to gain a true understanding of Earth and its place in the cosmos. Numbers of interest in science range over enormous magnitudes (figure 2.1). The number of protons contained in a single star, our Sun, is of order 1,000,000,000,000,000,000,000, 000,000,000,000,000,000,000,000,000,000,000,000; the size of an individual proton (itself made up of smaller elementary units) is of order 0.0000000000001 cm. (The term *of order* refers to how many powers of 10 a number contains, rather than the specific numerical value it has; hence 200 is of order 100, 40 is of order 10, etc.) These numbers are inconvenient to write down and manipulate in even the simplest mathematical expressions.

Hence *scientific notation* is universally used, where a number is expressed in terms of powers of 10. The number of protons in the Sun is of order 10^{57}; the size of a proton is of order 10^{-13}. To express the numerical value, in addition to the order of magnitude, one simply multiplies by the appropriate number. Hence, 5,000 is 5×10^3 and 0.004 is 4×10^{-3}. Any degree of precision can be handled readily; for example, 65,490 is 6.549×10^4 and 0.034256 is 3.4256×10^{-2}. Multiplication and division of such numbers is easy, the exponents in the power of 10 being added or subtracted, respectively, for the two operations.

The one drawback of scientific notation is that it dulls us to the enormous range of numbers that the scale of the universe demands. Somehow, writing 1.67×10^{-24} grams (the mass of a hydrogen atom) does not give us the same appreciation for the smallness of this number writing 0.00000000000000000000000167 gives. In contemplating the history of planet Earth and its place in the cosmos, it is too easy to manipulate such numbers without first considering the philosophical implications of their gigantic or minuscule quality!

2.3 MOTIONS OF EARTH IN THE COSMOS

We view a universe continually in motion. The most obvious movements, apparent to even the casual observer, are the paths of the Sun across the sky on a daily basis and the rising and setting of the Moon on an apparently slightly less reliable basis. The equivalent nocturnal rhythm of the rising and setting of the constellations also is easily discernible, though much less familiar to increasingly urban populations.

Those who are more interested watchers of the sky will notice two longer rhythms, the march of a changing Moon progressively through the day and night skies on a 28- to 29-day basis, and the annual ritual of the slow climb of the Sun toward a more northerly path in the sky during summer and toward a more southerly path during winter (readers in the Southern Hemisphere should reverse winter and summer in the description). At any given location the Moon occasionally wanders into a region of darkness, and

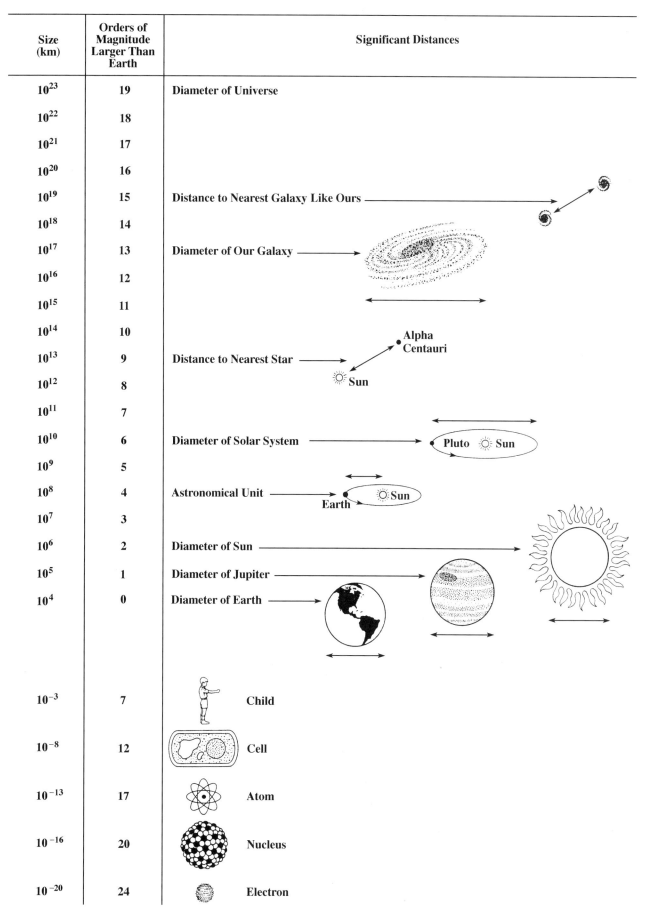

Figure 2.1. Sizes of various objects over the enormous range that the natural world encompasses. From Robbins and Jeffreys (1988) by permission of John Wiley and Sons.

reddens in what is called a lunar eclipse. The Sun's light is partially blocked once every few years from a given location, and totally blocked about once every few centuries at any given place, in a solar eclipse.

Even more subtle motions are available in the skies for those with the patience to watch. Five (eight, if a good telescope is used to enhance the eye's sensitivity) "stars" in the sky are seen to move against the background of the fixed stars on paths that execute peculiar back-and-forth dances; the speed with which these planets (from the Greek *planetes*, meaning wandering) move varies greatly, corresponding to timescales of months to centuries to orbit the Sun.

All of these motions are fully understandable on the basis of the Copernican model of a spinning earth, tipped modestly on its axis, orbiting about the Sun once each year, with other planets orbiting at greater or lesser distances from the Sun, and the Moon orbiting about Earth. We take this picture, quite appropriately, as fact, but few of us have paused to ponder the subtleties associated with working out such motions. Furthermore, slight changes in the shape of Earth's orbit have affected climate on cycles of tens of thousands of years, and the presence of the Moon in orbit about Earth apparently has prevented rather extreme swings in Earth's axial tilt, which could have led to very large climate instabilities in the past. Far from being a quaint part of the traditional curriculum of science in schools, the arrangement of the Sun, Earth, Moon, and other planets is in fact critical to understanding the stability of and variations in our climate on a range of timescales.

We discuss such climatic issues in Part III, but now we return to the basics of Earth's motions through the cosmos. The perception of movement of the Sun and constellations through the sky is akin to our experiences as children (or adults) on a carousel, watching people, trees, structures, and so on swing past us in regular, repetitive cycles. Because there is little sense of acceleration on the larger, slower (and hence grander) carousels, very quickly one can experience the illusion of being on a fixed world around which the external "universe" is moving.

The Moon's motion is somewhat more complicated; because it is orbiting Earth once every 28–29 days, it rises and sets at significantly different times each night. The analogy on our carousel is to watch a person who is walking briskly in the direction of the carousel's motion. Relative to fixed objects (standing adults, trees), our moving person will reappear later during each rotation of the carousel. Because our Moon is almost entirely illuminated by the distant Sun (some contribution from Earthlight is detectable on the otherwise unilluminated portion), different portions of the Moon are illuminated at different times of the month, creating *phases* (figure 2.2).

The orbit of the Moon is not aligned with Earth's orbit about the Sun, but rather is inclined from it by about 5 degrees. Because of this, during the time of the month when Earth, the Sun, and the Moon are all aligned in a given direction (the times of full and new Moon), the Moon generally appears on the sky significantly above or below the path of the Sun. Only when the time of full Moon coincides with the Moon crossing the plane defined by Earth's orbit around the Sun – the *ecliptic* – do we have true alignment. At this time, the full Moon gives way to a *lunar eclipse*, in which Earth's shadow obscures the Moon, or the new Moon is replaced by the dramatic *solar eclipse*, in which the disk of the Moon blocks out the light of the Sun (figure 2.2).

Eclipse prediction is not easy, because three motions are involved: the revolution of the Moon around Earth, the motion of Earth around the Sun, and the so-called *regression of nodes*, wherein the points at which the Moon crosses Earth's ecliptic rotate slowly in an 18.6-year cycle. This last motion can be visualized by imagining the orbit of the Moon as a circular glass sheet that cuts through Earth at a slight angle relative to the ecliptic. This sheet slowly revolves relative to Earth, completing one spin in 18.6 years. (The physical cause of the regression lies in the gravitational pull of the Sun, which exerts a torque because the lunar orbit is tilted or *inclined* relative to the ecliptic.)

These three motions are such that any particular sequence of eclipses recurs at an interval just over 18 years. The frequency of lunar eclipses is greater than the frequency of solar eclipses. Because Earth's shadow is much larger than the Moon when projected at the distance of the Moon from Earth, slight misses in crossing the node still produce a lunar eclipse. The lunar shadow is smaller and, coincidentally, the size of the Moon in the sky is just roughly that of the Sun. Thus the solar eclipse must occur very close to a node crossing for it to be total. Further, the orbit of the Moon around Earth is not a circle but an ellipse (see below); if the eclipse occurs when the Moon is farthest from Earth, the apparent size of the Moon is smaller than the Sun's disk, and a much less spectacular, *annular*, eclipse transpires.

Two remarkable cultures demonstrate both the subtlety and universality of tracking the rhythms of solar system objects. Stonehenge is a series of large rock monuments and circles laid out on the Salisbury plains of England. The earliest such construction, and least spectacular to the eye, is a large circle of *56 Aubrey* holes, spanning roughly 50 meters across, with a so-called heelstone off

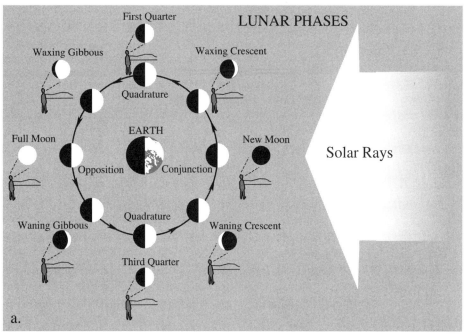

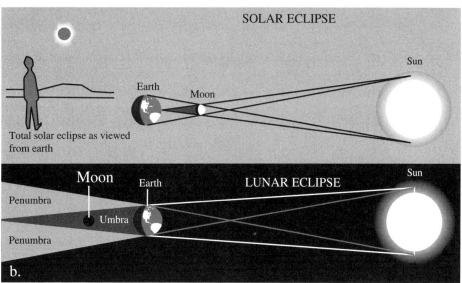

Figure 2.2. (a) Geometry of Earth, the Moon, and the Sun leading to the monthly cycle of phases; an earth-bound observer's view is shown next to each corresponding lunar position [adapted from Snow (1991, p. 31)]; (b) alignments of Earth, the Moon, and the Sun during total solar and lunar eclipses [after Hartmann (1983)].

to the northeast. This was set up by a Stone Age people about 4,800 years ago, perhaps a millennium before the spectacular large stone structures more familiar to tourists. Spurred by an initial suggestion by astronomer Gerald Hawkins, British astrophysicist Sir Fred Hoyle (1972) demonstrated that the 56 Aubrey holes could be used as an eclipse counter.

By moving stones representing the Sun and the Moon counterclockwise at certain prescribed rates (two holes every 13 days for the Sun and two holes each day for the Moon), one predicts the positions of the Sun and the Moon relative to the observer, on Earth, in the center of the ring. By moving two other stones, each 180 degrees apart, clockwise three holes each year to represent the precession of the lunar nodes, eclipses could be predicted reliably. When the Moon and the Sun are on opposite sides of the circle, and less than one or two Aubrey holes away from the node stones, a lunar eclipse would occur; when the Moon and Sun stones cross each other and are less than one or two Aubrey holes away from

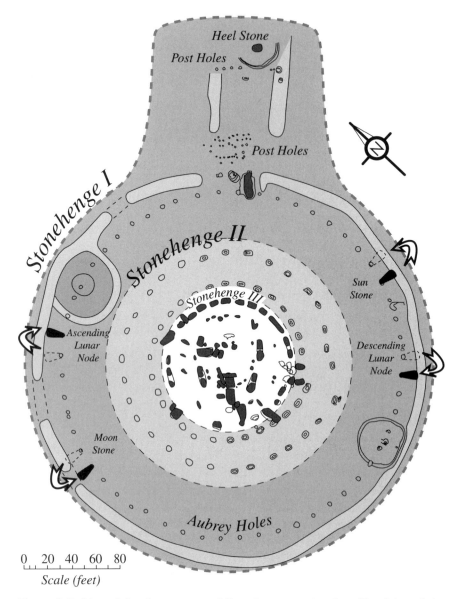

Figure 2.3. Map of the three stages of Stonehenge construction. The Aubrey holes and other sight points of Stonehenge I are identified. Adapted from Hoyle (1972 p. 22, Fig. 2.4) by permission of W.H. Freeman and Company.

a node stone, a solar eclipse is predicted to occur (figure 2.3). The counter scheme was not perfect, because about half of the predicted eclipses would not be visible in the skies above Stonehenge (the Aubrey circle representing the full 360 degrees of the sky including that beneath the horizon at Stonehenge); nonetheless, if correctly interpreted, it is a clever astronomical calculator.

Because none of the solar, lunar, or nodal cycles are exact multiples of the 56 holes, the counting rules are not exact. The marker positions would need resetting regularly by sighting the Sun and the Moon in the sky at key times of the year. The heelstone and nearby additional holes were used, according to Hoyle's model, for sighting and hence correcting the board positions.

Intriguing as the eclipse counter itself is, Hoyle brought up the significant issue of what the node stones would have meant to the people of early Stonehenge. The need for node stones to determine when full or new Moon points would have eclipses must have been derived empirically, because as invisible mathematical constructs one cannot see nodes in the sky. Given that the Sun and the Moon are common objects of worship in many cultures – even our own, as technologically advanced as it is – it is interesting to ask what the Stonehenge people thought their node stones really represented. Hoyle's proposal that the notion of an invisible all-powerful deity could have originated, at least for those people at that time, from the node stones on the eclipse counter is speculative but thought-provoking.

The Mayan people live in the Yucatan peninsula region of Mexico and Central America. From roughly 100 B.C. to A.D. 900, they produced large numbers of stone sculptures, or stelae, on which a complex system of calendar dates was engraved. The classical culture of organized city-states had several calendars, including one of 365 days and a 260-day religious calendar. This latter is close to, but not quite, the orbit period of Venus. Astronomer-archeologist Edward Krupp (1983) also has suggested that it might refer to the interval between passages of the Sun across the high point (zenith) of the sky at the latitude of important Mayan cities, occurring in May and August. There are other astronomical and biological cycles of significance close to 260, including the human gestation interval.

Most striking about the classical Maya was their sophisticated numbering system for precisely recording dates of major events in their history. The system allowed for extension of dates back in time, and some Mayan sculptures do so – back to arbitrarily large values. The longest date recorded on a Mayan stela corresponds to 1.4×10^{36} years, or 10^{26} times the age of the universe as determined by modern cosmology!

The classical Mayans regarded human history as one cycle embedded in nested sets of larger cycles. The Mayans established a "zero" date, prior to which events were played out by deities, which human events then mirrored. Hence history was already determined, in a sense, because it had been played out before, and time to the Mayans was cyclical. On the other hand, it was linear as well, in that the classical Mayan culture had a detailed chronological history of human events – battles, conquests, accessions – for which definite dates were assigned. Both significant human events and their mirrored supernatural events before the zero date often were pinned to particular points in the cycles of bodies in the sky, and the Mayans spent much time tracking and recording celestial movements so as to predict when significant events in human history might occur.

One might wonder whether this dual cyclical-linear concept of history arose out of the preoccupation of the classical Maya with calendar keeping, sky watching, and recording of dates, or vice versa. As with our own decimal system, where each digit placed to the left of preexisting digits represents a new power of 10 (and hence a larger supercycle of years, decades, centuries, millennia, etc.), the Mayan system of counting in twenties allowed cycles nested within cycles to be similarly expressed. It is interesting that our own Western concept of time also embodies both linear and cyclical elements: In our study of the history of Earth and its sister planets, we see this concept appear repeatedly.

2.4 COSMIC DISTANCES

2.4.1 The Planets

Distances to the planets could be gauged directly by sending spacecraft there, but in practice we need to know the answer to successfully reach the target. In fact, planetary distances began to be accurately determined hundreds of years ago, once the Sun-centered model of the solar system became fully developed. A German scientist, Johannes Kepler, around A.D. 1609 formulated a set of laws of planetary motion based on extensive observations by Tycho Brahe, a Danish astronomer. Kepler proposed that the planets moved around the Sun in *elliptical* orbits (figure 2.4), that a given orbit swept out equal areas in equal amounts of time, and that the square of the period of the orbit was proportional to the cube of the planet's mean distance from the Sun. (No understanding of *why* the planets obeyed these laws came out of Kepler's proposition, at least not immediately; the English mathematician and physicist Sir Isaac Newton decades later postulated the

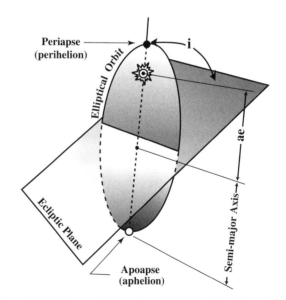

Figure 2.4. Characteristics of an elliptical orbit, where *a* is the semimajor axis and *e* is the eccentricity of the orbit. For a circular orbit, the product $a \times e$ is zero. *Periapse* is the closest point in the orbit to the central object providing the gravitational pull; *apoapse* is the farthest point. If the central body is the Sun, these points are called *perihelion* and *aphelion*. In our solar system, a reference plane is that of Earth's orbit, called the *ecliptic plane*; planetary inclinations, *i*, often are measured relative to this plane, as shown in the figure.

existence of an attractive force associated with the mass of an object, namely gravity. Kepler formulated his laws solely to fit observations; this is an excellent example of an *empirical* model.)

Given Kepler's laws and knowledge of the distance of Earth from the Sun, one can work out the distances to the other planets merely by determining the time it takes each to rotate once around the Sun, that is, the *period* of the orbit. Earth orbits the Sun once in one year. Jupiter orbits the Sun once in just under 12 years; taking the ratio of these periods, squaring it, and taking the cube root yields a mean distance from the Sun for Jupiter of about $5\frac{1}{4}$ times the Earth-Sun distance, or 5.25 AU. Pluto, the most distant planet, has an orbital period of 249 years and hence a mean distance of 39 AU. (We have seen only one-fifth of its orbital path around the Sun, but it was possible to fit an ellipse to its path and hence determine a period very soon after its discovery in 1930.)

However, there is one missing link: the distance of Earth from the Sun. We have expressed planetary distances in terms of Earth-Sun distance, but this is not very satisfying. The distance from Earth to the Sun was tackled by the Greek scientist Aristarchus of Samos, who worked out that when the Moon was exactly half full, the Sun-Moon-Earth would make a right triangle. The angle between the Moon and the Sun in the sky then yields, by simple trigonometry, the Earth-Sun distance – provided one knows the Earth-Moon distance! This distance, in turn, was found by comparing the size of the Moon to the size of Earth's shadow projected against the sky (and revealing itself during a lunar eclipse), yielding a size for the Moon roughly one third that of Earth. This then led to the lunar distance from Earth, and hence the Earth-Sun distance. Unfortunately, Aristarchus was unable to accurately measure the Moon-Sun angle in the sky, and did not get the right answer, but conceptually this is a correct procedure for getting the Earth-Sun distance, about 150 million km.

2.4.2 Nearby Stars

No stars that we see in the sky orbit the Sun. Instead, the Sun is 1 of 100 billion stars that orbit about a common center of gravity; this enormous collection of stars is called the Milky Way galaxy.

To measure the distance to stars relatively near our solar system, the optical effect of parallax can be used. To illustrate this effect, hold a pencil in front of your eyes and alternately close your left and right eye. The pencil is seen to shift against the background. The same effect is present when stars closest to us seem to shift the most during Earth's annual orbit around the Sun. The 300-million-km diameter of Earth's orbit serves as the equivalent of the separation between your eyes in the pencil experiment. By measuring how much stars shift against the background during observation (with highly sensitive telescopes) six months apart, absolute stellar distances are obtained. The nearest star, the Alpha Centauri multiple star system, is 4 light-years away. (A light-year is the distance light travels in a single year, about 10^{13} km.) Beyond a few hundred light-years from Earth, parallax shifts are too small to be measured and other distance techniques must be used.

The parallax technique itself has led to common use of a unit of stellar distance different from the light year: The *parsec* (from parallax-second) is the distance to an object that exhibits a parallax shift of 1 arc-second in the sky, which is 1/3,600 of a degree of angle, the full sky being 360 degrees around. Defined as it is for a baseline corresponding to the diameter of Earth's orbit, the parsec works out to be 3.26 light-years.

Aristarchus' model of an earth moving around the Sun was disputed by other Greeks because they could not see relative shifts in the position of stars from one side of Earth's path around the Sun to the other. We know now that the problem lay in the great distance to even the nearest star, which results in a parallax shift too small to be detected by the Greeks, who had no telescopes.

2.4.3 Nearby Galaxies

Beyond the distances accessible to parallax measurements one must use indirect techniques. If all stars were the same brightness, we could measure distances by comparing a star's brightness with that of one whose distance has been determined, for example, by parallax observations. The technique would be akin to looking out over a city from a hilltop and gauging distances to various streets by the apparent brightness of their streetlights. This is reasonably reliable because all streetlights of a given type are roughly the same *intrinsic* brightness. Light spreads out and dilutes in two dimensions as it moves away from its source, so that the apparent brightness of an object must decrease as the square of the distance from it. Precise measurement of the brightness, then, is a unique measure of distance as long as the intrinsic brightness is known, and there is no absorption of light by dust or gas between the observer and the source.

In fact, stars vary greatly in their brightness, depending primarily upon their age and mass (amount of material they contain). The brightness range for long-lived stars, so-called main sequence stars (see below), is 10 orders of magnitude; for stars in various early and late stages in which dynamic processes are occurring, the range can be much larger. Thus stellar brightness generally is not a useful measure of distance.

Luckily, there exists a group of stars whose intrinsic brightness is "labeled" in a readily decipherable way. These are the so-called *Cepheid*-variable stars, which pulsate or vary in brightness. The more rapidly a particular Cepheid pulsates, the dimmer it is. The relationship has been determined empirically for Cepheids that are close enough to Earth for the distance to be determined independently by parallax measurements, and hence for the star's intrinsic brightness to be worked out. This relationship is so similar for stars of this class that the intrinsic brightness of any given Cepheid is predictable from the pulsation period.

Cepheid pulsation periods can be measured out to great distances, limited only by the ability to detect the pulsations in very faint sources (faint because of the great distance). From the pulsation period, the star's intrinsic brightness thus can be determined. With large ground-based telescopes, the technique has been extendable out to the neighboring galaxies, some millions of light-years distant. The Hubble Space Telescope, positioned above Earth's distorting atmosphere, has been used to observe Cepheids even farther away. The extent of our own Milky Way galaxy is determined from this technique to be of order 100,000 light-years.

2.4.4 Beyond the Galactic Neighborhood

For more distant galaxies, Cepheids are too faint to be detectable and hence to have their pulsation periods measured. Distance determinations in the absence of Cepheid detections are much less precise. One common technique relies on the observed fact that, although stars overall vary greatly in brightness, there is a definite cut-off in the brightness of the very brightest of long-lived stars. These brightest stars are very blue, because their surfaces are extremely hot, and hence they can be recognized by their color. Intrinsic brightness is determined by looking at hot blue stars near enough to us that their distances can be determined independently.

Our urban analogy returns us to a hilltop above a city. As a resident of the city you might know that most of the streetlights are of such a design that they vary greatly in intrinsic brightness. This is true of the mercury-vapor type, the blue one, which dims with age. You then would not be able to use the observed brightnesses to determine distance from the hilltop. If, however, one type of streetlight, the one that gave off the bright orange glow, turned out to be reasonably constant, you could search for those as distance indicators. By comparing the bright-orange lamp brightnesses with one close by, the distance to which you already knew, you would have a means of estimating distance to the others.

Back in the cosmos at large, astronomers must identify blue stars not just by color but by the detailed spectra – that is, distribution of light with wavelength – that they emit (chapter 3). This is required because the apparent color of a star can be altered by dust intervening between it and the observer, making color an unreliable determinant of the hottest stars. Because the bluest stars are intrinsically brighter than Cepheids, they can extend the distance scale a bit farther out.

Certain stellar explosions, called *Type 1A supernovas*, seem to produce a characteristic peak brightness as the star explodes and then dims. By observing such supernovas in nearby galaxies for which Cepheid variables are measurable (to determine the galaxy's distance), the Type 1A supernova brightness can be calibrated. Because such explosions are enormously bright, millions of times that of a Cepheid variable, they allow the distance scale to be extended outward to 100 million light-years, well beyond the nearby galaxies.

Beyond this, or for galaxies in which no serendipitous supernova explosion is observed, the brightness of the whole galaxy must be used as a distance indicator. One might wonder whether this a reliable technique, given the wide variation in the brightness of different stars. However, it turns out that particular types of galaxies appear to have similar intrinsic brightnesses, at least to within a factor of 10, because they appear to have roughly the same number and distribution of stars. Galaxies come in spiral, elliptical, and irregular types, and such distance estimates are made using the spirals and ellipticals; the irregulars are too varied in their shapes and sizes to be useful in this regard.

2.4.5 To the Farthest Edge of the Universe

Hearing the horn of a passing car is an odd experience, if you remember that most car horns are designed to produce a sound of a single pitch. The pitch of a passing car horn is higher when the car is approaching and lower when it is receding. This phenomenon is known as the *Doppler shift*, and it applies equally to waves of light and to sound. Because light, like sound, travels at a finite speed, the relative motion between source and observer causes waves to bunch in the oncoming direction and to be stretched out in the receding direction.

We discuss the nature of light in chapter 3, but for now, it suffices to construct a mental picture of light as the movement of waves of electric and magnetic, or *electromagnetic*, energy through space. The distance between each crest of the wave determines the color of light as perceived by the eye or measured more precisely with an instrument called a *spectrometer*. (This is something of an oversimplification; light emitted by natural sources typically consists not of a single wavelength but a combination of many wavelengths which, overall, yields the perceived color of the light.) An observer moving toward a source of light will perceive the waves to be bunched, and hence the color of the light shifted to the blue. An observer moving away from the source will see a shift to the red in the color of the light. Because of the enormous (but finite) speed of light, 3×10^5 km per second (a billion kilometers per hour), blue and red shifts are not noticeable at speeds with which we are familiar.

Spectrometers can measure the color of galaxies very precisely. It has been found that more distant galaxies appear to be redder. There are three possible causes of the reddening: increasing amounts of dust absorbing blue light between the observer and the galaxy, very strong gravitational fields near the galaxy, or a high recessional velocity leading to Doppler shift.

The first possible cause is eliminated by measuring the positions of discrete lines in the spectrum (chapter 3); these are shifted toward the red by the Doppler or gravitational effects, but are unaffected by intervening dust. Gravitational fields as agents of red shift are a bit harder to eliminate, and may occur in some cases. However, in general, astronomers do not see other phenomena thought to be associated with strong gravitational fields when looking at most distant galaxies, and hence the bulk of galactic reddening should not be caused by strong gravity.

The third explanation seems to be the simplest and is supported by direct distance measurements to galaxies that are relatively near. It was American astronomer Edwin Hubble who first came to the sobering conclusion some half-century ago: The more distant the galaxy, the faster it is receding from us. In effect, the universe is flying apart from itself as if born in an enormous explosion.

The velocity distance relationship can be established using the cosmic distance-measurement techniques described above, and then can be extrapolated beyond those techniques to determine the distance to the farthest galaxies based on their red shifts (hence their recessional velocities). This relationship implies that the most distant galaxies are 10 billion to 20 billion light-years from our galaxy, and therefore from Earth. Because matter cannot move as fast as light, and the farthest galaxies are observed moving at a significant fraction of the speed of light, we are in fact seeing most of the universe and viewing back almost to the beginning.

Because the relationship between red shift and distance must be calibrated using nearby galaxies and more direct distance measures, it is sensitive to errors in the calibration techniques. Currently, strong disagreement exists between two such techniques. Hubble Telescope observations are sensitive enough to allow Cepheids to be observed in galaxies distant enough to yield a measurable red shift, but the result implies a universe only 8 billion to 12 billion years old. The Type 1a supernova observations seem to imply a much older universe, between 15 billion and 20 billion years old. The youngest age determined is a problem because not enough time is available to produce heavy chemical elements in stars (see chapter 4). Some very recent observations suggest that the traditional relationship between Cepheid period and brightness is in error and may need to be revised slightly; this could resolve the discrepancy in favor of the older universe.

Other evidence for cosmic birth in an enormous and ancient explosion lies in radio static which permeates all directions of the sky and can be heard on sensitive radio telescopes. This cosmic background radiation marks the horizon on the sky beyond which, during the very young universe, space was so hot and cluttered with dense matter that electromagnetic energy could not move freely through it. As the universe has expanded over billions of years, this horizon has receded; it is now so distant that the original energy of the explosion is red shifted into the radio part of the electromagnetic spectrum, which is defined and described in chapter 3. Recent mapping of this background radiation by a sensitive orbiting satellite experiment, *Cosmic Background Explorer*, indicates a remarkable uniformity that tightly constrains detailed models of how the cosmic explosion, or *Big Bang*, actually proceeded.

The expansion of the universe includes space itself in addition to matter. In consequence the red shift of the galaxies is not strictly a Doppler shift, which is the result of movement through a fixed medium. The galactic red shift is better thought of as a signature of the expansion of space itself, a phenomenon with no direct analog in our daily existence. Although further details of the Big Bang are a little outside the purview of our study of planet Earth, the consequences of the explosion are not. As we discuss in chapter 4, the Big Bang produced the two most abundant elements, hydrogen and helium, which was the start of a cosmic chemical evolution leading to the elements that comprise Earth and life. To understand the processes by which this took place we must understand something of the nature of matter at its very smallest scales: molecules, atoms, and subatomic particles.

2.5 MICROSCOPIC CONSTITUTION OF MATTER

All forms of matter with which we have direct familiarity are composed of a relatively small number of *chemical elements*. Of these, 111 such elements are known, of which roughly 90 occur in nature. The rest have been made in the laboratory; although some of these may occur in nature under extreme conditions (supernova explosions), they are too short-lived to be detectable.

Elements occur as chemically irreducible bits of matter called *atoms*; these are the smallest particles of matter that retain the chemical identity associated with elements. In our own lives we mostly encounter atoms combined into composites called *molecules*. Inside stars like the Sun, temperatures are high enough that atoms themselves are partly broken apart into negatively and positively charged pieces; the resulting form of matter is called a *plasma*. Very compact dense objects such as neutron stars (the collapsed remains of massive stars) contain matter under such extreme pressure that only subatomic particles called *neutrons* can exist. Finally, certain astrophysical observations have been interpreted as requiring very exotic or as yet unknown types of matter to exist in large amounts.

The search to understand the essence of matter, specifically whether it could be infinitely divisible or was reducible only to some definite elemental particle, began (in documented history) with the ancient Greeks. Democritus was a fifth-century B.C. Greek philosopher whose preference for an atomic model came largely from his views on human progress. If the material of the universe was built up of elementary particles, then the possibility existed that it was finite in complexity and hence understandable. The Roman poet Lucretius, in the first century B.C., elaborated on the philosophical aspects of atomism, arguing that, if atoms obeyed a set of natural laws, then everything in the universe obeyed such laws and the supernatural did not exist.

This did not sit well later with the medieval Catholic Church, which impugned both the poet and the philosophy. The Church's disapproval of atomism may have contributed in part to the labeling of scientists such as Roger Bacon (thirteenth century A.D.) as practitioners of black magic. Certainly, the unsuccessful attempts by the so-called alchemists to transform common metals such as lead into precious ones such as gold did little to advance understanding of the nature of matter. Ironically, however, these endeavors philosophically foreshadowed the discovery of nuclear processes, although they were hopeless at the energies available to the alchemists.

Laboratory evidence for a small number of different types of elements as the fundamental constituents of matter began to accumulate in the eighteenth century. Many common materials could be shown always to consist of irreducible proportions of other substances. Furthermore, when more than one sort of compound could be formed out of two elements, the ratio of the amounts of a particular element in one compound compared to the other could be expressed as small whole numbers. For example, the amount of oxygen in carbon dioxide is just twice that in carbon monoxide, for a fixed amount of carbon. These and other observations led Lavoisier in the eighteenth century, Dalton in the early nineteenth century, and others toward an understanding that the world was indeed comprised of a small number of elemental building blocks.

Experiments in the late nineteenth century involving electrical discharges in gases began to elucidate the nature of atoms as being composed of negatively charged *electrons* and positively charged *protons*. Experimentally, it was found that opposite charges (plus and minus) attract each other, whereas like charges repel. To ensure electrical neutrality, it was thought initially that these must be mixed uniformly in the atom. A very different distribution of these charges was revealed by Ernest Rutherford's famous experiment in the early twentieth century. Rutherford fired a narrow beam of α *particles*, positively charged fragments of atoms, at a very thin (4×10^{-5} cm) foil of gold. He then measured the various directions of scatter of the α particles, which are repulsed by the positively charged component of the atom. Most of the α particles were not deflected, but those that were scattered either nearly directly back or through very large angles. The results required that the positive charge of the gold atoms be concentrated in a very small volume, the *nucleus*, relative to the total volume of the atom, which is balanced by the negative charge of an equal number of electrons occupying a much larger volume.

It previously had been determined that the negatively charged electrons carried very little of the mass of the atom, and hence both the mass and the positive charge of the atom must reside in the very small nuclear space, worked out from experiment to be 10^{-12} of the volume of the atom itself. Furthermore, although it was found that the heavier elements had more protons in the nucleus, and correspondingly more electrons to ensure electrical neutrality, the mass of the elements did not increase linearly with the positive charge of the nucleus. The neutron, with zero electric charge, was postulated and discovered in the early 1930s. The proton and neutron have nearly the same mass, about 1.7×10^{-24} grams, and roughly 1,800 times the mass of the electron.

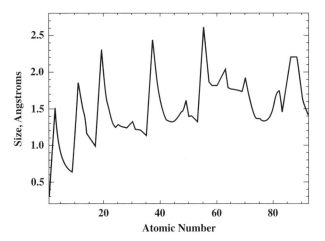

Figure 2.5. Size of elements as a function of their atomic number.

Elements were found to be defined by the number of protons, referred to as the *atomic number*. Atoms range in size, defined by the distance from the center of the nucleus to the outmost electron, from roughly 0.3 to slightly over 2.6 *angstroms*, where an angstrom is 10^{-8} cm. However, the elements do not increase linearly in size with increasing atomic number (figure 2.5). Instead, the atomic size zigzags in a fashion that is correlated with the chemical properties of the elements, or more specifically, the particular manner in which elements will bond with each other to make the enormous variety of materials in the world around us.

Patterns in chemical properties of the elements were recognized in the eighteenth century by French chemist Antoine Lavoisier. In the late nineteenth century the Russian scientist Dmitri Mendeleev constructed a so-called *Periodic Table* of the 60 or so elements known at the time, based on their experimental chemical properties. The modern version of this table, shown in figure 2.6, encapsulates the essential characteristics of the different types of atoms in the way they bond. The utility of the table was demonstrated repeatedly in the nineteenth century as properties of as yet undiscovered elements were predicted from the table.

Elements are arranged in the table in order of increasing atomic number, but in a periodic fashion such that elements that lie in the same column have similar chemical bonding properties. For example, hydrogen, lithium, sodium, and the other *alkali metals* down column IA all have a strong tendency to form two-atom (*diatomic*) molecules with fluorine, chlorine, bromine, and other *halogens* (or salt-formers) in column VIIA. Typical compounds are HF, HCl, NaCl, LiF, LiCl, etc. (chemical symbols for the elements are given in the periodic table). Two atoms in column IA will combine with one atom of the elements in VIA, the *chalcogens* (or ore-formers), so that H_2O, H_2S, Na_2S, and K_2S are all common *triatomic* compounds. The elements in column VIIIA, the *noble gases*, do not chemically bond or do so only weakly under relatively extreme physical conditions.

Note that, because a given element can bond with many other elements from different columns, there is a large degree of complexity in the number of molecules that can be formed. Furthermore, elements toward the middle of the table exhibit a tendency to combine in many different ways, even with a given second element, for example, carbon monoxide (CO) and carbon dioxide (CO_2). The origin of bonding patterns, and hence of the periodic table, lies in the particular number and configuration of electrons that an element possesses. Recalling that an element is defined by its atomic number, or number of protons, this also must be the number of electrons the element possesses to remain electrically neutral. Because the electrons move around in a volume that is much larger than the volume of the nucleus, it is logical that the interactions between the electrons determine the bonding between atoms.

An understanding of how electronic structure arises came primarily through the development of *quantum mechanics* during the early to mid-twentieth century. Quantum mechanics is a branch of physics that deals with the behavior of matter at very small spatial scales. Much of this understanding came through studying the light released or absorbed by electrons in the atom, a subject we take up in chapter 3. The key concept is that electrons possess definite values of energy as they move around the nucleus of the atom. The lowest energy level lies closest to the atom. Increasing energy levels are defined in terms of the pattern of electronic motion around the nucleus at a given energy level. Electrons have the property that they

Figure 2.6. Periodic table of the elements. The elements are arrayed according to their bonding tendencies, as described along the top of the table. The term *valence* refers to the electrons, usually the outermost ones in the atom, that most actively participate in the joining of elements to make molecules. The symbol, full name, atomic number, and atomic weight are given for each element. The atomic weight, by convention, is the average value for the stable (unchanging) isotopes of each element, weighted by abundance as described in the text. Some of the heavier elements have no stable isotopes, and so, approximate atomic weights for the longest-lived forms of these elements are given in parentheses. The Lanthanide and the Actinide elements represent series of chemically almost identical elements. Each of the series occupies a single place in the table as indicated by the asterisks. Names of the six heaviest elements are those officially adopted by the International Union of Pure and Applied Chemistry in September 1997.

PERIODIC TABLE OF THE ELEMENTS

Legend:
- Mn — Chemical Symbol
- 25 — Atomic Number
- Manganese — Element Name
- 54.94 — Atomic Weight

Strong tendency for outermost electrons to be lost to make full outer shell

Tendency to fill outer electron shell by electron sharing and gain or loss of electrons

Strong tendency to gain electrons to make full outer shell

Noble gases: outer shells filled; no tendency to gain or lose electrons

Transition elements: valence electrons not in outer shell

IA	IIA	IIIB	IVB	VB	VIB	VIIB	VIIIB	VIIIB	VIIIB	IB	IIB	IIIA	IVA	VA	VIA	VIIA	VIIIA
H 1 Hydrogen 1.01																	He 2 Helium 4.00
Li 3 Lithium 6.94	Be 4 Beryllium 9.01											B 5 Boron 10.81	C 6 Carbon 12.01	N 7 Nitrogen 14.01	O 8 Oxygen 16.00	F 9 Fluorine 19.00	Ne 10 Neon 20.18
Na 11 Sodium 22.99	Mg 12 Magnesium 24.31											Al 13 Aluminum 26.98	Si 14 Silicon 28.09	P 15 Phosphorus 30.97	S 16 Sulfur 32.06	Cl 17 Chlorine 35.45	Ar 18 Argon 39.95
K 19 Potassium 39.10	Ca 20 Calcium 40.08	Sc 21 Scandium 44.96	Ti 22 Titanium 47.90	V 23 Vanadium 50.94	Cr 24 Chromium 52.00	Mn 25 Manganese 54.94	Fe 26 Iron 55.85	Co 27 Cobalt 58.93	Ni 28 Nickel 58.70	Cu 29 Copper 63.55	Zn 30 Zinc 65.38	Ga 31 Gallium 69.72	Ge 32 Germanium 72.59	As 33 Arsenic 74.92	Se 34 Selenium 78.96	Br 35 Bromine 79.90	Kr 36 Krypton 83.80
Rb 37 Rubidium 85.47	Sr 38 Strontium 87.62	Y 39 Yttrium 88.91	Zr 40 Zirconium 91.22	Nb 41 Niobium 92.91	Mo 42 Molybdenum 95.94	Tc 43 Technetium (98)	Ru 44 Ruthenium 101.07	Rh 45 Rhodium 102.91	Pd 46 Palladium 106.4	Ag 47 Silver 107.87	Cd 48 Cadmium 112.41	In 49 Indium 114.82	Sn 50 Tin 118.69	Sb 51 Antimony 121.75	Te 52 Tellurium 127.60	I 53 Iodine 126.90	Xe 54 Xenon 131.30
Cs 55 Cesium 132.91	Ba 56 Barium 137.33	* (see below)	Hf 72 Hafnium 178.49	Ta 73 Tantalum 180.95	W 74 Tungsten 183.85	Re 75 Rhenium 186.21	Os 76 Osmium 190.2	Ir 77 Iridium 192.22	Pt 78 Platinum 195.09	Au 79 Gold 196.97	Hg 80 Mercury 200.59	Tl 81 Thallium 204.37	Pb 82 Lead 207.2	Bi 83 Bismuth 208.98	Po 84 Polonium (209)	At 85 Astatine (210)	Rn 86 Radon (222)
Fr 87 Francium (223)	Ra 88 Radium 226.03	** (see below)	Rf 104 Rutherfordium 261.11	Db 105 Dubnium 262.11	Sg 106 Seaborgium (263)	Bh 107 Bohrium (262)	Hs 108 Hassium (264)	Mt 109 Meitnerium (266)									

Lanthanide (Rare Earth) Elements

La *57 Lanthanum 138.91	Ce 58 Cerium 140.12	Pr 59 Praseodymium 140.91	Nd 60 Neodymium 144.24	Pm 61 Promethium (145)	Sm 62 Samarium 150.4	Eu 63 Europium 151.96	Gd 64 Gadolinium 157.25	Tb 65 Terbium 158.93	Dy 66 Dysprosium 162.50	Ho 67 Holmium 164.93	Er 68 Erbium 167.26	Tm 69 Thulium 168.93	Yb 70 Ytterbium 173.04	Lu 71 Lutetium 174.97

Actinide Elements

Ac **89 Actinium 227.03	Th 90 Thorium 232.04	Pa 91 Protactinium 231.04	U 92 Uranium 238.03	Np 93 Neptunium 237.05	Pu 94 Plutonium (244)	Am 95 Americium (243)	Cm 96 Curium (247)	Bk 97 Berkelium (247)	Cf 98 Californium (251)	Es 99 Einsteinium (252)	Fm 100 Fermium (257)	Md 101 Mendelevium (258)	No 102 Nobelium (259)	Lr 103 Lawrencium (260)

cannot exist identically in the same energy level with another electron. Two electrons can occupy one energy level only if a certain intrinsic property, called spin, is oppositely directed in the two electrons.

Certain preferred numbers of electrons exist at different energy levels, but most elements either have a deficit of electrons relative to the preferred number or have one or more excess electrons. There is a tendency then for elements to bond with each other in such a way as to produce the "right" number of electrons in each energy level. Direct transfer of electrons (*ionic bonding*) may occur, or the elements may simply associate closely in space so as to share electrons (*covalent bonding*). The different columns in the periodic table group the elements that have a certain excess or deficit of electrons relative to the preferred number for given energy levels. The table thus is a guide to how different elements are likely to bond. Elements on the left tend to donate electrons; those on the right need to acquire electrons. The rightmost column consists of elements that have the preferred number of electrons in all energy levels; these elements are chemically nonreactive and are called *noble gases*.

Elements in the middle columns of the table can either donate or acquire electrons with nearly equal likelihood. This leads to the many kinds of chemical bonds between these elements and others. Carbon, for example, can bond in many different ways with other elements. It is this versatility that is part of the reason why carbon is the most ubiquitous element in biological systems, playing a number of crucial roles. Overall, the variety of different propensities for bonding among the elements leads to the rich diversity of material properties in the universe (figure 2.7).

The neutral particles in the nucleus, the neutrons, do not affect the chemical properties of an element in a primary way. However, the same element can possess different numbers of neutrons, and these different varieties of the same element exhibit modestly different chemical properties. The total number of protons plus neutrons in a given atom is called its *atomic weight*. Atoms of the same element that have different atomic weights are called *isotopes* of that element. The average atomic weight of a given element in nature is listed in the periodic table of figure 2.6. The fact that this is a fractional number reflects the mix of different isotopes in a natural sample of that element.

Considering hydrogen as an example, the primary isotope, sometimes called *protium*, has no neutrons and one proton, for an atomic weight of 1 (the small mass of the electron, by convention, is not included). The next isotope of hydrogen, called *deuterium*, has one neutron and one proton for an atomic weight of 2. Tritium is next, with two neutrons and an atomic weight of 3. Taking the abundances of the three isotopes found commonly on Earth yields the average atomic weight for hydrogen given in figure 2.6 (1.00797). Remember, however, that *each individual atom has an integral atomic weight*, the mass of the electron not being included. No other elements have separate names reserved for their different isotopes. Instead, the atomic weight is attached as a superscript, so that protium (hereinafter referred to as hydrogen), deuterium, and tritium are ^{1}H, ^{2}H, and ^{3}H. The presence of a weight variation in the nucleus of the atom causes a small perturbation in the electron shell energies, leading to a subtle effect on the chemical and physical properties. Also, in the presence of gravity, natural processes tend to separate out isotopes of various types, an important effect in understanding aspects of Earth's history.

Isotopes of a given element may be stable, meaning that they have no tendency to change over time, or they may be unstable. An unstable isotope loses a portion of its nucleus (*radioactively decays*) through emission of particles of various types; very unstable isotopes may split apart. Some unstable isotopes last billions of years before they decay; others decay so rapidly that they are hard to study in the laboratory. The forces associated with the stability of the nucleus are discussed in chapter 3; radioactive decay as a means of forming elements and dating cosmic events is discussed in chapters 4 and 5, respectively.

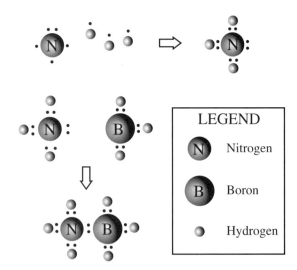

Figure 2.7. Simple example of how elements transfer or share electrons to achieve long-lived states: bonding of nitrogen and hydrogen to form ammonia, and bonding of ammonia and borane. Electrons that may be shared are indicated by dots; other electrons (not shown) are in energy levels much more tightly bound to the nucleus and do not participate in a significant way.

Figure 2.8. Image, using scanning tunneling microscopy, of an electron trapped in a ring of iron atoms. The electron is not merely a particle confined by the corral of atoms, but is also the waves seen traveling outward to and through the corral. Image produced by Crommie et al. (1993) and reproduced from Collins (1993) by permission of the American Institute of Physics.

In the discussion of chemical bonding and sharing of electrons, the reader may be left with a significant degree of dissatisfaction. How do electrons interact, and why do they preferentially move in certain patterns around atoms? To understand this requires that we free ourselves of the simple picture of elemental matter as particles. The behavior of microscopic atomic and subatomic particles displays attributes that have no real analog in the macroscopic world.

An electron is a wave pattern, partly localized in space and energy. An electron around an atom will have a particular wave pattern or *wavefunction*, which is altered when another electron is introduced to complete the energy level. Through the extraordinary insights that led to quantum mechanics, such wave patterns can be calculated mathematically to understand how electrons and nuclei will interact to form atomic and molecular associations. However, intuition has yet to catch up: One simply must imagine that, at smaller and smaller scales, the discrete nature of matter finally loses its meaning in a sea of wave packets interfering one with another.

Technology today is allowing us to see the wave-particle duality of nature. Electrons can be manipulated to image individual atoms in a technique called *scanning tunneling microscopy*. In the image shown in figure 2.8, IBM scientists arranged iron atoms in a circular pattern on a copper substrate and put a single electron in the center of this "corral." The single electron is seen not as a discrete particle trapped by the barrier of atoms but as a complex set of waves, extending beyond the corral.

Microscopic objects simultaneously exhibit particle-like and wave-like attributes, neither of which is a completely useful description of the ultimate behavior of matter. Yet, we are made of these things, and in detail this is the way the universe seems to work. The older discipline of classical mechanics is a useful approximation at scales larger than microscopic (and a necessary one, because of the great numbers of particles making up the macroscopic world). However, there is much in the story of Earth that requires taking the microscopic, quantum view: in particular, understanding the origin of the elements of which we are made, and the source of the light from the Sun which has driven much of the evolution of our atmosphere and life. We take up these subjects in the next two chapters.

2.6 QUESTIONS

a. Construct a mental picture of the distances within the solar system by scaling the diameter of the Sun to the size of a soccer ball. What then would the distance from the Sun to Earth be? From the Sun to Jupiter? From the Sun to the nearest star?

b. What in the appearance of a crescent Moon, particularly in the evening or early morning sky, might be a clue to the fact that the Moon is spherical?

2.7 READINGS

2.7.1 General Reading

Arny, T.T. 1994. *Explorations: An Introduction to Astronomy.* Mosby, St. Louis.

Hartmann, W.K. 1983. *Moons and Planets.* Wadsworth, Belmont, CA.

Krupp, E.C. 1983. *Echoes of the Ancient Skies: The Astronomy of Lost Civilizations.* Harper and Row, New York.

2.7.2 References

Collins, G.P. 1993. STM rounds up electron waves at the QM corral. *Physics Today,* Vol. 46 (11), 17–19.

Crommie, M.F., Lutz, C.P., and Eigler, D.M. 1993. *Science* **262**, 218.

Hartmann, W.K. 1983. *Moons and Planets.* Wadsworth, Belmont, CA.

Hoyle, F. 1972. *From Stonehenge to Modern Cosmology.* W.H. Freeman, San Francisco.

Mason, S.F. 1991. *Chemical Evolution.* Clarendon Press, Oxford.

Robbins, R.R., and Jeffreys, W.H. 1988. *Discovering Astronomy.* John Wiley and Sons, New York.

Schele, L., and Miller, M.E. 1986. *The Blood of Kings: Dynasty and Ritual in Maya Art.* George Brazziller, Inc., New York.

Snow, T.P. 1991. *The Dynamic Universe: An Introduction to Astronomy.* West Publishing, St. Paul, MN.

Taylor, M.D. 1960. *First Principles of Chemistry.* D. Van Nostrand, Princeton, NJ.

Chemical elements. In *Van Nostrand's Scientific Encyclopedia* (D.M. Considine, ed.). Van Nostrand Reinhold, New York, 1983, pp. 595–616.

3 FORCES AND ENERGY

3.1 INTRODUCTION

The previous chapters have touched on the scale of the universe and the nature of the smallest pieces of matter. The structure of the universe is determined not just by the matter contained within it, but by the forces that both bind matter together and compel it to move apart. These forces, which act at the macroscopic and microscopic levels, are thought to be carried by certain types of subatomic particles. In the case of electromagnetism the force-bearing particle is called the *photon*.

We have learned most of what we know of the universe around us by studying the light coming from objects; our most information-filled sense is that of vision, and we have augmented it through the use of devices that can measure in detail the energy distribution of the light. This energy distribution from celestial bodies reveals much about their chemical composition and physical condition. Light from one such self-luminous body, the Sun, is the primary power source for Earth's climate and for life on the planet. The light by which the Sun and other stars shine is not generated by chemical reactions, but by reactions involving the nuclei of atoms at enormous pressures and temperatures deep within these gaseous objects' interiors; these are called *nuclear reactions*.

The nuclear reactions powering stars have, over time, generated essentially all of the natural elements except hydrogen, the most abundant element, and some of the helium (the remainder having been made from hydrogen in the primordial Big Bang). Thus the elements that make up life today (carbon, nitrogen, oxygen, phosphorous, etc.), with the exception of hydrogen, were manufactured by the very same process that today provides the energy source sustaining life on the planet. This chapter sets us on an evolutionary course that joins up eventually with the history of Earth and life, as we consider the processes by which elements are made.

3.2 FORCES OF NATURE

Our lives are lived under the continual action of four forces that act in different ways upon matter. Two of these forces were deduced from the observation of everyday experiences; the other two act upon subatomic particles and were discovered and explored through laboratory experiments.

To discuss the nature of forces it is necessary first to define what a force is. This is surprisingly difficult, because we live constantly under the influence of forces (particularly gravity) that affect the paths of motions of objects. Thus, we are used to seeing a thrown baseball follow a parabolic trajectory under the influence of gravity, but in the absence of forces the ball would move with uniform velocity, that is, constant speed *and* direction, after it leaves the hand of the thrower. Thus, as first expressed by the seventeenth century English scientist, Sir Isaac Newton, in his extraordinary masterpiece *Principia*, every body continues in its state of rest, or of uniform motion in a straight line, unless it is compelled to change that state by forces impressed on it. An operational definition of *force*, then, is an action that causes a change in velocity (which could be a change in either speed or direction or both) of an object.

Closely related to force is acceleration, which is defined as the rate at which velocity changes. The reader may have inferred that velocity is a quantity that contains both speed and direction of a moving object. Thus, a car making a turn at a constant speed is accelerating, and its occupants feel a force on their bodies as surely as they do when the car is increasing or decreasing its speed in a constant direction. The force exerted by an object is acceleration multiplied by the object's mass.

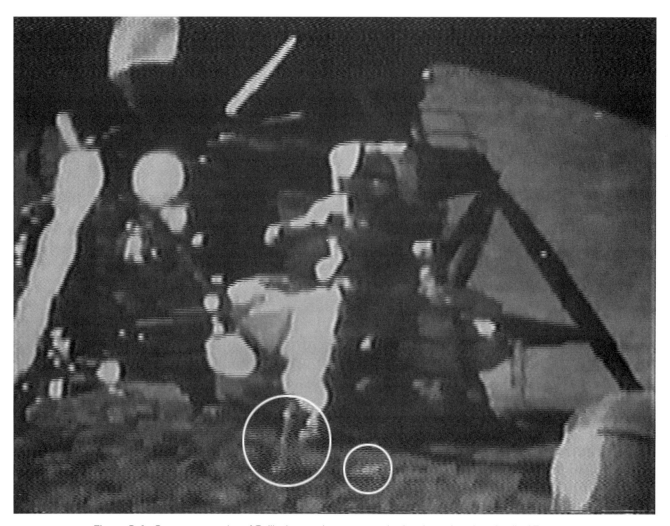

Figure 3.1. Space-age version of Galileo's experiment on gravitational acceleration. Apollo-15 mission commander David R. Scott drops a hammer (left circle) and a falcon's feather (right circle) simultaneously in the airless environment of the moon. With no drag, both the heavy and the light objects hit the lunar ground at the same time, demonstrating that gravitational acceleration is independent of mass. Image from NASA television.

The *gravitational force*, or gravity, is the attraction that all bodies exert on one another by virtue of their mass. The acceleration due to gravity is proportional to the mass of the attracting body. Because all objects have a gravitational force, one might say the attraction is mutual. However, for humans standing on Earth, the gravitational acceleration imparted to them by Earth is much greater than the acceleration they impart to Earth.

A careful reading of the above definition reveals that a cannonball and a feather will be accelerated by Earth's gravity at the same rate. This seems counterintuitive, but our experience is "contaminated" by the effect of atmospheric drag on the less-massive feather. One can minimize the effects of atmospheric drag by using two balls of the same size but of different weights, and dropping them both at the same time, but a far more dramatic demonstration was conducted on the airless Moon in 1971. Apollo-15 commander David Scott, space-suited against the lunar vacuum, dropped a massive rock hammer and a falcon's feather brought from Earth simultaneously; the television camera showed both reaching the ground at the same time (figure 3.1). Note that the gravitational *force*, however, is directly proportional to the mass of the object being accelerated. Thus, the hammer hit the lunar surface with much more force than the falcon's feather, in spite of the fact that they were being accelerated to the same extent by the lunar gravity.

Gravity is a so-called long-range force; it decreases according to the square of the distance between objects. On Earth we do not notice this, because the relevant distance is that to the center of the Earth. By moving up to the highest mountain (Everest), one moves only 0.15%

of Earth's radius above its surface; thus the gravitational attraction of Earth decreases by only 0.3%. However, the force of Earth's gravity at the distance of the Moon, some 380,000 km (or 60 Earth radii) away, is 3,600 times weaker than at Earth's surface.

This inverse-square property of gravity is responsible for the characteristics of the orbits of the planets around the Sun (and of natural satellites, or moons, around the planets). Kepler's laws, which describe the elliptical shape and the property that the planet's path sweeps out "equal area in equal time" along the orbit, are both consequences of this property. Orbital motion is a balance between the gravitational force exerted by the Sun and the force associated with the changing velocity of the planets (and likewise for the Moon's motion about Earth and that of other natural satellites about their parent planets). Artificial satellites are launched into orbits around Earth by imparting to them a velocity sufficient to achieve a similar balance.

Tides arise from a particular effect associated with the distance dependence of the gravitational force and the fact that macroscopic bodies therefore will experience different forces at slightly different points in their interiors. The resulting tidal distortion can lead, under some circumstances, to stresses in the interiors of the planet and its satellite, which produce frictional heating of the interior. In the case of Earth and the Moon, tidal interaction causes the oceans of the Earth to slosh back and forth, which we see as the rising and lowering of the ocean along shorelines during high and low tides, leading to energy dissipation that slows the rotation of Earth and causes the Moon to gradually spiral outward to a larger orbit. (The ocean tides are modulated by the Sun as well, which is more massive than the Moon but much farther away.) The implications of the lengthening day are discussed in part III.

The root cause of the gravitational force is poorly understood. In the context of *general relativity*, essentially a geometric theory of the origin and effect of gravity, the German physicist Albert Einstein visualized space as being distorted around objects, the extent of distortion being dependent upon the mass. Any physical object existing in space will have its path altered, or experience an acceleration, because of the distortion of space. Even the fundamental particles of light – photons – which have no mass, are predicted by this model to have their paths bent by gravity and this has been verified experimentally. Moreover, in relativity theory, time is a fourth dimension in the fabric of space-time: The theory predicts that the passage of time slows in the presence of a gravitational field, a prediction that also has been verified experimentally.

However, such a picture does not actually explain how matter interacts to produce the space-time distortions, and we must turn to a particle viewpoint: Forces (including gravity) are assumed to arise by the mediation of special particles. A mass-bearing (or "massive") particle (neutron, proton, electron, for example) emits a force-carrying particle. The resulting recoil changes the velocity of the emitting massive particle, and the collision of the force particle with a second massive particle causes a velocity change in the latter. The properties of the force-bearing particle and the emitting and absorbing mass particles determine the strength and attractive or repulsive nature of the force. For gravity, the force-bearing elemental particle is called a graviton. This particle, a theoretical construct, has never been observed; however, other types of force-bearing particles have, leading physicists to hope that such a particle can be found for gravity.

The *electromagnetic* force is the force of repulsion or attraction that bodies with net electric charges exert on each other. Unlike the gravitational force, which is entirely attractive, electric charges come in two varieties – positive and negative – hence allowing two directions to the force: Like charges repel; unlike charges attract. As with gravity, however, electromagnetism is a long-range force, decreasing as the square of the distance between bodies. Macroscopic objects, such as people, rocks, Earth, and the Sun, contain essentially equal numbers of positive and negative charges; hence we experience very little electromagnetic force. (Rub a balloon on a piece of fur, however, and a few charged dust particles accumulate on the balloon, allowing it to stick to walls.)

At the atomic level, where individual electrons are involved, the electromagnetic force dominates in chemical reactions (the sharing or exchanging of electrons) to form molecules. Furthermore, the physical properties of liquids and solids are dominated by the effects of the electromagnetic force associated with electronic attraction and repulsion. When we stand upon Earth, gravity pulls us to the center of the planet; we do not fall through the ground because the ground has solidity, and this in turn is due to the electromagnetic bonding of the atoms and molecules in the solid material.

The particle carrying the electromagnetic force is the photon. When a charged particle is accelerated, photons are emitted or absorbed. These photons carry electromagnetic energy through space, in a manner that is akin to waves traveling through a physical medium, such as sound waves through air. Electromagnetic waves, or trains of photons, can range in wavelength over arbitrarily large values. Those in the region of 5×10^{-5} cm, or 0.5 micron, stimulate the human eye and are known as *visible light*. The electromagnetic force includes both electric and magnetic fields. A changing electric field induces a magnetic

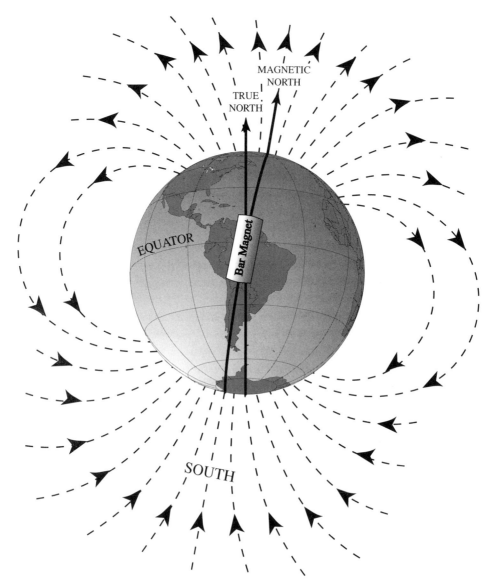

Figure 3.2. Visualization of the lines of magnetic force around Earth, generated by fluid motions deep inside our planet. Also shown is the misalignment between Earth's rotational axis and its magnetic axis. The shapes of the lines are valid close to Earth; farther away, the solar wind pushes the lines of force away from the Earth, creating a more complex magnetic structure. After Press and Siever (1978).

field, and vice versa. Earth and five other planets possess magnetic fields that apparently are generated by the motions of electrically conducting fluids in their interior (figure 3.2). We detect the direction of Earth's field using magnetized iron, in the common device known as a compass. A few elements such as iron and nickel possess the property that they can be magnetized permanently by virtue of the tendency for certain kinds of alignments of their electrons. Such elements are *ferromagnetic*.

The *strong nuclear* force acts to attract protons and neutrons and hence to bind them into a nucleus, overcoming the repulsion between the like-charged protons. It is a short-range force the effect of which increases very sharply (*exponentially*, see figure 3.3) as the distance between particles shrinks, but is negligible beyond about 10^{-13} cm. For this reason, nuclei tend to be less stable with increasing atomic number; some of the heaviest nuclei actually split apart or *fission*. The mediating force particle, called a gluon, is an exotic particle, evidence of which exists only in particle accelerator experiments.

To understand the nature of the strong nuclear force, however, requires delving into the structure of the neutrons and protons themselves. They are not truly elementary particles, but in fact are composites of particles called quarks,

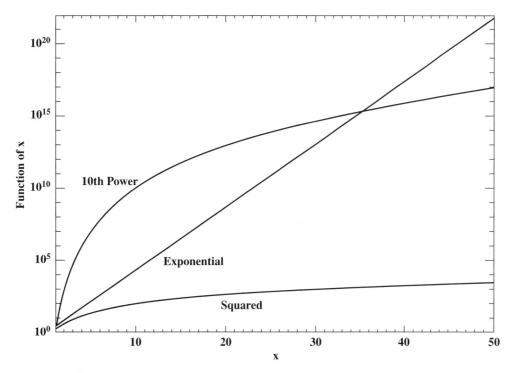

Figure 3.3. Functions that have an exponential dependence on some physical quantity increase (or decrease) much more quickly than those that depend only on some power of that physical quantity. Shown are three functions of a parameter x: x raised to the power 2, or x^2 (labeled "squared" on the figure); x raised to the power 10, or x^{10} (labeled "10th power"); and the "exponential" of x, or 2.7 raised to the power x. The exponential function is encountered again in chapter 5, where radioactive dating is explained, and in chapter 22 in the discussion of climate modeling. The exponential of x will always exceed x raised to any power for large enough values of x. On the vertical axis, values are plotted in scientific notation, and each tick mark represents a factor-of-10 increase from the tick mark below it.

which carry fractional electric charge. Three quarks are required to make up protons and neutrons, of two different types – "up" and "down." Protons are amalgams of two up and one down quark, whereas neutrons are two down and one up. (Knowing that the charges of the quarks must add up to zero for the neutron and plus one for the proton, the reader might have fun deducing the fractional charges on the up and down quarks.)

There are four other types of quarks, predicted by theory, which when compounded produce exotic massive particles normally found in particle accelerators (machines that collide subatomic particles at very high speeds) and extreme conditions in the cosmos; evidence for all six quarks has been found in accelerator experiments. The strong nuclear force, strictly speaking, binds quarks together, and in doing so creates a bound set of protons and neutrons that we call the *atomic nucleus*.

The trade-off between the influence of the strong force and the electromagnetic force provides a rationale for the number of protons and neutrons in naturally occurring stable isotopes. Sticking two or more protons together, in the absence of neutrons, is an inherently unstable exercise because the repulsive electromagnetic force between the like-charged protons overcomes the attractive strong nuclear force. Inserting uncharged neutrons, which add to the attractive strong force, stabilizes the nucleus. As one moves upward in atomic number, larger nuclei formed of more protons and neutrons are less efficiently bound because the volume of the nucleus begins to exceed the effective range of the strong force. Hence a higher proportion of neutrons relative to protons is required to stabilize the nucleus (figure 3.4) with increasing atomic number.

Eventually, beyond element 92 (uranium), the nucleus simply becomes so big that instability cannot be avoided. Heavier elements have been created by smashing nuclei together in nuclear reactors or particle accelerators. These *artificial elements* behave in exactly the same way as the naturally occurring elements; in particular, the electrons continue to systematically occupy higher energy levels (more distant from the nucleus) with increasing atomic number, as described in chapter 2. The artificial elements tend to fission into lighter elements on short timescales.

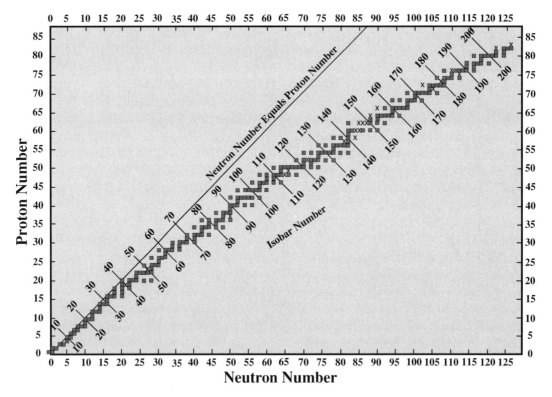

Figure 3.4. Distribution of stable atomic nuclei. The number of protons (atomic number) is plotted against the number of neutrons (atomic mass minus atomic number). For a given number of protons, and hence a given element, there are often several stable isotopes, and in some cases, many. Beyond the lightest elements, stable nuclei tend to have more neutrons than protons. Some isotopes are labeled with an X; these are not strictly stable, but change very slowly over billions of years. The short diagonal lines define nuclei that have the same mass; that is, they have the same *total* number of neutrons and protons. This *isobar* number is important in the discussion of radioactive decay in chapter 4. Redrawn from Broecker (1985).

The distinction between artifical and natural reflects an Earth-centered bias, because some energetic processes elsewhere in the cosmos produce small quantities of the so-called artificial elements.

It is possible that some ultraheavy elements are stable. Somewhat analogous to electrons, neutrons and protons can be visualized as being organized within the nucleus in a series of concentric energy levels. As with the electrons, particular stability is achieved when levels are completely filled. Beyond uranium, the next stable nucleus is at atomic number 114. In 1995, scientists in Germany succeeded in synthesizing elements 110 and 111, and nuclear physicists are now anticipating working up to element 114. How stable this element will be – and hence how long it will last – cannot be predicted reliably by nuclear theory, but will be tested soon by experiment.

The final of the four forces is the so-called *weak nuclear* force. It acts on electrons and more exotic atomic particles that are related to electrons. (These particles including the electron are not made of quarks, but are "elemental" on the same level that the quarks are.) The weak force also acts over a very short range, akin to the strong nuclear force. The weak force is manifest in an atomic nucleus when a neutron converts into a proton and an electron, with the electron leaving the atom. The result is that an unstable isotope of one element is converted into a different element with the same atomic weight as that of the decaying isotope. The mediating force particles are the so-called *massive vector bosons*.

3.3 RADIOACTIVITY

Radioactive decay refers to the spontaneous change of an atomic nucleus through emission of a particle, or splitting of the nucleus. Four types of radioactive decay can occur:

Alpha decay, or α decay, involves emission from the nucleus of two protons and two neutrons as an aggregate. Such an agglomeration, called an α particle for historical reasons, is in fact the nucleus of a helium atom, and is very stable. The original atom is left with a reduction of four in atomic mass, and two in atomic number, and hence is converted into a lighter element.

Beta decay, or *β* decay, involves conversion of a neutron in the nucleus into an electron and a proton. The proton stays behind, and the electron departs from the nucleus. This decay process, mediated by the weak nuclear force, leaves the atomic weight the same but advances the atomic number by one.

Gamma decay, or *γ* decay, does not alter either the atomic weight or the atomic number of the nucleus. A photon is emitted from a nucleus that has been put in an excited state (because of a collision or another decay process), a state in which the configuration of the protons and neutrons is at an elevated energy level. The loss of the photon decreases the energy of the nucleus, but the number of protons and neutrons remains unaltered.

Fission is the splitting of a massive atomic nucleus into two less massive pieces, forming two new elements of lower atomic number and weight than the original decaying element. Spontaneous fission involves release of energy as the nucleus splits. Fusion, the opposite process, involves the combining of lighter nuclei to form a heavier one.

Radioactive decay plays an important role over the history of the solar system in providing heat sources for planetary interiors, and natural chronometers in rocks through which the ages of important planetary and cosmic events can be dated. We discuss these further in chapter 5. We return to fusion in chapter 4 as the primary source of energy coming from stars, including the Sun.

3.4 CONSERVATION OF ENERGY, AND THERMODYNAMICS

A very important and universal concept of physics is *conservation of energy*. Simply put, energy is neither created nor destroyed, but only transferred from one form to another. Energy can be divided roughly into two forms: energy of motion, or *kinetic* energy, and energy stored in some fashion, called *potential energy*. Kinetic energy is straightforward to visualize; a moving car or falling stone both possess it. Kinetic energy is computed readily as half the mass of the object multiplied by the square of its velocity. Thus, when a car doubles its speed, it is quadrupling its kinetic energy; this is why the destructiveness of automobile accidents increases so dramatically with speed.

Potential energy is stored energy. The storage medium might be certain chemical compounds that, under the right conditions, tend to react in such a way as to release heat or exert pressure on a container; gasoline is an example. Storage of energy also can involve placement of material in a field that can induce movement; a weight suspended above the ground possesses potential energy, which becomes kinetic energy if the string is cut. A battery contains potential energy that can be released by creating a circuit of conducting wires connecting the two terminals.

Heat illustrates that energy can be considered as both kinetic and potential, depending on the context. Heat itself is in fact the random motions of molecules or atoms of a substance; each molecule can be thought of microscopically as having some kinetic energy by virtue of its motion. On the other hand, from the macroscopic point of view, a hot substance can be used to drive an engine, when there is a colder reservoir available to provide a direction of heat flow. In this situation the substance can be thought of as possessing potential energy by virtue of its heat content. Most important, though, is the concept that energy can be changed from one form to another and hence made to do useful work. Work itself usually is defined in terms of a change in the state of a system, for example, movement, increase in volume, change in chemical composition. The rate at which energy is expended, for example in doing work, is defined as *power*. The basic metric unit of energy is the joule; a joule per second (power) is a watt.

Energy and matter are interchangeable; the sum of energy and the energy equivalent of matter is conserved. (This is a more general statement of energy conservation than the one given above.) Einstein's famous formula refers to the conversion of matter into an equivalent amount of energy equal to the original mass times the square of the speed of light in vacuum. A hydrogen bomb converts large amounts of mass into energy, by converting four hydrogen atoms into helium; the mass of each helium atom is somewhat less than that of the four hydrogens, and the "missing mass" goes into the energy of the explosion.

One fundamental concept related to heat is *temperature*. Temperature does not directly measure the heat content of a body, which also requires knowing the heat capacity, or how much heat can be stored in some object. For example, a very tenuous gas at high temperature may have less heat content than massive adobe walls at a much lower temperature. In fact, temperature relates to the average kinetic energy, and hence speed, of random collisions between the atoms of an object. Temperature is measured in a number of ways, the mercury thermometer being just one.

Temperature scales are important in this book. The Fahrenheit scale is commonly used but is awkward scientifically because 0°F doesn't correspond to *absolute zero*, or minimal motion in the atoms of an object. The *Celsius* scale, used most places in the world, has the same problem. To get a Fahrenheit temperature, multiply the Celsius temperature by 9/5 and add 32. Thus, water boils at 100°C, or 212°F. A more rational scale, called the *Kelvin* scale, just slides the Celsius scale so that 0 Kelvin is absolute

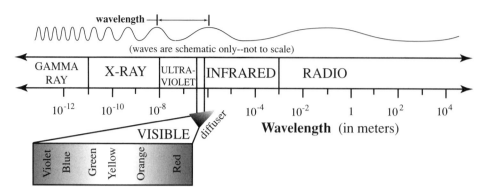

Figure 3.5. Schematic of electromagnetic spectrum, showing the names assigned to the various wavelength ranges. The visible part of the spectrum is expanded below to allow colors to be labeled. Wavelengths in meters are shown using scientific notation.

zero. To do this, just add 273° to the Celsius temperature. Thus water, which boils at 100°C, does so at 373 K (the degree sign is not used in the Kelvin scale). Room temperature is roughly 300 K. The freezing point of water, at 32°F or 0°C, lies at 273 K.

3.5 ELECTROMAGNETIC SPECTRUM

What we perceive with our eyes as light is a particular form of electromagnetic radiation, as are radio waves, x rays, and ultraviolet light. Electromagnetic energy propagates through space in the form of massless particles called photons, which we first introduced as the particles that mediate the electromagnetic force. Because photons are not fully localized in space, they also have the properties of waves of electromagnetic energy, which move through space as alternating electric and magnetic fields. Photons are created when free electric charges (such as electrons) are accelerated, or when bound electrons shift from one energy level in an atom to a lower one. Because of the wavelike properties of photons, electromagnetic radiation can be characterized by its *frequency*, the rate at which the waves pass a definite point, expressed as hertz or waves per second, and *wavelength*, the distance from one crest to another of the wave. The energy content of a photon is just proportional to its frequency.

Electromagnetic radiation travels through a vacuum at a definite velocity, 300,000 kilometers each second (3×10^5 km/s). In a material medium, electromagnetic radiation slows down, by an amount dependent on the frequency of the radiation. Devices called *spectrometers* therefore can be constructed to make the path traveled by light a function of wavelength, such that the various wavelengths are detected within different portions of the spectrometer, and the intensity (or number of photons) at each wavelength can be measured.

Figure 3.5 shows the names assigned to the different portions of the electromagnetic spectrum. It must be emphasized that these names are for convenience, are sometimes historical, but in no case imply a fundamental difference in the nature of photons from one end of the spectrum to the other. Photons at different wavelengths do interact with matter in very different ways, as is apparent throughout the book; nonetheless, the essential nature of photons as massless carriers of electromagnetic energy is the same throughout the spectrum.

Photons, in interacting with matter, can produce characteristic patterns or distributions of electromagnetic energy that reveal the physical and chemical nature of the matter. These *spectra* are most conveniently divided into three types: *continuum, emission,* and *absorption*. Continuum radiation is the broad distribution of photons characteristic of the temperature of a material, which in turn is just a measure of the mean speed of the atoms as they collide with each other. A material that is dense enough, or otherwise has the right properties, to absorb photons effectively and re-emit them will have a precise relationship between its temperature and the number of photons emitted at each wavelength. Such a material, called a *blackbody* if it is a perfect absorber of radiation, will exhibit a definite pattern of emission of radiation, called its *Planck function* after the German physicist Max Planck, who in 1918 received the Nobel Prize for his work on blackbody radiation.

Figure 3.6 shows the pattern, or continuum spectrum, of radiation emitted by objects of different temperature. Light in the middle of the visible part of the spectrum is emitted by objects of temperature roughly 6,000 K, such as the Sun. Objects in a typically heated room, at 300 K, emit in the infrared. Microwaves are the peak of the Planck function for very cool objects (10 K), and gamma rays require objects in the million-degree range.

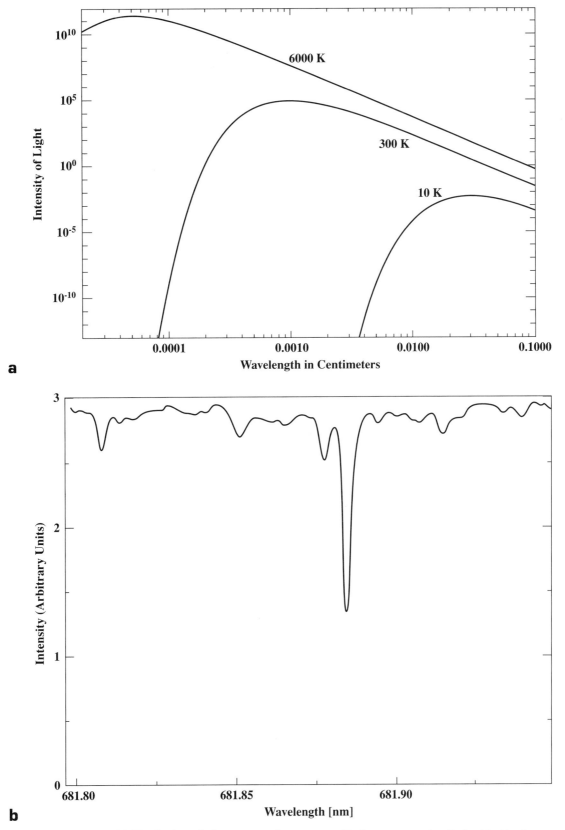

Figure 3.6. (a) Distribution of electromagnetic energy emitted versus wavelength for blackbodies of temperatures 10 K, 300 K, and 6,000 K. The intensity of the photon radiation is expressed as energy emitted (in units of joules) every second at each wavelength, per square meter, per solid angle. The reader can appreciate, without worrying about the units, the fact that cooler objects radiate at longer wavelengths than warmer objects, and emit far less energy as well. (b) Example of a spectrum from the molecule methane, in wavelengths corresponding to red light, as measured in the laser spectroscopy laboratory of G. Atkinson at the University of Arizona. The wavelengths in this panel are given in nanometers (nm), or billionths of a meter (1 nm is 10 angstroms).

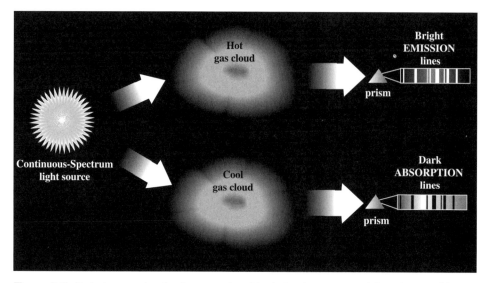

Figure 3.7. Emission spectra (top) are produced by hot substances emitting photons. Absorption spectra (bottom) occur when a flow of photons is intercepted, and absorbed at discrete wavelengths, by a cooler material. After Snow (1991).

Note that only very hot objects create photons in the visible part of the spectrum, where we see with our eyes. Common objects around us, including ourselves, are visible because such photons bounce off, or are reflected from, such objects and then are available to stimulate the retinas of our eyes. *Reflection* is simply the redirection of existing photons coming from a luminous source. We see colors different from that of the source of illumination because physical materials selectively absorb photons at certain wavelengths and hence reflect only a subset of those incident upon them. The textures that we see in objects relate to how photons are reflected or scattered in different directions. Without the photons emitted from luminous sources such as the Sun, we would have no photons available by which to sense material objects around us. Moreover, as we discuss in later chapters, these photons provide the energy for warming Earth's surface and atmosphere, and the energy for most of the living systems existing on Earth.

Most materials are not perfect blackbodies, and even those that are have this property only over limited wavelength ranges. Because of the electronic structure in atoms and molecules, with fairly well-defined energy levels between which electrons can move, electromagnetic energy tends to be absorbed or emitted at fairly definite wavelengths. If an electron drops from a higher energy level to a lower one, a photon is created, that is, emitted from the atom or molecule, at a wavelength or frequency characteristic of the difference between the two energy levels. A discrete, bright line will be seen at that wavelength if the light from the object is passed through a spectrograph.

The lines, or spectrum, emitted by an object that is undergoing electronic changes in energy level can be as intricate as fingerprints on a human hand, and just as diagnostic of the composition of that material (i.e., of what elements it is made); spectra taken in the visible and ultraviolet parts of the spectrum reveal such lines. Furthermore, in molecules, shifts of electrons from one major energy level into another are divided further into shifts between sublevels, associated with how the atoms in the molecule are vibrating relative to each other, and how the molecule is rotating. (Recall that rotation is an acceleration, because the object is not moving in a straight line; hence photons may be released or absorbed during rotation.) Spectra taken from the near infrared through to the microwave can reveal the identity of molecules through these complex transitions (figure 3.6).

In very-high-temperature objects, such as the interiors of stars, collisions are large enough (and photons energetic enough) that atoms are partly or completely stripped of electrons. These *ions* then are positively charged particles, and the soup of positively charged ions and negatively charged electrons is a plasma. Ionization occurs near the surfaces of hotter stars, and emission spectra from such ions is characteristic of how many electrons each atom has lost.

Complementary to emission spectra are absorption spectra, where photons are absorbed by atoms or molecules with associated increases in the energy levels of the electrons, or increases in the vibrational or rotational energy of the molecule. Such absorption spectra are, for our purposes, just the complement of emission spectra. Whether one finds absorption or emission spectra depends on whether a hot material is emitting photons into a colder region (*emission* spectra) or a colder object is absorbing photons from a warmer environment (*absorption* spectra) (figure 3.7). In understanding Earth and the other planets, both kinds of spectra are important. What is key here is that the pattern of emission or absorption lines, even from a material composed of many kinds of molecules, can be deciphered to determine what molecules are present. Additionally, absorption or emission lines are more prominent when there are more atoms or molecules to absorb or emit photons at characteristic wavelengths; hence spectra provide information on the amount of atoms or molecules in a certain material.

3.6 ABUNDANCES IN THE SUN

One of the most striking first accomplishments of modern spectroscopy was the identification of a previously unknown element in the Sun. Beginning with the German scientist Josef Fraunhofer in the early nineteenth century, scientists had mapped dark absorption lines coming from the bright surface, or *photosphere*, of the Sun, and identified most of these lines with elements known at the time by measuring their spectra in the laboratory. A prominent line escaped identification until 1895 when the British chemist William Ramsey isolated a gas from a uranium-bearing mineral *cleveite*. Heating the gas in the laboratory produced the previously unidentified line in the solar spectrum, and the newly discovered element was named *helium* after the Greek word for the Sun (Helios).

Since then, a thorough identification of abundances of elements in the Sun has been pursued, in part for its significance with regard to solar system objects. If the Sun and the planets had a common origin (chapter 11), then, because the Sun has retained essentially all of its gas from that time, it should reflect the original, primordial mix of material from which the planets formed. There are difficulties in deriving a complete inventory from spectroscopy,

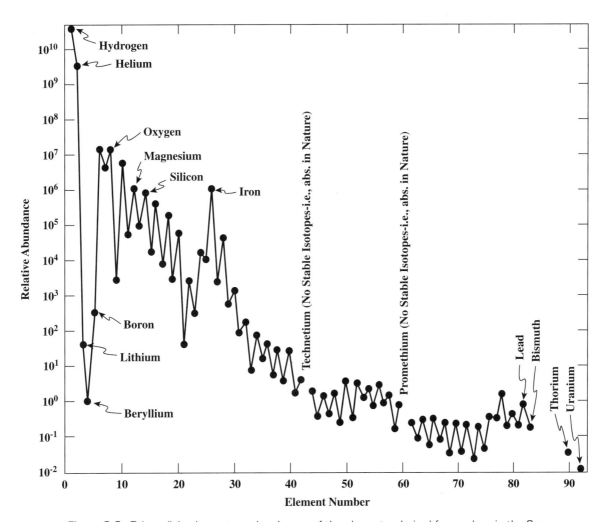

Figure 3.8. Primordial solar system abundances of the elements, derived from values in the Sun and primitive meteorites (see chapter 5). Abundances are plotted on a *logarithmic* scale, so that each tick mark on the vertical axis means an increase or decrease by a factor of 10. Reproduced by permission from Broecker (1985).

however. One only sees the surface of the Sun, and to assume that the interior abundances are equal to those on the surface requires the notion that the Sun is thoroughly mixed. This is not entirely true, and corrections must be made. In addition, the nuclear reactions that power the Sun (chapter 4) convert some elements to others, and corrections must be made for this also in deriving an inventory of "original" planetary material.

Abundances of elements in the Sun are summarized in figure 3.8. Most are derived from spectra of the Sun's surface, but additional information is folded in. This includes direct sampling by spacecraft of the *solar wind* (a stream of charged atoms emanating from the Sun) and chemical analysis of a class of meteorites (rocks that originally were in orbit around the Sun and collided with the Earth) that are thought to be relatively unaltered since the birth of the solar system (chapter 5) and hence have some elements in their original relative abundances.

The resulting graph is a guide to answering some important questions about how the planets, including Earth, have evolved over time, and such problems are discussed in later chapters. Perhaps more profound, however, is that the pattern of elemental abundances reveals something about *how* these elements were formed in the first place. The decline in abundance toward higher proton number, the peak near iron, and then a further decline, and the zigzag nature of the abundances for odd and even proton numbers reflect the superposition of a number of natural nuclear reactions that have taken place since the birth of the universe as a whole, and to which we turn in chapter 4.

3.7 QUESTIONS

a. What are the various conversions from one form of energy to another that take place as an automobile engine is started and then engaged to propel the automobile along a road?

b. What is the original source of energy for the gasoline that is used to power the automobile? (Hint: Check chapter 23.)

3.8 READINGS

Arons, A.B. 1990. *A Guide to Introductory Physics Teaching*. John Wiley and Sons, New York.

Boorstein, D.J. 1983. *The Discoverers*. Vintage Books, New York.

Broecker, W.S. 1985. *How to Build a Habitable Planet*. Eldigio Press, Palisades, NY.

Cookson, C. 1995. A smashing new atom. *Financial Times London*, March 25–26, 1995.

Hawking, S.W. 1988. *A Brief History of Time*. Bantam, New York.

Mason, S.F. 1991. *Chemical Evolution*. Clarendon Press, Oxford.

Pippard, A.B. 1957. *The Elements of Classical Thermodynamics*. Cambridge University Press, Cambridge, UK.

Poppy, W.J., and Wilson, L.L. 1965. *Exploring the Physical Sciences*. Prentice-Hall, Englewood Cliffs, NJ.

Press, F., and Siever, R. 1978. *Earth*. W. H. Freeman, San Francisco.

Snow, T.P. 1991. *The Dynamic Universe: An Introduction to Astronomy*, West Publishing, St. Paul, MN.

4 FUSION, FISSION, SUNLIGHT, AND ELEMENT FORMATION

4.1 INTRODUCTION

The understanding of the origin of sunlight (and starlight in general) was a nineteenth and early twentieth century development that culminated in the release of nuclear energy in human-made devices on Earth. Beyond the implications (both negative and positive) of such developments, however, lies the profound perspective gained in the latter half of the twentieth century regarding the origin of the elements of the periodic table. The existence and abundances of the 90-odd elements that make up Earth, the planets, the solar system, and the universe beyond have an explanation that lies in natural nuclear reactions that have taken place in the several generations of stars preceding the formation of the Sun and the solar system.

4.2 STARS AND NUCLEAR FUSION

The observable cosmos around us is, by and large, made of stars. Stars are spheres primarily of hydrogen and helium gas in balance between the attractive force of gravity pulling everything inward and the pressure forces associated with the high temperatures of stars' interiors, tending to push the material outward. Most stars eventually evolve, through nuclear processes described below, into dense spheres of carbon, oxygen, or exotic neutrons; some collapse into the mysterious and incredibly dense *black holes*.

The copious amounts of photons coming out of stars, including the Sun, are a signature of the enormous temperatures in their interiors. The origin of these high temperatures, and hence of sunlight or starlight, was a matter of debate throughout the nineteenth century. A hypothesis by the British physicist Lord Kelvin, that the Sun was radiating away the energy associated with its initial collapse from clouds of interstellar gas and dust, met with a timescale problem: The Sun would cool in several tens of millions of years, but various lines of evidence suggested that terrestrial rocks were older by at least a factor of 10. However, the essential and simple concept that the infall of material by gravity toward a common center, forming a star or planet, would generate heat is essential to understanding the heat budget of Earth, as we discuss in chapter 11. Another possible source, radioactivity of heavy elements, was advanced around the same time, but the spectroscopic determination that the Sun is mostly nonradioactive hydrogen and helium made this hypothesis also untenable.

By the 1930s, physicists began to grasp the essential workings of the atomic nucleus, including the fact that with sufficient force, one could overcome the repulsive barriers between the nuclei of atoms and induce lighter nuclei to combine to form heavy nuclei, in a process called fusion. In the case of four hydrogen nuclei (each of which is just a single proton) combining together, the most stable resulting nucleus requires that two of the protons transform to neutrons. This is accomplished only through a modestly complex series of steps, outlined below, but the important point is that the resulting nucleus *has less mass than the original four protons*. The missing mass ΔM has been converted to energy ΔE, according to the Einstein formula $\Delta E = \Delta M c^2$.

Analogous to electrons, certain numbers of protons and neutrons assembled as nuclei represent especially stable structures. In general, the stability of the nonradioactive nuclei increases as the atomic number increases toward iron; beyond iron, the stability tends to decrease. Therefore, fusion reactions tend to produce energy as heavier nuclei are assembled, only up to iron (figure 4.1). Nonetheless, this does not mean that it is easy to fuse two nuclei together; sufficient pressure (or collisional force, and hence temperature) is required to overcome first electronic

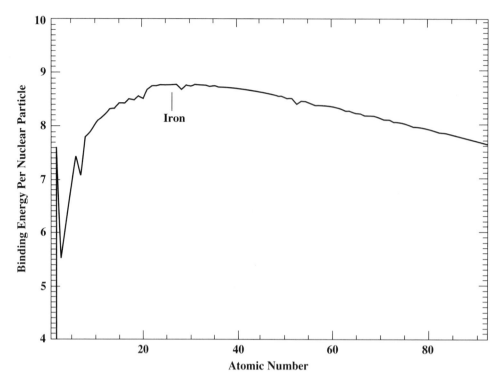

Figure 4.1. Binding energy of the nucleus as a function of atomic number. The higher the energy, the more stable is the nucleus against fragmentation or other decay. Note that stability is highest around the atomic number corresponding to iron. The binding energy is expressed relative to the number of protons and neutrons in the nucleus, and the units are millions of *electron volts*. One electron volt is 1.6×10^{-19} joules, and is a convenient unit for energies on the small scale of atoms. The curve was computed from a model that roughly fits the measured value for most elements, but there are small deviations from the experimental values.

repulsion and then repulsion associated with the two colliding nuclei.

Production of heavier elements from lighter ones by fusion in stars appears to be a process of fundamental importance to the evolution of the cosmos and in particular to the existence of solid planets. It is therefore worth getting a flavor for the kinds of reactions that take place. We focus on the fusion of hydrogen to heavier elements. The simplest and most basic fusion process is called the *p-p chain*, or proton-proton chain, and requires that only hydrogen and helium be present.

The simplest of the p-p chains, often called ppI, involves three separate *reactions*, as sketched in figure 4.2. A reaction is defined as a discrete step in the process in which one or more atomic nuclei are fused to form certain products. In ppI, step 1 involves two hydrogen nuclei (protons) colliding to form a deuterium nucleus (one proton and one neutron) and two atomic fragments. One such fragment is identical to an electron in mass, but of opposite charge, and is called a *positron*. Also released is an exotic particle with little or no mass and a propensity for passing easily through matter. Such *neutrinos* have been detected experimentally.

In step 2, the deuterium nucleus collides with another hydrogen nucleus, i.e., proton. The net result is the release of light, or photons, and the generates of the two nuclei into a light isotope of helium, consisting of two protons and one neutron. This isotope, helium-3, or ^{3}He, is quite rare in the cosmos, because it is destroyed easily by further fusion reactions. However, some of it escapes from the Sun in the solar wind, and has been detected.

Finally, two helium-3 nuclei collide and form the most stable isotope of helium, helium-4, ^{4}He, consisting of two protons and two neutrons. This nucleus stays intact under present conditions in the Sun's interior but undergoes further fusion in more massive stars. In addition to the helium-4, two protons (i.e., the hydrogen nuclei) are produced.

If we take all of the reacting nuclei and the product nuclei from the three stages of the ppI chain, we see that, in net form, four protons have been converted into one helium nucleus, which consists of two protons and

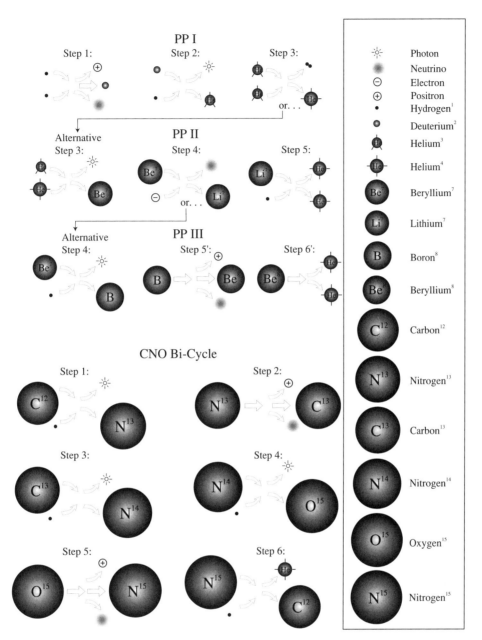

Figure 4.2. Steps involved in four kinds of fusion reactions in stars, all of which convert hydrogen to helium.

two neutrons. A net loss of mass has occurred, and that lost mass (a small fraction of the total) is liberated as energy, mostly as photons, and generates the light we see coming from the stars.

The ppI chain is not the only proton-proton fusion reaction chain that occurs in stars. Indeed, it is possible for helium-3 to collide with a helium-4 nucleus to form beryllium, and from there lithium (ppII chain) and, in a fraction of reactions, boron (ppIII chain), but in the end these heavier elements are destroyed in favor of helium-4 again.

Beryllium and lithium act as *catalysts* in the nuclear reaction; they control the speed of the reaction sequence, in this case by being good targets for electrons and protons, without being consumed in the process. We see examples of catalysts in *chemical* reactions (that is, involving whole atoms as opposed to just the bare nuclei) in chapter 13 but the principle is the same as with the nuclear reactions discussed here.

The energy liberated per helium nucleus produced is 4×10^{-12} joules, enough to power a 40-watt (40-joule per second) lightbulb for only 10^{-13} seconds. However, the Sun contains enough hydrogen to produce 10^{56} helium atoms; if all of the hydrogen were so converted, the amount of energy released would be 4×10^{44} joules. The Sun is observed to emit photon energy, its main form of removal of the energy, at a rate of 4×10^{26} watts; therefore, the pp-chain could sustain this process for 10^{17} seconds, or 30 billion years. This calculation is a bit off because (a) much of the hydrogen is too far from the Sun's center to experience high enough temperatures to undergo fusion and (b) the Sun's brightness (luminosity, or power) has varied with time. More careful calculations yield roughly 12 billion years of steady hydrogen fusion for the Sun, of which 4.5 billion years has already transpired.

The p-p chain is only one of two cycles converting hydrogen to helium in stars. The carbon-nitrogen-oxygen, or *CNO cycle*, requires that the three heavier elements, so familiar to us on Earth, be present in the region of nuclear burning. In this cycle, carbon acts as a catalyst to facilitate, through the intermediate formation and destruction of nitrogen and oxygen, the creation of the helium-4 nucleus from four protons. The sequence is potentially much faster than the p-p chain because, in the latter chain, two hydrogen nuclei (protons) must collide to initiate the process, and this is inefficient because of the small size of the protons. In the CNO cycle, all collisions are between protons and larger nuclei such as the carbon nucleus (six protons and six neutrons). However, there is so little carbon in the center of the Sun that the CNO cycle is currently less important than the p-p cycle. As fusion proceeds in the Sun and helium builds up, the interior temperature of the Sun will increase; as the temperature increases, the CNO cycle will gain in importance and eventually dominate.

Fusion requires very high temperatures to provide nuclei with enough velocity to overcome the repulsive electric force between the protons. In stars, the high temperatures are achieved through the enormous pressure associated with the mass of the star: Our Sun is 1,000 times more massive than Jupiter and 300,000 times more massive than Earth. Most stars, however, are smaller than the Sun, and in the interiors of the smallest stars, or *red dwarfs*, nuclear reactions barely proceed, and are much slower than in the Sun. These stars are cooler, and hence appear red rather than yellow, but they are much longer-lived because they burn hydrogen more slowly. Stars more massive than the Sun undergo hydrogen fusion much more rapidly, are much brighter and bluer than the Sun, but are far shorter-lived. During the time over which stars undergo hydrogen fusion, their brightness and size change only slowly; this stable portion of their evolution is referred to as the stellar *main sequence*.

The lifetime and luminosity of main sequence stars, sorted according to their mass, have important implications for the habitability of orbiting planets and the chance that life will have enough time to evolve into complex forms before these stars become unstable. Figure 4.3 shows the time for stable hydrogen fusion in stars as a function of their mass; this is the main sequence lifetime. Stars several times more massive than our Sun do not last long enough to give complex life a foothold on any planets orbiting them, if the timing of evolution of life on our planet is a fair guide (chapter 12).

Like normal hydrogen, deuterium is being depleted today by fusion processes in stars. In fact, deuterium can undergo fusion at lower temperatures than can hydrogen. The reaction is simple: It is the second step of the ppI chain in which a deuterium nucleus and hydrogen nucleus collide to form a helium-3 nucleus with liberation of energy in the form of a photon. The reaction of two protons, the first step of the ppI chain, requires much higher collision velocities and hence limits hydrogen fusion to objects more massive than those that just undergo deuterium burning. The threshold for hydrogen fusion in stars of solar composition is 85 times the mass of Jupiter; for deuterium burning it is only 13 times Jupiter's mass. Because of the very small abundance of deuterium in the cosmos – 50 parts per million relative to hydrogen – deuterium fusion can only power a star for a few million years at most, compared to the billions of years that stars such as the Sun shine by hydrogen fusion.

4.3 ELEMENT PRODUCTION IN THE BIG BANG

Hydrogen fusion produces helium, which builds up as a kind of thermonuclear "ash" in the interiors of main sequence stars. Stellar explosions, which we discuss below, can deliver helium to the gas in interstellar space, the *interstellar medium*. This production of helium from hydrogen is just one example of *stellar nucleosynthesis*, or the production of elements within stars. It is a somewhat special case, however, because, unlike most of the elements, much of the helium present today in the cosmos is thought to be primordial, like hydrogen. The origin of the primordial material is presumed to lie in the initial explosion that started the universe, that is, the Big Bang.

Evidence for an initial explosion of matter to create the cosmos exists primarily in the observed expansion of groups of galaxies away from each other, and in the pervasive background static, mentioned in chapter 2, which can be heard at radio wavelengths. This background static is produced over a range of wavelengths, and the energy

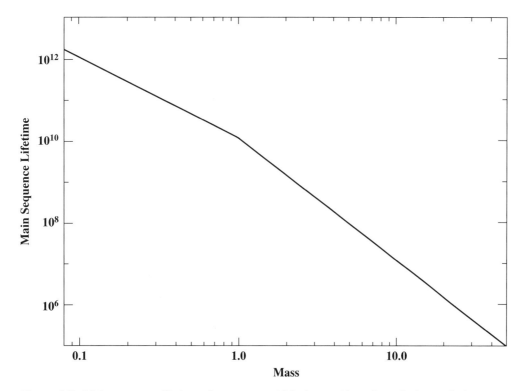

Figure 4.3. Main sequence lifetime of stars over which they stably undergo hydrogen fusion, as a function of mass of the star expressed in solar masses. The Sun's mass on this scale is 1. The lifetime is given in years. Note that the lifetime and mass must be expressed in powers of 10 because of the broad range of stellar masses and ages.

is distributed so as to be a nearly perfect blackbody with a temperature of 2.7 K. The most straightforward interpretation of this cosmic static is that it is the last light from the initial explosion, red-shifted by its great distance from us, marking a transition from a universe that in its first moments was suffused with photon radiation scattering off of a dense gas of subatomic matter.

During the initial phases of the expansion after the Big Bang, the universe consisted mostly of neutrons compressed to an extremely high density, much like the present-day interiors of *neutron stars*, the remnants of stellar collapse. Very quickly, however, β decay created a population of electrons and hydrogen nuclei, that is, protons. Helium was formed in this dense soup of matter through capture of a neutron by hydrogen to form its heavy isotope, deuterium, followed by collision between two deuteriums to make tritium (the next heavy, and unstable, isotope of hydrogen) plus hydrogen, and terminating with a collision between tritium and deuterium to make helium-4. A branch-off in the step involving two colliding deuterium nuclei produces helium-3, some of which survives today as a primordial remnant.

The Big Bang production of helium was a different process than the p-p chain in stars, emphasizing that different ambient conditions (in this case, the much higher densities in the Big Bang than are obtained in stars) force the nuclear reaction sequence to be different. In addition to most of the present-day helium coming from Big Bang nucleosynthesis, it is thought that most lithium available today was made at that time.

4.4 ELEMENT PRODUCTION DURING NUCLEAR FUSION IN STARS

Fusion reactions beyond hydrogen burning in stars require increasingly higher temperatures because the nuclei, up through iron, are progressively more stable (figure 4.1). Consider the Sun as a typical star of intermediate mass and age. The temperature near the Sun's center is computed from physical models to be roughly 1.5×10^7 K, fully adequate for hydrogen fusion, but a temperature of 10^8 K is required to initiate the next stage in which helium fusion takes place. However, as more hydrogen is consumed, leaving a core of helium, the density increases and the temperature rises. Computer models suggest that temperatures near the center of the Sun are already 10% higher than they were at the time hydrogen fusion was initiated. The continued slow increase in the

Sun's internal temperature leads to increasing luminosity, which will have profound consequences for Earth's habitability well before the end of our star's main sequence lifetime.

As the Sun approaches the end of its main sequence life some 7 billion years hence, hydrogen fusion will become progressively concentrated in a shell around the then-helium core. This core no longer will be supported by the heat from fusion reactions, and will begin to contract rapidly, heating up to the threshold temperature for helium fusion. Two helium nuclei, that is, α particles, each composed of two neutrons and two protons, will collide to form ^{8}Be. This is an unstable isotope of beryllium, with a large cross section, and hence will easily capture another helium nucleus, producing the most abundant carbon isotope, carbon-12 (^{12}C).

Although the energy production of helium fusion is small compared with that of hydrogen fusion, the sudden pulse of ignition will force the Sun to expand dramatically and become a *red giant*, which will extend out beyond the orbit of Venus, almost to Earth. The luminosity of the Sun will increase sufficiently to bake the Earth and melt the water ice of Jupiter's Galilean moons.

Helium burning produces heavier elements by capture of additional α particles, each succeeding element having a mass number 4 higher than the previous element. Thus, α-particle capture by carbon produces oxygen-16; capture by oxygen-16 produces neon-20, and magnesium-24 then is produced from neon. In addition to this rather simple production by helium nuclei, which explains the high abundance of elements with mass numbers divisible by 4, other reactions are going on. The CNO cycle during hydrogen fusion has produced other elements such as nitrogen-14. Addition of α particles leads to heavy isotopes of oxygen (^{18}O), neon (^{22}Ne), and magnesium (^{25}Mg). Other products of the CNO cycle are converted to ^{13}C, ^{15}N, and ^{17}O.

Further cycles of nucleosynthesis occur as helium is exhausted and the star again goes through core collapse and resulting heating. At a temperature of about 1 billion Kelvin, carbon fusion is initiated. Two carbons combine, releasing a helium nucleus and producing ^{20}Ne or releasing a proton and yielding heavy neon (^{23}Ne). Isotopes of magnesium, sodium, aluminum, and silicon are produced in the carbon fusion process as well. Oxygen nuclei also undergo fusion to produce isotopes of silicon, phosphorus, sulfur, magnesium, and aluminum.

Finally, temperatures exceeding 4 billion K permit silicon fusion to take place, such that iron and elements nearby in the periodic table are produced. This is the end of element production by fusion in stars because, beyond iron, the nuclei decrease in stability; fusion thus requires energy input and is not self-sustaining. Only the more massive stars make it all the way to silicon burning; the Sun and smaller stars will only progress through helium fusion, after which final collapse will produce a small *white dwarf star*. For stars massive enough to produce iron by silicon fusion (nine or more times more massive than the Sun), the end is more dramatic: The termination of silicon fusion and core collapse produce a violent *supernova* explosion.

The fusion process accounts for the existence and abundances of only some of the elements and their isotopes in nature, and in particular cannot produce directly elements heavier than iron. To understand production of the rest we must look more closely at what happens in the interior of stars, as well as in the violent processes of stellar expansion and explosions.

4.5 PRODUCTION OF OTHER ELEMENTS IN STARS: *S*, *R*, AND *P* PROCESSES

The deep interiors of stars undergoing fusion reactions are dense fluids of protons, neutrons, electrons, and heavier nuclei (composites of protons and neutrons). In addition to the direct fusion reactions considered above, the capture of protons and neutrons by nuclei can build up the atomic mass (and in the case of proton capture, the atomic number) in ways distinct from the main fusion reaction sequences. Neutron capture is much more likely than proton capture, because electrostatic repulsion does not have to be overcome. Sources of free neutrons become important in the helium burning stage of a star's life. The production of ^{25}Mg from ^{22}Ne and an α particle, and the conversion of ^{13}C and an α particle to ^{16}O both liberate neutrons and are thought to be their primary sources.

The process of neutron capture in a stable stellar interior is the *s*, or slow, process. It is so defined because the flux of neutrons is such that the time between successive captures of neutrons by a nucleus may range from 10 to 10^5 years. An understanding of neutron capture is greatly aided by the sort of diagram shown in figure 4.4. The graph shows the various elements and their isotopes, collectively called the *nuclides*, plotted as the number of neutrons on the horizontal axis versus the number of protons, defined earlier as the atomic number, on the vertical axis. Thus, a horizontal movement on the graph is from one isotope to another of the same element; a vertical movement is from one element to another. The atomic mass of a species is given by the sum of the neutron number and the proton number. *Isobars*, or species of the same atomic weight, lie on a diagonal line from lower right to upper left on the chart.

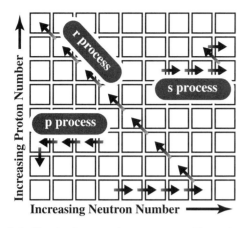

Figure 4.4. Graph of s, r, and p processes. Elements and isotopes exist in squares defined by a proton number (vertical axis) and a neutron number (horizontal axis). Straight horizontal arrows to the right are neutron captures; diagonal arrows represent β decays. Because of its complex nature, the p process cannot be shown directly but is equivalent to the horizontal line moving to the left accompanied by a downward vertical step. Adapted from Broecker (1985).

Capture of a free neutron moves an isotope horizontally along the graph, converting it to a heavier isotope of the same element. Eventually, the isotope reaches a neutron number which is not stable, and decays radioactively. Because of the long time between neutron captures in the s process, such an unstable nucleus will undergo radioactive decay before the next capture, and the relevant radioactive process is β decay, in which a neutron converts to a proton and an electron, but the atomic mass of the nucleus (number of protons and neutrons) is preserved. On the graph, such an event moves the isotope diagonally up and to the left, along the isobaric line. β decay continues until a stable nucleus is reached, and then continued neutron capture moves the isotope, now a different element, horizontally to the right again.

The resulting abundances of elements and isotopes are determined both by the neutron flux and the relative cross sections of the various nuclei created. As mentioned earlier, the stability of nuclei depends separately on the numbers of both neutrons and protons. Certain of these numbers, as with electrons, are particularly stable whereas others are not. This is in addition to the unstable situation of having too many neutrons relative to the number of protons, leading to β decay. Very stable nuclei have small cross sections for capturing neutrons and hence tend not to be converted to heavier isotopes or isobars. The limited rate of neutron addition relative to β decay forces the pattern of diagonal movement along an isobar as soon as an unstable isotope is reached. Thus, although the s process is important in making many elements and isotopes above iron, it cannot produce the more neutron-rich isotopes.

The question of which stellar environments are most important contributors of s-process elements is a continuing debate. Presumably, the s process goes on in all stars undergoing fusion beyond the hydrogen stage, but we are interested in stars from which material eventually is expelled in sufficient quantities that it is an important contributor to the interstellar medium and, eventually, to new generations of stars and planets. *Asymptotic Giant Branch* (AGB) stars swell in the late stages of nuclear burning and consist of a core of carbon and oxygen that is not undergoing fusion, surrounded by a shell undergoing helium fusion and a final, outer hydrogen layer. These stars appear to be abundant and hence are important sites for s-element production.

Not all heavy elements can be made by the s process. Some neutron-rich isotopes require that neutron capture proceed quite far to the right, through the unstable isotopes, before β decay takes over. Rapid addition of neutrons, or an *r process*, is required. Here, capture of neutrons is rapid enough that very neutron-rich nuclei are produced, until the binding of additional neutrons becomes so unfavorable that the net capture rate is no longer competitive with β decay, and a cascade of β decays moves the neutron-heavy elements diagonally to the left in figure 4.4 until a stable nuclide is reached.

Once one understands the stability of the various nuclides, charting their production by the s and r processes becomes a kind of board game in which the pieces are moved according to rules determined by nuclide stability, neutron fluxes, and the ambient physical conditions, elucidated through laboratory experiments and computer models. But what environment could be so neutron-rich as to enable the r process to occur? Stars several times more massive than the Sun that have completed fusion cycles up through production of the iron-group elements explode as supernovas. Various neutron-rich environments within supernovas have been invoked as possible sites for production of elements by the r process, but none seems capable of producing the full mix of r-process elements seen in the galaxy. The problem is an intricate one because not only must conditions be right for r-process element production, but the material then must be ejected into interstellar space without being further altered significantly.

A promising site for the r process has been identified by Clemson University astrophysicist Bradford Meyer and colleagues (1994) in the winds coming from neutron stars. After the explosion of a star as a supernova, the remnant cinder collapses with no further prospect of fusion reactions to halt the collapse. If the star is massive enough, collapse will continue "forever" and a black hole will be formed. Most supernova remnants, however, stop collapsing when the pressures are high enough that all electrons and protons are squeezed together to make neutrons. This incredibly dense neutron star is only a few kilometers across, yet it contains potentially as much mass as the Sun. For the first 10 seconds or so of its existence, an intense wind of neutrons flows from the neutron star, and it is in this exotic cosmic breeze that many or most of the r-process nuclides might be produced. The rate of neutron star births is thought to be high enough to make these winds a primary source. The reader should regard this model not as the last word, but as an illustration that the search for the birth sites of the elements is tied closely to an understanding of the exotic processes by which stars evolve and die.

Some nuclides in figure 4.4 are relatively proton-rich and are shielded from s- and r-process production by other stable nuclides. Some 35 nuclides out of the hundreds of stable and near-stable nuclides known to exist are in this state. For some time it was thought that a p process to produce such material must involve addition of protons. This is difficult because high temperatures are required to produce sufficiently energetic collisions for protons to overcome the electrostatic repulsion of other protons. Appropriate environments for proton addition within stars were difficult to find.

An alternative mechanism that enriches protons in a nucleus is removal of neutrons, as sketched in figure 4.4. To make the p-process nuclides, the removal would have to occur from stable nuclides, ones for which β decay will not operate. Exposing nuclides to very high temperatures for short periods of time is one possibility, because the neutrons will "drip" off of the nuclides first, followed by protons; if the process is truncated early enough, the net result is relatively proton-rich nuclides. Certain regions of the interiors of supernovas have been identified as providing the right environment for the p process, in which the supernova shock itself provides a short high-temperature burst. At least two different kinds of supernovas appear to be required to produce the right mix of the p-process nuclides, and it is clear that much more work will be required to fully understand how these are formed.

4.6 NONSTELLAR ELEMENT PRODUCTION

Once expelled into interstellar space by supernova explosions or the more quiescent ejection of envelopes around lower-mass stars, element production and evolution are not terminated. Most nuclei that are ejected from supernovas have initial velocities an appreciable fraction of the speed of light. Nuclei that intersect our solar system and hit Earth are called high-energy *cosmic rays*. Collisions between ambient interstellar hydrogen and the high-energy nuclei cause spallation or splintering of portions of the heavy nuclei. This l *process* is a primary one in the production of lithium, beryllium, and boron.

Additionally, once produced, isotopes not fully stable begin to radioactively decay, which is another kind of element and isotope production process. Decay times range over large values, from seconds through billions of years. As described in chapter 5, the abundance of decay products of some of these isotopes, trapped in rocks, provides a wealth of information ranging from the age of the solar system to the timing of geologic events on Earth.

4.7 ELEMENT PRODUCTION AND LIFE

Figure 4.5 summarizes the mechanisms that produce the elements occurring in nature. The extent to which it is possible to understand the sources of elements and their isotopes is remarkable, given that only a century ago scientists were still struggling with the concept of the nature of elements and the underlying structure of the atom. Today we have a glimpse of the wide range of processes – from the Big Bang through stellar fusion and supernova explosions – responsible for the mix of elements present today in the cosmos.

It is particularly intriguing to examine the elemental abundances and notice that the fundamental building blocks of life – carbon, hydrogen, nitrogen, and oxygen – are quite abundant relative to most other elements. Except for hydrogen, which is the primordial element, these others are abundant because they are direct products of the fusion reactions powering stars.

The high abundances of silicon and iron-group elements have planetary implications. Silicon is the last of the source materials for main fusion reactions, the products being iron and elements close to it. These elements of moderate atomic weight are the basic building blocks, with oxygen, of Earth and its sister terrestrial planets; the compounds of such elements are loosely referred to as rocks and metals.

Go out into the dark skies of a moonless country night and gaze at the multitude of stars. Let your eyes run from the seven sisters of the Pleiades to the red giant Betelgeuse in the constellation Orion. In this visual sweep, one captures the alpha and omega of element production: young stars just beginning their conversion of hydrogen to helium

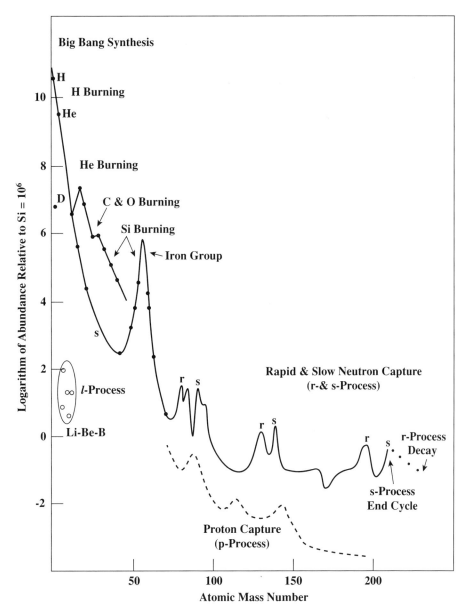

Figure 4.5. Summary of mechanisms by which elements are produced, plotted as the logarithm of the element's abundance versus the element's atomic weight. On the vertical scale $2 = 10^2$, $4 = 10^4$, etc. The mechanisms themselves are discussed in this chapter. Modified from Mason (1991) by permission of Clarendon Press.

by fusion, and the red giant going through its terminal stages of fusion before the frenetic final neutron production of heavy elements. There in the sky are the cosmic factories making the elements that, in the distant future, might become part of some strange biology on an as yet unformed world.

4.8 QUESTIONS

a. Given the story of element production described in this chapter, would you expect life to have been possible during the very first generation of stars after the Big Bang?

b. Why might one not expect to encounter intelligent life on a planet orbiting a star twice the mass of the Sun?

4.9 READINGS

4.9.1 General Reading

Broecker, W. 1985. *How to Build a Habitable Planet*. Eldigio Press, New York.

4.9.2 References

Aldridge, B.G. 1990. The natural logarithm. *Quantum* **1**(2), 26–29.

Cloud, P. 1988. *Oasis in Space: Earth History from the Beginning*. W.W. Norton, New York.

Clayton, D.D. 1968. *Principles of Stellar Evolution and Nucleosynthesis*. McGraw-Hill, New York.

Mason, S.F. 1991. *Chemical Evolution*. Clarendon Press, Oxford.

Meyer, B. 1994. The r-, s- and p-processes in nucleosynthesis. *Annual Review of Astronomy and Astrophysics* **32**, 153–190.

Sackman, J., Sackman, I-J., Bootnroyo, A.I., and Kraemer, K.E. 1993. Our Sun III. Present and Future. *Astrophys. J.* **418**, 457–468.

Wilford, J.N. 1992. Scientists report profound insight on how time began. *New York Times*, April 24 **CXLI**(48, 946), p. 1.

Wilson, T.L., and Reid, R.T. 1994. Abundances in the interstellar medium. *Annual Review of Astronomy and Astrophysics* **32**, 191–226.

II THE MEASURABLE PLANET: TOOLS TO DISCERN THE HISTORY OF EARTH AND THE PLANETS

5 | DETERMINATION OF COSMIC AND TERRESTRIAL AGES

5.1 OVERVIEW OF AGE DATING

To understand the history of Earth in the cosmos, we must be able to establish ages of physical evidence and timescales over which processes have occurred. The task is daunting because of the enormous spans of time over which the physical universe and Earth have existed, and several different approaches must be used. In chapter 2, we discussed observations leading to the conclusion that the universe is in an overall state of expansion, which began some 10 billion to 20 billion years ago. In this chapter we discuss rather precise techniques that enable us to determine the age of Earth and other solid matter in the solar system very confidently: Some 4.56 billion years ago, the planet we live on began to take shape.

It is useful to distinguish between two kinds of chronologies that are constructed in regard to Earth's history, because the techniques and uncertainties are quite different. A *relative chronology* is derived by observing the relative position in which the remains of an event lie. In sediments on Earth, older layers of soil, sand, and rock are deposited first, and then overlain by subsequent layers. Geologic processes might turn a whole stack of layers upside down, but fossils present in the layers, which can be compared to those in other layers worldwide, enable us to determine the age progression of the layers. We discuss relative geologic dating in chapter 8.

Similar relative records of events can be read from the surfaces of planets; on the Moon we find evidence, discussed in chapter 7, of an early period of frequent impacts on the surface to form craters, followed by extensive volcanic flooding to make the lunar mare. On Mars, dried-up river channels are seen to be overlain by impact craters in some places, but cut through pre-existing craters in others. Such photographic evidence allows a relative chronology of events to be constructed. In cases in which the average rate of physical processes can be estimated, relative ages can be assigned rough absolute values; however, this is an approach fraught with potential error.

Absolute chronologies, our main concern here, contain information on the actual times at which events took place. To construct such chronologies requires a natural and well-calibrated clock, with markers indicating when the "ticking" began. On a macroscopic level, biological growth effects such as rings in trees or seasonal events such as thawing of lake water provide such clocks; we encounter these much later. Certain microscopic processes, atomic or nuclear, have the simplicity and predictability required to act as very precise clocks over enormous time spans. Radioactive decay in particular provides both the regularity and the markers required for such measurements, and as we see below, there is a broad range of radioactive nuclides characterized by varying longevity that occur in natural materials. Scientists have applied these to problems ranging from the age of ancient settlements (carbon-14 dating) to the time when element formation first began in our galaxy (using long-lived uranium and thorium isotopes, among others).

5.2 THE CONCEPT OF HALF-LIFE

To understand how radioactive isotopes, introduced in chapter 3, can be used to date the materials within which they are found, we must delve into a little physics and mathematics. We have talked in chapter 2 about quantum mechanics and the consequent *probabilistic* nature of atomic processes. Imagine a single atom that is radioactive. Although one might know, from its particular identity as a radioactive isotope of a given element, whether it is likely to decay sooner rather than later, it

Table 5.1 Half-Lives of Important Radioactive Elements

Parent	Daughter	Half-Life (Year)
^{235}U	^{207}Pb	0.70 billion
^{238}U	^{206}Pb	4.5 billion
^{232}Th	^{208}Pb	14.0 billion
^{40}K	^{40}Ar	12.0 billion
^{40}K	^{40}Ca	1.4 billion
^{87}Rb	^{87}Sr	49.0 billion
^{147}Sm	^{143}Nd	106.0 billion
^{26}Al	^{26}Mg	700,000
^{14}C	^{14}N	5,730

is not possible to predict how long before it will decay, even approximately. This would seem to contradict the notion that radioactivity provides a precise clock for dating events.

However, because decay is a probabilistic event, precision is achieved through considering an ensemble of a large number of atoms of the same isotopic species at once. This is easy to do, because macroscopic materials contain enormous numbers of atoms. The rate of decay of a large number of atoms of a given radioactive species can be measured quite precisely. One way to express this rate is in terms of the *half-life*, which is simply the time it will take half of a sample of radioactive atoms to decay when the number of atoms is very large. The half-life is a measurable and dependable characteristic of a particular radioactive isotope (table 5.1).

Another important aspect of the radioactive decay process is that the number of decays in a given time is just proportional to the number of radioactive atoms present. This makes sense because, for a decay to take place, there must be radioactive atoms present, and the more present, the more decays that are likely to take place in a given amount of time. In fact, over a very short time interval δt, where δ indicates a discrete change in a quantity, a simple algebraic equation describes the change δN in the number N of radioactive atoms of a particular element present:

$$\delta N = -NR\delta t.$$

Here we use R to represent the rate at which the radioactive decay occurs. R is the reciprocal of the mean lifetime of the radioactive atoms, which is 1.4 times the half-life. Note also that the three quantities on the right-hand side – the number of atoms, the rate, and the time interval – are to be multiplied together. As is common in scientific writing, we do not put multiplication signs (×) between the symbols. The minus sign is needed to indicate that the decay process decreases the number of atoms over time.

The equation only tells us the change in isotope number over time. If we want to know what the number N is as time passes, for example, over 10 days, we could add the increments δN over the time increments δt that total 10 days. Because N changes at each time step, we must make our steps sufficiently small that we accurately track N. There is a mathematical procedure, called *integration*, that we can perform to find N:

$$N(t) = N(0)e^{-Rt}.$$

In this equation we introduce two new types of symbols. $N(t)$ and $N(0)$ are simply shorthand for the number of radioactive atoms at times that we label t and 0, respectively. Our time 0 is arbitrary; depending on the situation that we are calculating, $t = 0$ might be last Wednesday at 9 A.M. or it might be the moment that earth began. But, it makes sense that, to compute the number of radioactive atoms at time t, we must know what that number is at some earlier time.

More peculiar is the symbol e. It is shorthand for exponential, and it is a special function that produces a unique number for every value of $-Rt$. The value of e^1, or just e, is $2.71828\ldots$, with the ellipses indicating that the number is not exact as written. Then, e^2 is e multiplied by e, or $7.38904\ldots$. For negative numbers, we just take reciprocals: e^{-2} is $1/e^2$, or $0.135335\ldots$. Things get a bit hairy when we have fractional powers for the exponent, that is, $e^{-3.45}$ for example, but this represents a number also, which can be worked out on a scientific calculator, computer, or using printed tables. An exponential curve, e^x, which rises very steeply as x increases, is displayed in figure 3.3.

Why do we end up with this exponential function describing radioactive decay? It is because the number of atoms decaying is proportional to the number of radioactive atoms present. If this were not the case – if, for example, the number decayed δN per time interval δt were just proportional to R – then the decay law would be simple: $N(t) = N(0) - Rt$. However, many physical processes – radioactive decay, growth of bacteria (because each bacterium present splits into two), initial growth of a fertilized egg, etc. – operate in such a way that the change in a quantity depends on how much of that thing is available. Such processes are referred to as *exponential* in their growth, or *inverse exponential* if there is a negative sign in the power, as in radioactive decay.

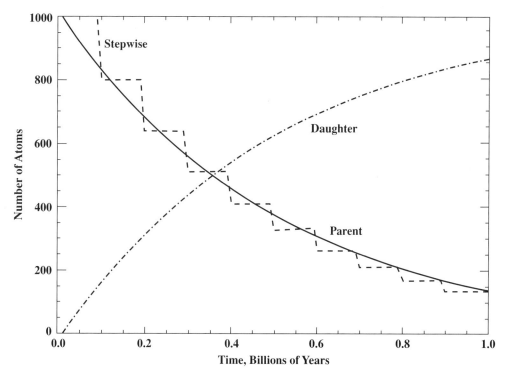

Figure 5.1. Radioactive decay law. A sample of 1,000 atoms (a very small number compared to real samples analyzed) is assumed, which have, for example, a half-life of 355 million years. The numbers of parent atoms remaining, and daughter atoms produced, are shown as a function of time. The half-life can be read from the graph as the time corresponding to the crossing of the parent and daughter curves. The dashed line is the number of parent atoms remaining based on simply adding the increments δn; this stairstep pattern only roughly approximates the real decay law.

Figure 5.1 shows the progressive, inverse exponential decline of radioactive atoms during a decay process. It also shows when the half-life is reached – after half of the initial atoms have decayed. The radioactive decay law is fundamental to what follows in this chapter, though it is by no means the whole story, as we shall see. Nonetheless, the predictability of the decay of an ensemble of a large number of atoms is at the crux of the use of this process as a clock.

In the remainder of the chapter we consider two different approaches to dating materials by radioactivity, distinguished by what actually can be measured in the system. The first of these is *radiocarbon dating*, limited because of the short half-life of radioactive carbon-14 to organic remains of living things that died less than about 70,000 years ago. The second technique is applied to radioactive isotopes with much longer half-lives, such that both the amount of the original radioactive isotope, hereinafter the *parent*, and the product, hereinafter the *daughter*, species can be measured or inferred. This technique is used in the dating of terrestrial and extraterrestrial rocks many millions or billions of years old.

5.3 CARBON-14 DATING

The stable isotope carbon-12 (^{12}C) is one of the more abundant atoms in the cosmos, and a foundation for biology on Earth. Carbon-13 (^{13}C) also is present as a stable isotope in all natural carbon-bearing systems, but at much lower abundance. The next heavier isotope, carbon-14 (^{14}C) is continually produced in Earth's atmosphere as the most abundant nitrogen isotope, ^{14}N, absorbs neutrons produced from an influx of atomic fragments – the *cosmic rays* from energetic sources in the galaxy. The absorption of the neutron leads to ejection of another neutron or a proton, but primarily the latter. When the proton is ejected, the atomic number decreases by one but the mass stays at 14, and hence ^{14}C is manufactured (figure 5.2).

As table 5.1 indicates, ^{14}C decays with a half-life of approximately 5,730 years. The production of ^{14}C by neutron bombardment and its decay lead to a roughly constant, but small, abundance in the atmosphere. Because it is virtually chemically identical to ^{12}C (the higher mass creating only small differences), ^{14}C combines with oxygen to make heavy carbon dioxide, ^{14}C^{16}O$_2$, and then finds its way into plants through photosynthesis, and thence

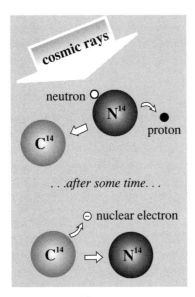

Figure 5.2. Production of ^{14}C from nitrogen and cosmic rays, and its decay. After Cloud (1988, p. 84).

through the food chain to the rest of the biological world. Living organisms continually exchange their carbon with the atmosphere via photosynthesis or respiration and food consumption, and live a short time, except for some very ancient trees, compared with the half-life of ^{14}C. Thus, in the majority of living things, the ratio of ^{14}C to ^{12}C is a constant.

When an organism dies, exchange stops or, actually slows, because bacterial and nonbiological processes still move materials in and out of the dead organisms, but at very low rates. The carbon-14 within the dead organism decreases over time according to the radioactive decay law. Biological materials that are less than roughly 70,000 years old have enough remaining ^{14}C that the electrons resulting from the decay can be directly counted in the laboratory, thus sensitively measuring the amount of ^{14}C remaining relative to the total carbon (masses 12, 13, and 14) in the sample. By comparing this number with the amount of ^{14}C relative to total carbon in the biosphere, and knowing the decay rate, the age is determined. The daughter, nitrogen-14 (^{14}N), is of no help, because it is identical to the rest of the ^{14}N that dominates our atmosphere and hence carries no signature of a radioactive origin.

In chapter 21 carbon-14 dating figures prominently in the construction of a chronology of climate change in the latter part of the last Ice Age and the more recent, postglacial, period. It also has been a critical tool in dating abandoned structures built, for example, in the arid southwestern United States by cultures now long gone, the wooden beams and remains of fire pits being of key importance. What, however, are the uncertainties associated with the technique?

In any scientific measurement, errors crop up associated both with the act of measurement and with the assumptions behind the interpretation. Measurement errors are familiar to anyone who has had to measure and build something. In using a ruler to determine what length of a beam to cut, for example, one might measure several times, or cut several beams to the same length. Repeated measurements or cuts reveal *random* errors, caused by small changes in positioning the ruler or the cutting tool. These errors generally can be estimated with some reliability, based on the sensitivity of the measurement and other factors. *Systematic* errors can be more insidious: In our example, perhaps the ruler is faulty, our carpenter has astigmatic eyesight, or the cutting tool is out of alignment in some fashion. These errors can be difficult to detect but can ruin measurements. Experimental work in science is successful only so far as the errors can be reliably estimated and controlled.

In the interpretation of carbon-14 measurements, an obvious and crucial assumption has to do with the amount and rate of carbon-14 produced in the atmosphere over time. The manufacture of carbon-14 depends on the cosmic-ray flux, and this is known to vary as the strength of Earth's magnetic field changes, and as the Sun varies in its level of activity. Carbon-14 also may vary with changes in ocean circulation, which brings up varying amounts of carbon dioxide stored in deep water, or with other climatic or geologic events. Uncertainties in ages determined by carbon-14 dating must fold in these possible variations.

Cross-correlation, where possible, with independent dating techniques, for example, tree rings (chapter 21), is essential for calibration: It reveals that the ^{14}C level may have differed from recent values prior to about 3,500 years ago, necessitating a revision in some earlier dates from ^{14}C data. Tree-ring studies cannot go back over the tens of thousands of years accessible to carbon-14 dating, however, and so the earliest dates have larger uncertainties. Attempts to account for variations in the Sun, Earth's magnetic field, and ocean circulation lead to the calibration shown in figure 5.3. Dates obtained with carbon-14 are younger than the actual age of the sample, and this discrepancy increases with age. At 20,000 years before present (B.P.), the discrepancy is 3,000 years; it is worse for more distant ages. At present, industrial and military activities are increasing the production rate of ^{14}C in the atmosphere, so that the archaeologists of the future will need to correct for the increased amount of the isotope in organisms living at this time.

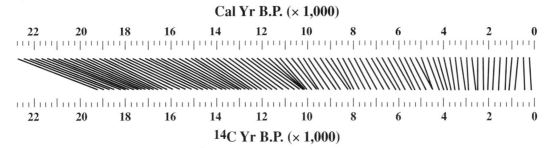

Figure 5.3. Relationship between actual age of a sample (cal yr before present; B.P.) and the age from carbon-14 dating (^{14}C yr B.P.). Both axes are in units of thousands of years, and so, the figure goes back to 22,000 years ago. From Bartlein et al. (1995) by permission of Academic Press.

5.4 MEASUREMENT OF PARENTS AND DAUGHTERS: RUBIDIUM-STRONTIUM

The plausibility of dating very ancient events, such as the formation of the Earth and planets, by *radiogenic* (produced by radioactive decay) nuclides lies in the fact that these atoms are produced in stars in calculable amounts, expelled through supernova explosions, and decay in a regular fashion after formation. As elements, including the radioactive ones, became trapped in the solid material around our newly forming solar system, the initial abundances were modified through radioactive decay. The chemical affinity that particular elements have for certain rock phases provides the means for determining how much radioactive isotope was originally incorporated in the rock, and then the age since trapping via measurement of the present abundance of the radioactive isotope.

The main difficulty that confronts radioactive dating in which the decay process cannot be detected directly is the ambiguity involved in knowing the actual initial abundances of the radioactive species and the decay products of the species. If 0.1 gram of the parent is present now in a sample, was there 1 gram when the rock formed? 2 grams? 10 grams? We could simply measure the amount of daughter product in the rock, but not all of the daughter product came from the decay of the parent. There might have been some daughter atoms present in the rock when it was formed. There is no guarantee that the rock was formed without any initial daughter atoms, and so, the only way to find the initial amount of daughter element is to find another isotope of the same element, not formed by radioactive decay, that acts as a chemical tracer of the daughter element.

Rubidium-87 (^{87}Rb) decays to strontium-87 (^{87}Sr) with a half-life of 49 billion years. There is a stable isotope of strontium, ^{86}Sr, formed directly in supernovas. Because isotopes of the same element behave nearly the same chemically, ^{86}Sr and ^{87}Sr would have tended to be trapped together in the same portions of the grains that form the rock under analysis. The ^{87}Rb would be locked elsewhere in the grains and, as it decayed, would have produced a local region of the rock enriched in ^{87}Sr.

Mathematically, the measured ratio of ^{87}Sr to ^{86}Sr in a particular grain equals the initial ratio at grain formation, plus the measured ratio of ^{87}Rb to ^{86}Sr multiplied by the number of mean-lifetimes that have elapsed. (This expression is approximate – the measured ratio of ^{87}Rb to ^{86}Sr is actually multiplied by ($e^{Rt} - 1$). However, for the long half-life of ^{87}Rb our approximate expression suffices.) If we could find a part of our rock that contains no rubidium, that is, has only strontium, we could determine the initial strontium abundance and hence the age of the rock.

In real rocks, there is a range of strontium and rubidium abundances in different grains, and so, we can't isolate pure strontium. However, we can make a plot of the abundance of ^{87}Sr in each grain versus the abundance of ^{87}Rb, all relative to the ^{86}Sr abundance. Ideally, all grains in a rock of a given age should form a straight line on such a graph (figure 5.4).

Where the straight line crosses (intercepts) the vertical axis is the initial amount of ^{87}Sr in the rock. This is the amount of ^{87}Sr that formed before the rock grains condensed from the interstellar gas. The age since the rock formed is just given by the slope of the line: The slope is the age divided by the mean lifetime of ^{87}Rb. Since the mean-lifetime is measured in the lab, we can read the age in years from the slope of the plotted line. The older the rock, the less ^{87}Rb relative to ^{87}Sr there is in the sample today, and the steeper the slope of the line. The younger the rock, the more rubidium there is relative to strontium, and the less steep the slope.

Suppose that, after the rock formed, there was heating or chemical contamination that added some extra ^{87}Sr, leached out some ^{87}Sr, or did similar sorts of things to the ^{87}Rb. The affected grains would be skewed away from the ideal, straight line that represents a well-determined, single

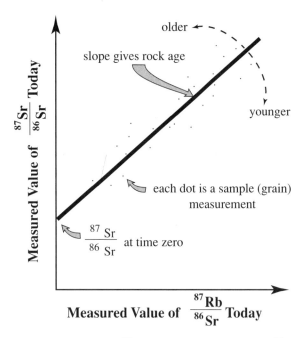

Figure 5.4. Amount of ^{87}Sr relative to the amount of ^{87}Rb at present, both shown as ratios to the ^{86}Sr abundance. Labels on the figure show how various quantities are determined.

age since formation for the sample grains. Such a rock would not be suitable for dating, and this is immediately evident from the graph. The isochron technique therefore is self-correcting: A rock that has been severely altered will not have grains whose ^{87}Sr to ^{87}Rb analysis falls on a straight line (figure 5.5).

But are not all rocks altered to some extent? To determine the age of the solar system, we cannot use rocks that are now part of a planet – they likely have been melted at least once during and after the planet came together. Instead, the most primitive-appearing, least-altered meteorites are chosen to determine the age of the solar system. The meteorites with the most primitive composition that we can find, which have an elemental composition as close to the Sun as possible, are the *carbonaceous chondrites*.

The results of an actual analysis on a suite of meteorites are shown in figure 5.6. A number of isotopic parent-daughter systems are used to determine the age of the meteorites and solar system bodies; the pioneering work yielding essentially the currently accepted age was done by Claire Patterson of the California Institute of Technology in the 1950s, using the uranium-lead system. The ages of the most primitive meteorites center rather precisely on 4.56 billion years old. The oldest Moon rock brought back by the Apollo astronauts was found to be 4.4 billion to 4.5 billion years old using samarium-neodymium and lead isotopic systems. On Earth, one cannot find rocks older than 4.0 billion years, suggesting that no rocks in the upper parts of Earth escaped heavy alteration early in our planet's history. However, events earlier in Earth's history, such as formation of the core, can be determined using isotopic dating as described in chapter 11.

So, the oldest solid samples in our solar system cluster around 4.5 billion to 4.6 billion years old, and therefore this is the age of the solar system; that is, 4.5 billion years or so have elapsed from the time solids began to form what ultimately would be planets to the present.

Corroborating this determination are models of the structure of the Sun as a function of age. Models and observations of other stars show that a star expands slowly with time as it fuses hydrogen farther and farther from its central core. The radius and luminosity of the Sun correspond to the value expected at an age of roughly 4.5 billion years.

5.5 CAVEAT EMPTOR

This chapter has tried to explain simply and logically how scientists date geologic events and timescales using isotopes of elements that decay at measurable rates. These rates are determined in the laboratory and both theory and laboratory evidence point firmly to the concept that, even for half-lives of billions of years, the rates are constant and a function only of which particular isotope we are considering. Nonetheless, these techniques must be done carefully and require checks and cross-checks among different isotopic systems and painstaking error analysis.

The age determinations are not easy, but the reproducibility that has been achieved from sample to sample and system to system makes the determinations of cosmic

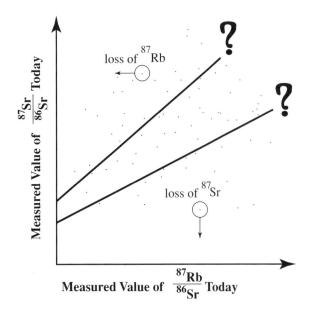

Figure 5.5. Possible effects on the age curve of chemical alteration of rock samples, in which either rubidium or strontium might be lost.

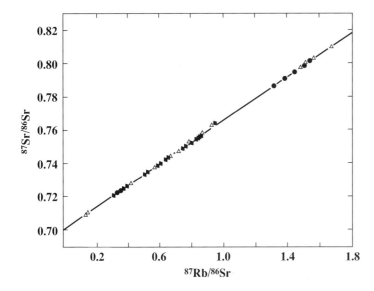

Figure 5.6. Determination of rubidium and strontium abundances for a number of meteorites whose appearance and chemical composition suggest that they have not been altered in planets. The data points fall very tightly on a curve that yields an age of 4.56 billion years. Modified from Minster et al. (1982) by permission of Macmillan Magazines Limited.

ages highly convincing. Particularly firm is the age of the solar system, and hence the beginning of the formation of Earth and its planetary neighbors, as 4.56 billion years ago. Radioactive isotopes have been used to date the time since element formation began in stars. This age, 12 billion to 16 billion years, is much less certain but still a useful direct constraint on an important event in the early history of the universe. Isotopic systems are also the foundation for dating geologic events on Earth itself, forming the backbone of the history described in part III.

5.6 QUESTIONS

a. What possible sources of contamination might a geochemist have to guard against while radioisotopically dating rock samples?
b. What arguments would you make to justify the assertion that the radioactive decay rates of the elements have not been altered over the age of the universe?

5.7 READINGS

5.7.1 General Reading

Broecker, W. 1985. *How to Build a Habitable Planet.* Eldigio Press, New York.

5.7.2 References

Allègre, C.J., Manhès, G., and Göpel, C. 1995. The age of the Earth. *Geochimica et Cosmochimica. Acta* 59, 1445–1456.

Bartlein, P.J., Edwards, M.E., Shafer, S.L., and Barker, E.D., Jr. 1995. Calibration of radiocarbon ages and the interpretation of paleoenvironmental records. *Quaternary Research* 44, 425–433.

Cameron, A.G.W. 1993. Nucleosynthesis and star formation. In *Protostars and Planets III*, (E.H. Levy and J.I. Lunine, eds.). University of Arizona Press, Tucson, pp. 47–73.

Cloud, P. 1988. *Oasis in Space: Earth History from the Beginning.* W.W. Norton, New York.

Minster, J.F., Birck, J-L., and Allègre, C.J. 1982. Absolute ages of formation of chondrules studied by the ^{87}Rb-^{87}Sr method. *Nature* 300, 414–419.

Patterson, C. 1956. Age of meteorites and the Earth. *Geochim, Cosmochim. Acta* 10, 230–237.

Radioactivity and other dating techniques. In *Van Nostrand's Scientific Encyclopedia* (D.M. Considine, ed.). Van Nostrand Reinhold, New York, 1983, pp. 2387–2389.

Swindle, T.D. 1993. Extinct radionuclides and evolutionary timescales. In *Protostars and Planets III* (E.H. Levy and J.I. Lunine, eds.) University of Arizona Press, Tucson, pp. 867–881.

6 OTHER USES OF ISOTOPES FOR EARTH HISTORY

6.1 INTRODUCTION

In addition to the dating of rocks by measuring amounts of radioactive isotopes and their decay products, isotopes can be useful as indicators of climate variations on Earth over its long history. Here, the key is to use stable isotopes of the same element. The difference in mass between the isotopes leads to separation, called fractionation, of the isotopes in natural systems; the separations in some cases are a function of the climate, specifically temperature.

To use isotopes as climate indicators, four key features are required:

1. availability of stable isotopes of the same element whose separation depends on temperature;
2. incorporation of the fractionated isotope mixture in some storage medium that is preserved for a long time;
3. ability to measure accurately the ratio of the various isotopes;
4. a means to date, in an absolute or a relative sense, the age of the stored isotope data.

6.2 STABLE ISOTOPES, SEAFLOOR SEDIMENTS, AND CLIMATE

6.2.1 Carbon

Three important elements for tracking climate changes are carbon, oxygen, and hydrogen. Consider the carbon first. Carbon has two stable isotopes, ^{13}C and ^{12}C. Recall that ^{14}C is radioactive and used for dating relatively recent events. Certain biological processes distinguish mass differences in isotopes. We cannot survive on deuterated water (HDO or $^1H^2HO$). Likewise, plants are observed to preferentially take up ^{12}C in carbon dioxide (CO_2), and hence preferentially enrich the atmosphere in ^{13}C. The more temperate the climate, the more land area that is available for plants, and the more ^{12}C that is taken up. In ice ages, global plant activity is reduced, and so, less ^{12}C is taken up.

How is the excess or deficit of ^{13}C in the atmosphere recorded? Certain single-celled sea organisms produce plates, called *coccoliths*, that serve as protection. These coccoliths are composed of calcite, $CaCO_3$, where Ca is the element calcium. The carbon in the calcite comes from the carbon dioxide in the atmosphere, and the organisms are observed to take up ^{13}C and ^{12}C with equal propensity. Therefore, the calcite shells or coccoliths of tiny sea organisms record the atmospheric ratio of ^{13}C to ^{12}C. When the organisms die, the shells fall to the ocean floor, where they are buried over time. The calcite shells are a biologically produced mineral that can remain intact on the seafloor for enormous lengths of time.

Ancient seafloor sediments often have been uplifted in more recent mountain chains through geologic processes. As the ancient sediments, now hardened as rock, are raised and then exposed to view by erosion, the calcite contained within can be analyzed for the ^{13}C to ^{12}C ratio. The higher the ^{13}C amount, the warmer the climate was at the time the shell-forming creature lived. If the sediments can be dated, by means of radioactive isotopes or by relative techniques involving fossils (chapter 8), the temperature as a function of time thus is determined through this proxy measure.

6.2.2 Oxygen

The oxygen bound up as water in the oceans also produces a record of Earth's surface temperatures. Water on Earth is a mixture of the stable isotopes $H_2^{16}O$, $H_2^{17}O$, and $H_2^{18}O$, with the ^{16}O being by far the more abundant. From laboratory measurements, $H_2^{16}O$ is known to be preferentially evaporated from the oceans to form

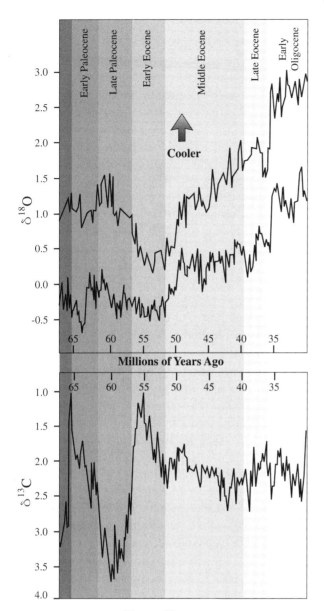

Figure 6.1. Records of ^{18}O and ^{13}C enhancements from fossils of ocean creatures in seafloor sediments, ranging in age from 65 million to 30 million years ago. The horizontal axis is time before present; the corresponding geological epochs (chapter 8) are marked on the top. In the oxygen graph, the upper curve corresponds to bottom water and the lower curve to the near-surface part of the ocean. The carbon determinations are for shallow water only. Larger amounts of ^{18}O and smaller amounts of ^{13}C correspond to colder climate conditions, as indicated by the direction of the arrow on the graph. However, the carbon isotopic record in particular is strongly affected by processes other than climate changes. Redrawn from McGowran (1990).

clouds. During cold periods on earth – Ice Ages – the polar caps were built up and spread outward in the form of ice sheets. The sources of the ice sheets are storm systems that dump large amounts of moisture on high-latitude continents in the form of snow. Therefore, water actually was lost from the oceans during ice ages and stored as ice in great continental ice sheets. Because it is $H_2{}^{16}O$ that is preferentially evaporated from oceans, during Ice Ages the oceans should be enriched in ^{18}O. The same kind of enrichment occurs for ^{17}O, but its mass difference from ^{16}O is half that of ^{18}O, and so, climate effects are smaller. For that reason, we focus only on the $^{18}O/^{16}O$ record because it is read more easily.

Again, a storage medium for the oxygen is required and, as before, tiny sea creatures play the role. To make the calcite of coccoliths requires oxygen as well as carbon ($CaCO_3$). As with carbon, the oxygen is taken up without regard for the isotope number, ^{18}O and ^{16}O equally. The oxygen source is water in the ocean. Hence these microscopic shells record the ratio of ^{18}O to ^{16}O in the ocean at the time of their formation. Unlike the carbon, the measurement for oxygen is available from organisms at both the ocean's surface and its depths, because oxygen extracted by deep-sea shellmakers will record the local $^{18}O/^{16}O$ ratio in the ocean water.

As the organisms die, the calcite shells are deposited on the ocean floor, buried by progressive sedimentation and become part of the rock record. Ancient ocean sediments exposed by geologic events allow the calcitic oxygen to be extracted, the $^{18}O/^{16}O$ ratio measured, and the global temperature determined. To put these data into a climate chronological sequence, the sediments or surrounding rock layers then must be dated.

Figure 6.1 shows an example of ocean temperatures for the time period from 65 million to 30 million years ago, which covers the demise of the dinosaurs through the golden age of mammals (see chapter 19). Remember that the ^{18}O signature is available for both bottom water and near-surface water by selecting the mineralized remnants of benthic foraminifera and planktonic coccoliths, respectively. In the case of the carbon isotopic record, only material from photosynthesizing plankton is available for this time. The reader should judge the quality of the correspondence between the surface oxygen and carbon records of the temperature. Because the two isotopes are affected by different oceanic and biological processes, they disagree on the pattern of warm versus cold times. In particular, the carbon record is affected by geologic processes to a significant extent and provides a poorer indicator of climate than does the oxygen record.

6.2.3 Hydrogen

About 150 parts per million (ppm) of water on Earth contains heavy hydrogen, or deuterium, primarily paired with a light hydrogen to make HDO (as opposed to normal water, H_2O). As with the heavier isotopes of oxygen, deuterated water tends to be preferentially left behind during evaporation of ocean water near the equator. Thus air masses moving away from the equator and hence toward colder latitudes are slightly enriched in normal water; that is, they possess somewhat less than the 150 ppm of HDO that is typical for the ocean. As rain and snow form in the air mass, the deuterated water is preferentially and progressively removed from the storm system in the precipitation, so that storms near the poles drop snow that is significantly depleted in deuterium. The depletion is exaggerated during colder climate episodes relative to warmer, because the drop in temperature from equator to pole is larger during colder times. (This is checked by mapping the distribution of plant species during warm times versus ice ages.) Also, the tendency of deuterium to fall out in the rain and snow is exaggerated at lower temperatures (figure 6.2).

The resulting *deuterium fractionation* has been used to study the most recent epochs of glacial climate and warm episodes in between, by sampling the ice sheets that cover Antarctica, Greenland, and other very cold places. The record in such ice sheets extends back less than 300,000 years in Earth history, but it is much more detailed than that in the more ancient seafloor sediment record of carbon and oxygen isotopes. Colder times are characterized by more exaggerated deuterium fractionation and hence greater deuterium depletion in the ice laid down at that time. Warmer episodes show less deuterium depletion. Because the ice also contains the oxygen isotopes discussed above, the deuterium and oxygen-18 depletions can be compared to help build confidence in the paleotemperatures (ancient temperatures) derived from the core. Chapter 21 discusses the application of stable isotopes to understanding the nature of interglacial warm periods such as that in which we now live.

6.3 A POSSIBLE TEMPERATURE HISTORY OF EARTH FROM CHERTS

Oxygen isotopic exchange potentially provides a temperature indicator back through 80% of Earth's history. *Cherts* are hard rocks, composed largely of very-fine-grained silica. Silica is silicon and oxygen: SiO_2. It can occur as bands in limestones, as nodules, or in other physical forms. Cherts form in a wide range of environments, precipitating directly out of rivers or ocean waters, or forming from rocks that are subjected to mild increases in temperature and pressure. *Biogenic* chert, that is, chert made by organisms such as sponges or radiolaria that secrete silica, is probably the most abundant.

Of interest to us here is that the oxygen isotopic content of the chert bears a definite relationship to that of the environment in which it is made. If precipitation occurs in an ocean environment, then the ^{18}O content of the silica decreases with increasing temperature in a manner that can be quantified in the laboratory. Essentially, the chert, which preserves very well as a sediment through time, acts to record the ambient water temperature through the oxygen isotopic enhancement during its formation.

Unfortunately, using cherts as indicators of the surface temperature of Earth is extremely complicated because cherts form in so many different environments and the ^{18}O values may be altered in ways that have nothing to do with the surface temperature. Geochemists Paul Knauth at Arizona State University and Donald Lowe at Louisiana State attempt to use cherts to determine ancient ocean temperatures in spite of these difficulties. They argue that, for most (but not all) types of chert, processes during or after formation will tend to lower the ^{18}O enhancement in cherts relative to the value obtained during precipitation from ocean waters. Therefore, for a collection of cherts of a given age, the cherts with the highest ^{18}O values should most nearly reflect equilibration with ocean waters during formation. Hence, the cherts with the highest ^{18}O value at a given time provide a measure of Earth's ocean temperature.

A temperature history of Earth from cherts is shown in figure 6.3 updated by Knauth (1992) from his original work with Lowe. Some qualifications must be applied to this analysis. First, the chert samples were formed over a range of latitudes, leading to the concern that one is mixing latitudinal and time variations in temperature. As we discuss in chapter 19, however, warmer ice-free climates, which have dominated over most of Earth's history, experienced much less variation of temperature with latitude than we experience today. The second issue is more of a complication, and has to do with whether the oceanic value of ^{18}O has really been constant over time. One source of variation is episodes of massive glaciation interspersed throughout Earth's history (figure 19.2). Formation of glaciers alters the baseline ^{18}O value in the oceans. Further, the baseline ^{18}O abundance, relative to ^{16}O, may have been lower than today's during the first quarter of Earth's history, based on the chemistry of the most ancient cherts. Knauth attempts to correct for the last of these effects, and a second temperature scale is shown on the graph, which may be more appropriate for samples older than 3.4 billion years.

Reading the temperatures from the upper envelope of the chert data, one sees a general decrease over Earth's

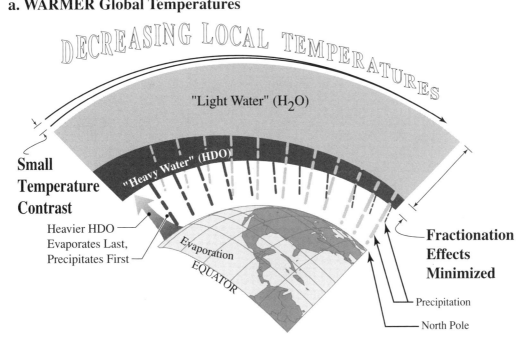

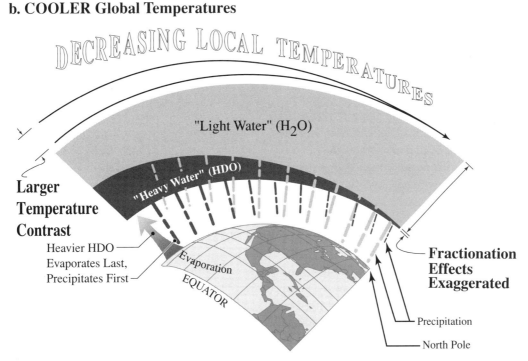

Figure 6.2. Progressive depletion of deuterium-bearing water from equatorial ocean to the polar ice sheets, during (a) warmer and (b) colder times. The steeper equator-to-pole temperature drop under cold climate conditions exaggerates the fractionation relative to that in warmer times. HDO is shown in black, H_2O in grey.

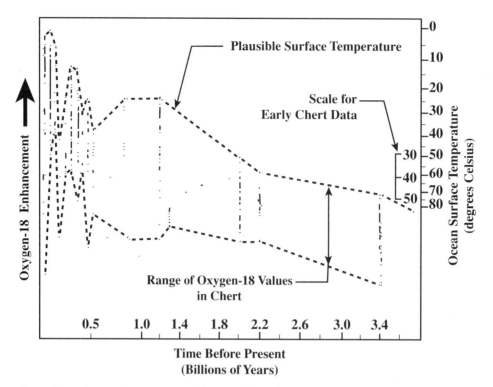

Figure 6.3. A possible temperature history of Earth from cherts collected in Australia. These preserve ratios of ^{18}O to ^{16}O back to 3.5 billion years ago. Plotted is ^{18}O as a function of time, in billions of years. On the right side is a corresponding temperature scale, based on ^{18}O and assuming a single oceanic value for that isotope through time. However, for very early times, possible changes in bulk ^{18}O content of Earth's hydrosphere (ocean and atmosphere) might have occurred, necessitating a temperature correction; this alternative scale also is shown and should be used for earliest times. As discussed in the text, the temperature record is most likely given by the upper envelope of the various data points compiled from the chert samples. The vertical range at each time reflects chemical and physical alteration of the sediments. Modified from Knauth (1992) by permission of Springer-Verlag.

history, such that average surface temperatures recorded in the oldest cherts, just a billion years after the planet's formation, are some 40° to 60°C (80° to 120°F) higher than today! If the data are being interpreted correctly, early Earth was suitable only for the most heat-tolerant bacterial forms.

One must extract conclusions from a single type of data with an enormous amount of caution and skepticism. Unfortunately, other evidence for Earth's temperatures in the first quarter of its history is extremely sketchy. The intriguing conclusion from the chert data should not be dismissed, but it must be recognized that the technique is regarded as highly controversial, and even invalid, by many geologists. It is consistent, at least, with the extensive geologic evidence that liquid water was stable on Earth back at least 3.8 billion years ago. Although this may not seem surprising, it is of some significance because stellar models argue that the Sun was much less luminous at that time than it is today. We discuss this *faint early Sun* problem in chapter 14. For now, it suffices to note that Knauth's isotopic data and interpretation create a problem because they require early Earth to have been significantly warmer than at present. Spacecraft images of Mars say the same about that planet, and this dual-planet dilemma regarding climate change and the early Sun represents a major puzzle that we tackle later in the book.

6.4 QUESTIONS

a. Why is it important to use more than one isotopic system to determine the history of Earth's surface temperature?

b. In using cherts to determine global temperatures in the past, how would you address the question of whether the oceanic ^{18}O value has been constant over Earth's history?

6.5 READINGS

Jouzel, J., and Merlivat, L. 1984. Deuterium and oxygen 18 in precipitation: Modeling of the isotopic effects during snow formation. *Journal of Geophysical Research* **89**, 11,749–11,757.

Knauth, L.P. 1992. Origin and diagenesis of cherts: An isotopic perspective. In *Isotopic Signatures and Sedimentary Records* (N. Clauer and S. Chandhuri, eds.). Springer-Verlag, Berlin, pp. 123–152.

Knauth, L.P., and Lowe, D.R. 1978. Oxygen isotope geochemistry of cherts from the Onverwachte group (3.4 billion years), Transvaal, South Africa, with implications for secular variations in the isotopic composition of cherts. *Earth and Planetary Science Letters* **41**, 209–222.

McGowran, B. 1990. Fifty million years ago. *American Scientist* **78(1)**, 30–39.

Robert, F., Michel-Rejou, A., and Javoy, M. 1992. Oxygen isotopic homogeneity of the Earth: New evidence. *Earth and Planetary Science Letters* **108**, 1–9.

Shackleton, N.J. 1986. Paleogene stable isotope events. *Paleogeography, Paleoclimatology, Paleoecology* **57**, 91–102.

Vostok Project Members. 1995. International effort helps decipher mysteries of paleoclimate from Antarctic ice cores. *EOS* **76**, 169.

Wilson, A.T. 1980. Isotopic evidence for past climate and environmental change. *Journal of Interdisciplinary History* **10**, 795–812.

Cherts. In *Van Nostrand's Scientific Encyclopedia* (D.M. Considine, ed.). Van Nostrand Reinhold, New York, 1983, p. 624.

7 RELATIVE AGE DATING OF COSMIC AND TERRESTRIAL EVENTS: THE CRATERING RECORD

7.1 INTRODUCTION

The absolute dating techniques of chapter 5 rely on very precise laboratory analyses of rock samples. For Earth, an abundance of accessible samples exists. However, with respect to the rest of the solar system, only meteorites, small bits of asteroidal and cometary debris – interplanetary dust particles (IDP), and samples from the Moon have been delivered to terrestrial laboratories for age analyses. One class of meteorites, the Shergottites – Nakhlites – Chassigny (SNC), may have been ejected from Mars by collision with one or several asteroids. Aside from these cases, we have no known samples of material from large bodies in the solar system and thus cannot date major geologic events on the surfaces of the bodies in an absolute fashion.

Instead, scientists use *relative* dating techniques to infer time histories of the moons and planets in the solar system, and they rely primarily on the record of bombardment, or *cratering*, of the surfaces of these bodies. We describe this technique and the physics of cratering in the present chapter. In addition to providing a foundation for inferring key aspects of the solar system's history, this discussion provides a good foundation for the presentation in chapter 8 of relative age dating on Earth, which relies on geologic processes other than cratering but for which the principles are much the same.

7.2 PROCESS OF IMPACT CRATERING

Impact cratering is a process in which a high-speed projectile collides with a solid surface, forming an excavated region called a crater. Impact craters, and the closely related form of craters caused by massive explosions, such as nuclear detonations, can be distinguished from those produced by other processes, such as volcanism or collapse due to groundwater withdrawal, by their distinctive appearance (figure 7.1).

Projectiles impact with velocities imparted by virtue of their orbital motion and the gravitational pull of the target planet. Impact speeds vary depending on the target planet's distance from the Sun and the strength of its gravitational field. Typical impact speeds onto the Moon, due to the free-fall velocity of projectiles at 150 million kilometers from the Sun, are 40 kilometers per second, or just short of 100,000 miles per hour.

An automobile hitting a surface at 100,000 miles per hour delivers an impact energy a million times higher than if it were in a head-on collision with another vehicle at 50 miles per hour, that is, 100 miles per hour relative velocity, because impact energy scales as the square of the velocity. However, it also scales with the mass, and the bigger craters on planetary surfaces are formed by impactors that are kilometers in size. The energy released by just one such impactor, kilometers across, is equivalent to the release of the world's entire nuclear arsenal – at the peak prior to current disarmament – many dozens of times over! Such an enormous release of energy on a habitable planet has the capability to transform oceans and atmospheres, and to destroy life on a planetary scale.

At impact with the ground, the projectile plows into the surface, its energy of motion rapidly converted into heat, and the impact itself sends *shock* waves, familiar examples of which are thunder and sonic booms, into the ground. The ground itself is compressed and shattered by the enormous temperatures and pressures of the shock wave. The projectile also is shocked and shattered. The shock waves travel outward and downward in the ground in a hemispherical pattern. Nearest the impact, rock is vaporized or melted; farther away it is pulverized. As the shock waves travel away from the impact, the ground begins to rebound toward the center of the hemispherical cavity or crater,

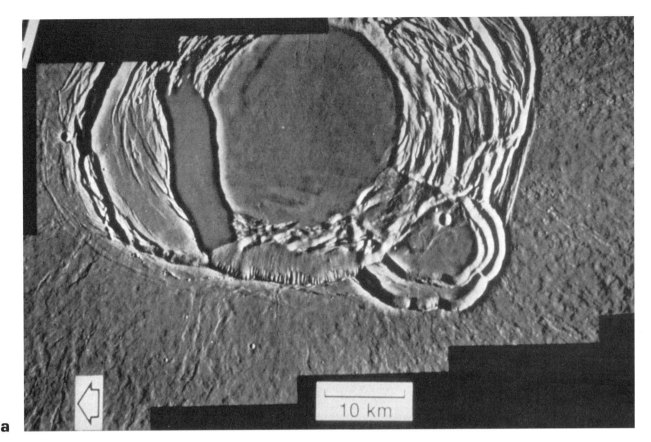

Figure 7.1. Examples of craters formed by different processes: (a) Caldera at the summit of the Martian volcano Aescraeus Mons (Mosaic of NASA Viking images generated by J. Zimbelman at Lunar and Planetary Institute); (b) sinkhole near Montevallo, Alabama, 120 meters across, formed by the action of groundwater (U.S. Geological Servey photo).

Figure 7.1. (Continued) Examples of craters formed by different processes: (c) explosion crater about 250 meters across in Nevada, generated by a 30-kiloton nuclear warhead detonated underground (U.S. Department of Energy); (d) Meteor crater, Arizona, a small (1-km-diameter) impact crater.

forming a central peak in the case of moderate-sized to large impact craters. The central peak can form only because the rock is in a temporary state of being partly molten and partly solid, the solid part being so weak that the shock waves moving back and forth can readily push the material. As the shock waves dissipate, the central peak remains intact.

The shock waves also raise a rim around the crater as well as eject material off the sides, this material (ejecta) shooting into the air as hot molten (liquid) rock, traveling many times the size of the crater away from the center, forming lines of smaller *secondary* craters as well as streaks or *rays* of material as it strikes the ground. The impactor itself is obliterated and becomes a small part of the ejecta.

Figure 7.2 shows the stages of crater formation and the final shapes of typical small and large craters. There are many variations: Small craters do not have well-developed central peaks. Extremely large impactors send shock waves through deeper parts of the target's interior, where the warmer rock or ice can flow more easily, creating large-scale wave patterns that are preserved as *multiring basins*. Mare Oriental on the surface of Earth's Moon, Valhalla on Jupiter's moon Callisto, and Gilgamesh on Jupiter's moon Ganymede are examples. Craters also may take on different forms depending on whether ground ice is present, and the strength of the planetary crust: Weak crusts will cause crater topography to disappear over time, leaving only ghostly outlines. Finally, erosion by water and subduction of crust (chapter 9) have removed most of the craters on Earth, and left many others barely discernible (figures 7.3a–f).

7.3 USING CRATERS TO DATE PLANETARY SURFACES

Craters can be used to determine how old one surface is relative to another because the rate of impacts over time is thought to have declined slowly over the past three-quarters of solar system history, having decreased quickly prior to that from a much larger initial rate. Surfaces that are young, that is, which have been renewed through lava flows, mountain building, erosion by water, and other geologic processes, will show fewer craters than surfaces that are much less active, or older. Because of this we can use the abundance of craters on various surfaces of a planetary body to determine, in a relative sense, when certain kinds of geologic processes occurred relative to others. The freshness of craters, that is, how bright their debris or ejecta blankets appear and how sharp their features are, provides an additional refinement to the relative dating process. It is also possible to compare relative ages from one planet or moon to another, provided one can calibrate the rate of impacts in one part of the solar system relative to another.

An example from the Moon provides a classic illustration. The bright areas of the surface of the Moon are very heavily cratered regions called the *lunar highlands*, as revealed by telescopes and images from lunar orbiting spacecraft (figure 7.4a). In these regions the density of

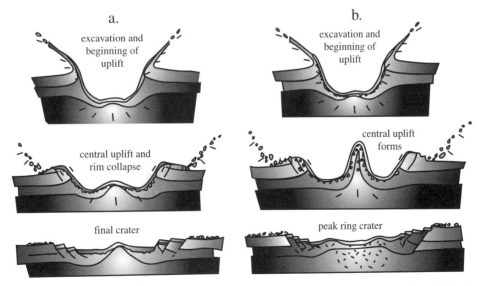

Figure 7.2. Stages in the formation of (a) small and (b) large craters. Modified from Melosh (1989, p. 142) by permission of Oxford University Press.

craters is so great that craters overlap with and are superimposed on each other; down to the limit of resolution on the image, one sees a scene filled with craters.

The dark portions of the Moon, on the other hand, consist of areas that are smooth and relatively devoid of craters. These *mare* (Latin for seas, the seventeenth and eighteenth century interpretation from telescopic views) show ample evidence of craters that have been partly covered or obliterated by the material that makes up the smooth, dark surfaces (figure 7.4b). The simplest interpretation is that the mare are lowland basins that were flooded by lavas sometime after an early, heavy bombardment of the Moon occurred. The flooding obliterated most of the craters, leaving a fresh surface on which some remains of old craters can be seen, and a few small, new craters were formed by impacts after the lava solidified.

Explorations by the Apollo astronauts from 1969 to 1972 returned hundreds of pounds of lunar samples from mare and highland regions. Radioisotopic techniques, described in chapter 5, were used to provide absolute dates for the solidification of these rocks from original molten materials. The lunar highlands are old, with rocks dating as old as 4.5 billion years. The mare deposits typically are 4.2 billion to 4.3 billion years old, significantly younger than the highlands.

The age estimates based on the cratering density on the lunar surface are confirmed by the absolute dating of mare and highlands provided by rock samples. It then would appear possible to use crater densities on other worlds, not accessible for sample collection at present, to construct chronologies as well. The most straightforward chronology involves determining relative ages of events on a surface, that is, which event preceded another. This simply requires counting craters as well as looking for evidence of craters partly obliterated by geologic processes. More difficult is to try to assign actual dates, which requires

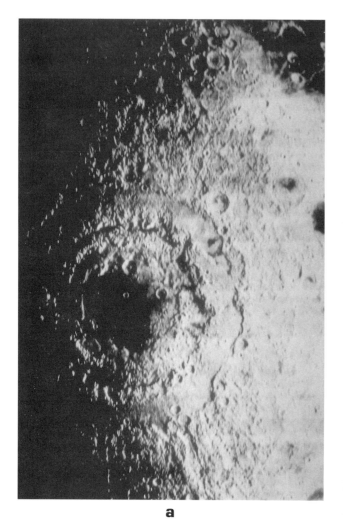
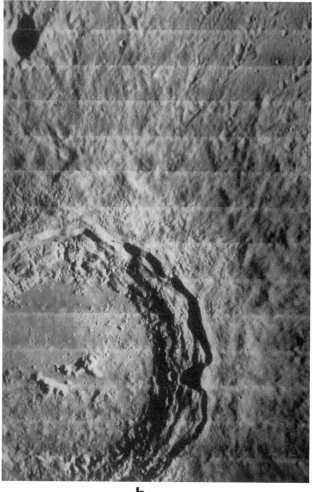

a **b**

Figure 7.3. Varieties of impact craters: (a) Large multiring basin, Mare Oriental, on the Moon; (b) classic large crater, Copernicus, with central peak, on the Moon. Photos (a) through (e) are courtesy of NASA.

c

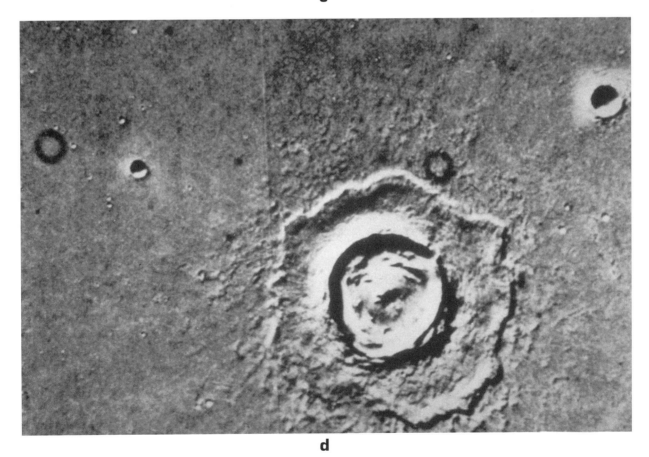

d

Figure 7.3. (Continued) Varieties of impact craters: (c) smaller lunar craters without peaks; (d) pedestal crater on Mars, formed by melting of ground ice during impact. Photos (a) through (f) are courtesy of NASA.

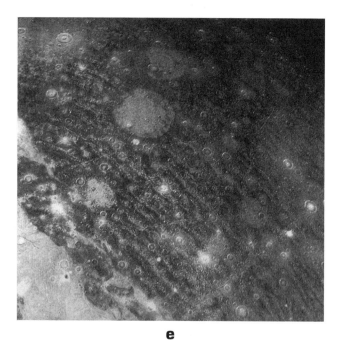
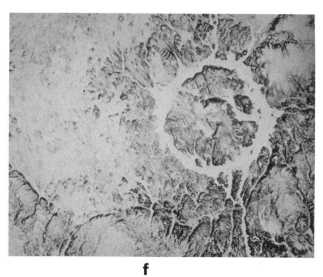

Figure 7.3. (Continued) Varieties of impact craters: (e) relaxed craters, or *palimpsests* on Ganymede (Voyager image); (f) eroded crater on Earth, now comprising Lake Manicouagan, Quebec, Canada. Photos (a) through (f) are courtesy of NASA.

assuming that the lunar crater density and ages based on Moon rocks can be transferred directly to other bodies in the solar system.

7.3.1 Relative Ages of Events on a Planetary Surface

Impact craters can be used to determine the relative ages of geologic features on a planetary surface. Two examples of this are shown, one from Mars and one from Jupiter's moon Ganymede, using images from Viking and Galileo missions in figures 7.5a and 7.5b, respectively. In the case of Mars, the geologic features of interest are channels that clearly were cut by water, but today are dry along with the rest of the planet. Are the features young or ancient? Was the climate wet up through recent times, such that life might have evolved to an advanced stage?

Examination of the Viking images such as that in figure 7.5a reveals that the Martian channels typically are overlain by impact craters, some fairly substantial in size. Other regions of the Martian surface have far fewer craters, and hence we can say that the channels are, relatively speaking, ancient. Determining a more exact age requires tying the cratering rate to some absolute timescale. At the same time, we know that the channels are not among the oldest features, either, because many cut through craters that must therefore be older than the channels. A chronology can be assembled in which channel formation occurs after formation of the oldest Martian terrains but before a number of other geologic events that are recorded in the surface.

Ganymede, the third of Jupiter's four giant moons (Io is closest to Jupiter, and then Europa, Ganymede, and Callisto), shows lines in its spectrum typical of water ice. However, the mass of the planet is too heavy given its volume (mass over volume is density) to be pure water ice. The best guess, based on models of solar system formation, is that the heavier component is a *silicate*, or common rocky material containing silicon, oxygen, magnesium, some iron, and other elements. Therefore, unlike our own Moon, Earth, Venus, Mercury, and Mars, which are made up mostly of silicon-bearing rock and metal, Ganymede is half-rock, half-ice. This is true for Callisto, as well, but not Europa and Io: They are both mostly rock, although Europa has an outer veneer of ice and possibly liquid water.

One might expect, from terrestrial experience, that the ice might behave differently in an impact than rock. In fact, the pressures and temperatures in the hypervelocity impacts we have been describing are so large that there is little difference. Furthermore, temperatures in the distant outer solar system, where Jupiter and its moons reside, are so low that ice behaves much like rock as a material making up the solid *crusts*, or outer layers, of Ganymede

Figure 7.4. Two very different terrains on the Moon: (a) The lunar highlands show craters of all sizes filling all available surface space. (b) The lunar mare regions are smooth, dark plains with a few fresh craters and remains of large craters, in various states of preservation, which were present when lavas flooded the lunar surface.

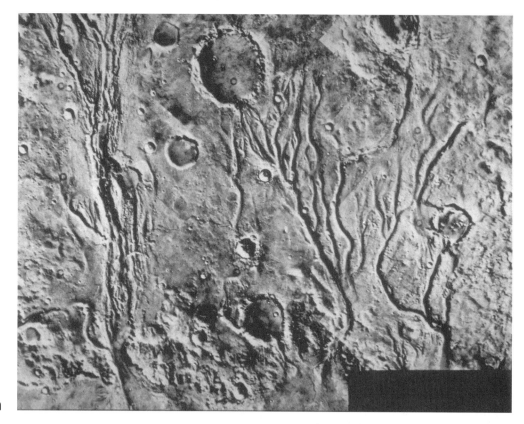

Figure 7.5. (a) NASA/Viking image of Martian surface cut by channels. (b) NASA/Galileo image of dark and light terrains on Jupiter's moon, Ganymede.

and Callisto. The surface temperature near the equator of these moons is typically 165 K, very far below the ice melting point of 273 K.

Images of Ganymede reveal two types of surfaces: a dark, heavily cratered terrain, and a bright, lightly cratered terrain. The paucity of craters on the latter surface immediately suggests that it is a younger feature, perhaps ice that has been extruded from the interior along cracks and flowed outward. In places, it is possible to see where a crater on the older dark terrain has been partially obliterated by the new material. It is also possible to tell something about how the cracks and new material formed by looking at distortions in partially preserved craters along the edges of the bright terrain. The difference in brightness between the dark and light terrains remains a matter of speculation; silicates and perhaps some carbon-bearing materials are well mixed with the water ice, perhaps dating back to the original formation of Ganymede.

The ability to learn something about the sequence of events on a surface by looking at crater densities is a tool of primary importance in solar system studies. It is a new development of a much older technique applied to Earth geology to look at *superposition* of layers to assemble a history of a given region. On Earth, water and geologic activity have effectively erased the cratering record, so that the use of craters as a geologic tool was a novel idea that did not come into its own until planetary exploration began some three decades ago.

7.3.2 Absolute Chronology of Solar System Events

Relative age dating is limited in the amount of information derived. Ideally, one wants to assign ages to events on the surfaces of planets and moons so as to understand their history and ultimately that of the solar system. Imagine how limited our own understanding of the history of human cultures would be if we only knew the order of events, but not their antiquity or duration.

In the case of Earth's *geologic* history, even before radioisotopic dating provided reliable dates, estimates of ages could be made on the basis of notions of the accumulation rate of *sediments*, debris brought from high to low places by the action of water. Early work tended to overestimate the rates of sedimentation and hence produced a compressed timescale relative to what is accepted today based on radioisotopic determinations. With the help of radioisotopic dating, the rates of geologic processes are now better understood and calibrated, such that indirect dating techniques such as sedimentation are enhanced as tools in assembling the history of Earth.

The situation for an absolute chronology from planetary cratering is similar in that radioisotopic dating has been used to construct a chronology for Earth's Moon, which then has been applied, with caveats, to other solar system bodies. The Moon is the only body for which radioisotopic dating of terrains of varying crater density can be performed; Earth's cratering record is too sparse. (It is not possible to determine unequivocally from what part of the Martian surface the SNC meteorites were derived; hence they cannot help calibrate the cratering record on Mars.)

The oldest parts of the Moon, the highlands, have by far the largest number of craters; the younger mare possess the least. This is consistent with the decreasing population over time of debris in orbits around the Sun. Theories of planet formation, which we discuss in chapter 10, hold that the planets were assembled from smaller pieces of rock and ice through relatively low-speed collisions which allowed the pieces to stick together. In the final phases of this process, most of this *protoplanetary* material was perturbed by close encounters with the planets into highly elliptical orbits, guaranteeing that any subsequent collisions with the planets would be at high speeds, producing craters. Over time this remnant debris of planet formation was swept up by the planets, so that the available impactor population has decreased dramatically from the beginning to the present-day.

A simple law governing the rate of impacts over time, consistent with the sweep-up picture described above, and with the lunar cratering record, has the *inverse exponential* form shown in figure 7.6. The curve is characterized by a very steep decrease initially, as large amounts of material are swept up by the nearly fully grown planets, followed by a transition to a slowly decreasing rate of impacts. The cratering record on the Moon tells us when the transition occurs between these two regimes. Further, it provides information about the tail-off in impacts at later times, though with limited capability because of the paucity of new craters. More difficult to discern is the precise steepness of the early curve, because the cratering rate was so high that lunar highland surfaces are completely covered with craters: New impacts simply obliterate all or part of old ones and only a lower limit on the ancient cratering rate is accessible.

The dating of Moon rocks fixes the transition in the cratering curve at roughly 3.8 billion to 4.0 billion years before present; the period of intense cratering before that is called the *late heavy bombardment*, referring to the tail end of the planet-formation (*accretion*) process. Interestingly, the oldest whole rock samples on Earth date back to roughly the same time. We know that this does not

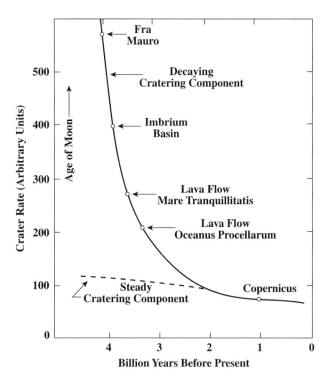

Figure 7.6. Number of impacts versus time on the surface of the Moon; the curve is labeled with ages of rocks collected at the Apollo landing sites, and an estimate for the age of the large crater Copernicus. The dashed line shows what the cratering rate would look like if the recent cratering rate were extrapolated along a straight line back to the beginning. After Lang and Whitney (1991).

represent the age of Earth because the rocks are rather evolved, showing the action of liquid water on their chemistry and texture; additionally, meteorites record much earlier dates back to 4.56 billion years before present. Instead, Earth was simply too active geologically at earlier times to preserve older rocks and, as we see in later chapters, had little or no continental land mass on which such rocks could be preserved.

An additional piece of information on the bombardment history of the Moon, one crucial for calibrating the impact history of other solar system bodies, is the distribution of crater sizes. Crater sizes are related fairly directly to those of the original impactors, and hence a model of crater formation can yield the original impactor size distribution.

A typical crater population will exhibit the evolution sketched in figure 7.7. On a plot showing the number of craters below a given size, versus size, the data tend to fall on a broken line with two different slopes. The steeper slope, occurring for larger crater diameters, directly reflects the size distribution of the original impacting population that produced the craters. At smaller crater sizes, *saturation* effects tend to reduce the number of observed craters: Smaller craters are so numerous that they readily fill a surface until new craters simply obliterate the old ones. Over time, as shown in the figure, the breakpoint at which saturation takes over moves to larger crater sizes

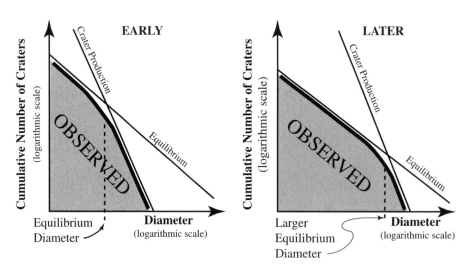

Figure 7.7. How a population of craters evolves over time. Shown in each graph is the number of craters with a diameter less than *D*, as a function of *D*. The thinner lines are guides to two idealized populations of craters. The steeper line, "Production," is what is produced directly from a particular size distribution of the impacting bodies. The shallower "Equilibrium" line is the result of saturation, i.e., obliteration of craters by newer impacts in a very crowded crater field. The right-hand panel represents the situation at a time later than the left-hand panel, showing that, as a surface gets older, the effects of saturation extend to larger and larger crater diameters. Redrawn from Melosh (1989, p. 192).

as the surface is increasingly filled. Determination of the breakpoint on such a plot for a cratered region of a planetary surface provides a measure of its age when correlated against surfaces that are absolutely dated, such as those of the Moon.

Mercury and Mars show heavily cratered terrains with distributions similar to those on the Moon. Not only does this allow us to determine how ancient the various terrains are, it also leads us to conclude that the impactor populations on the Moon, Mars, and Mercury are similar. This strongly suggests that the population of impactors that have struck the Moon over time originate from beyond Earth orbit, and in fact are in orbits around the Sun that take them well beyond Mars into the outer solar system. Most of these impactors may have been comets derived from reservoirs of debris left over from planetary formation. Additional impact debris likely is derived from the asteroid belt between Mars and Jupiter. Examples of crater size-frequency distributions for the inner planets are given in figure 7.8.

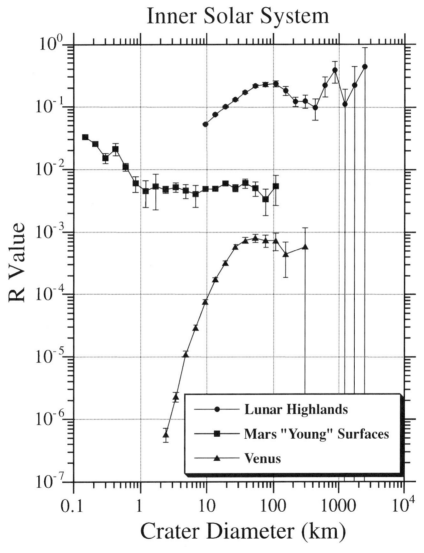

Figure 7.8. Crater size-frequency distribution for the Moon, Mars, and Venus. The vertical axis is a measure of the abundance of craters of a given size, and this is plotted against the crater diameter. The lunar highlands represent terrain that has not been geologically active since close to the beginning of the Moon's history. The young terrain on Mars has been subject to tectonic forces that erased the ancient crater record. The steep falloff in small-crater abundance for Venus is a result of that planet's thick atmosphere, which shields the surface from the smaller impactors. Courtesy of Prof. Robert Strom, University of Arizona.

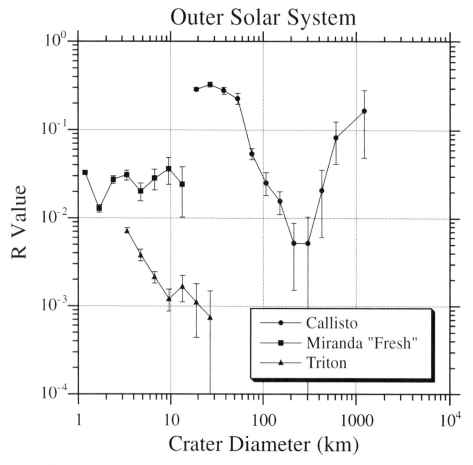

Figure 7.9. Crater size-frequency distribution for three objects in the outer solar system: Callisto, Triton, and younger surfaces on Uranus' moon Miranda. The strong turn-up of the curves for smaller craters is a characteristic not seen on inner solar system bodies. Courtesy of Prof. Robert Strom, University of Arizona.

The crater distributions on the moons of the giant planets are similar neither to those of the Moon, Mars, and Mercury nor to each other (figure 7.9). Each giant planet seems to have defined a unique population of impactors for its moons, with the only general resemblance that the population has decreased sharply with time. The cratering size-frequency distribution for the moons of the giant planets probably can be understood best by invoking two populations of impactors: those in solar orbit, perhaps the same as those that had peppered the inner solar system with craters, and a unique population of debris orbiting each of the giant planets. This local debris, likely the leftovers from the formation of the giant planets and their moons, has had a different history for each of Jupiter, Saturn, Uranus, and Neptune. In the case of Saturn, there is even evidence in the crater record for the breakup of a large moon late in solar system history, perhaps in the orbit now occupied by the irregularly shaped moon Hyperion.

The cratering record in the outer solar system is a useful tool for assembling a rough history of this region but, as yet, it is too difficult to retrieve samples of icy moons for radioisotopic dating. From studies of craters we now know that Callisto's heavily cratered surface probably dates back close to the beginning of the solar system; the heavily cratered terrain on Ganymede might be slightly younger but certainly predates the bright lightly-cratered regions on that moon. Saturn's small moon, Enceladus, exhibits smooth regions bordered by areas where craters have been partly covered by bright flows; clearly this moon has been active enough that fresh water ice and other compounds poured out onto the surface recently. Uranus' satellites differ in their crater density and hence bespeak varying levels of geologic activity the nature of which is otherwise poorly understood. Finally, Jupiter's Io is devoid of craters and the same is nearly true of Jupiter's Europa and Neptune's Triton, satellites that possess a history whose levels of geologic activity rival or exceed that of our Earth in erasing the cratering record.

7.4 CRATERING ON PLANETARY BODIES WITH ATMOSPHERES

In our discussion of the inner solar system, the reader might have noticed that we omitted mention of Venus. That planet's surface lies beneath an enormous atmosphere with a surface pressure 90 times that of Earth. The geology was finally mapped with completeness by a radar imager aboard the U.S. *Magellan* spacecraft beginning in 1989. Impact craters are sparse, indicating a relatively youthful surface renewed by volcanism, likely similar to the basaltic types of volcanism on the Earth, a near twin in terms of size and mass. However, most striking is the predominance of large impact craters relative to the distributions seen on airless bodies (figure 7.8).

The thick atmosphere of Venus acts as a shield, preventing smaller impactors from reaching the surface intact. The pressure of the atmosphere, close to the surface where it is densest, exerts a stress on the incoming material. Most comets and even asteroids are relatively weak; that is, their interiors have been fractured and hence are not strongly glued together. The effect of aerodynamic forces is to crush the impactor; the individual fragments then spread out. Very small impactors may spread out well above the surface, the individual fragments burning up in the atmosphere; this is the likely fate of the small asteroid that exploded above the Tunguska River in Siberia in 1908, knocking down people 60 km from the impact site but leaving no crater. Very large bolides are hardly affected by the atmosphere and leave a normal crater. Intermediate-size impactors disrupt and partially disperse, leaving an oddly shaped crater or a field of smaller craters clustered close together.

The size thresholds defining the different results of an impact into an atmosphere depend to some extent on the strength of the impactor, but much more importantly on the thickness of the planet's atmosphere. This leads to the intriguing possibility that one could use the crater distribution on ancient terrains to learn what the atmosphere of a planet was like a long time ago. If Venus, for example, had a much thinner atmosphere in its past, ancient terrains should show a crater size distribution akin to that of the Moon. Likewise, Mars may have had a thick atmosphere in the past, and perhaps the most ancient terrains record evidence for this. Unfortunately in the case of Venus, the surface has been so geologically active that no ancient terrains appear to be present, making it impossible to use the crater record as a probe of the early atmosphere.

It may be possible to apply this approach to Saturn's moon, Titan, which is larger than the planet Mercury and has an atmosphere at its surface four times denser than the Earth's air at sea level. In consequence, Titan's atmosphere is intermediate in thickness between Earth's and that of Venus, and filters out small impactors rather effectively. Some models of the evolution of this moon suggest that its atmosphere might have been much thinner several billion years ago when the Sun was less luminous (see chapter 14); if this turns out to be the case, craters too small to have been produced during the current thick-atmosphere epoch should be present. Our first detailed glimpse of the surface, hidden behind a global haze, will come from the European *Huygens* probe as it descends through the atmosphere in 2004, part of the joint U.S.-European Cassini–Huygens mission to the Saturn system. The probe and a Saturn orbiter will map Titan's surface, including the distribution of craters.

7.5 IMPACTORS THROUGH TIME

In our discussion of the dating of events through the cratering record, we noted that underlying the impactor flux is the assumption that collisions are winding down as the solar system is slowly cleared of debris. Although this general concept is a useful one, it is important to consider that sources of collisional debris for the inner solar system continue to be supplied over time as comets are perturbed out of their orbits in the Kuiper Belt and Oort cloud and travel inward to possible collisions with the planets and their moons.

It has been suggested that, occasionally, large numbers of such comets may be perturbed into the realm of the terrestrial planets, temporarily increasing the impact rate and producing comet showers. Although the evidence for such major events is tenuous (and the idea comes in and out of vogue), the more general notion that the flux of comets through the inner solar system is variable is sensible. The consequences of a collision of a comet-sized body with Earth are profound for the geology, climate, and biology of this planet. We explore these consequences in chapter 18.

7.6 QUESTIONS

a. Could the cratering record on Titan be used to determine whether that moon of Saturn ever had an atmosphere much thicker than it does today? If so, how?

b. Why is it necessary to use lunar samples to determine the cratering rate as a function of time in the inner solar system? Is there a reliable way to do so without physical samples?

7.7 READINGS

Bloom, A.L. 1978. *Geomorphology: A Systematic Analysis of Late Cenozoic Landforms.* Prentice-Hall, Englewood Cliffs, NJ.

Engel, S., Lunine, J.I., and Hartmann, W. 1995. Cratering on Titan and implications for Titan's atmospheric history. *Planet Space Science* **43**, 1059–1066.

Lang, K.R., and Whitney, C.A. 1991. *Wanderers in Space: Exploration and Discovery in the Solar System.* Cambridge University Press, Cambridge, UK.

Melosh, H.J. 1989. *Impact Cratering: A Geologic Process.* Oxford University Press, New York.

Renner, M. 1983. Preparing for peace. In *State of the World: A Worldwatch Institute Report on Progress Toward a Sustainable Society* (L. Stark, ed.). W.W. Norton, New York, pp. 139–157.

Spudis, P. 1992. Moon, geology. In *Van Nostrand's Scientific Encyclopedia* (D.M. Considine, ed.). Van Nostrand Reinhold, New York, 1983, pp. 452–455.

8 RELATIVE AGE DATING OF TERRESTRIAL EVENTS: GEOLOGIC LAYERING AND GEOLOGIC TIME

8.1 INTRODUCTION

Prior to the invention of radioisotopic techniques for dating rock samples, geologists determined relative ages for rocks using simple principles of how rocks and their fragments are deposited, and using remains or records of extinct life to correlate samples from different locations. When combined later with the dating of rocks by radioisotopic techniques, a detailed history of Earth could be developed. We work with this history repeatedly throughout the rest of the book. This chapter serves as an introduction to the techniques used to assemble such a record.

8.2 CATASTROPHISM VERSUS UNIFORMITARIANISM

When we look at Earth's landforms, we are viewing a snapshot, a moment in a vast span of time during which mountains rise and fall, seas expand over land areas and contract again, and continents shift their positions and grow slowly from new rock added by volcanoes. These processes all require vast amounts of time for their completion, but most do not proceed in a smooth, gradual manner. Instead, geologic processes are a combination of gradual effects and sudden catastrophes. The earthquakes that shake California represent sudden failures of rock after the buildup of stresses over time as one portion of California slowly glides past the other, as we discuss in chapter 9.

The realization that Earth changes in this hybrid fashion was long in coming. Much of the history of the development of geology was a battle between those who argued in favor of *uniformitarianism*, and hence gradual change over enormous spans of time, and those who claimed that Earth was young and shaped by catastrophic processes. As estimates for the age of Earth climbed, it appeared the uniformitarians were right. In the past few decades, though, the importance of catastrophic change on Earth has become clear, in large measure through study of other planets. However, radioisotopic dating (chapter 5) confirmed a large age for Earth, and so, both camps were right.

The long history of this debate has a literature all its own, and it is appropriate only to touch on some aspects here, to illustrate the ways people attempted to gauge Earth's age, and as an introduction to geologic processes.

8.3 ESTIMATING THE AGE OF EARTH, WITHOUT RADIOISOTOPES

One of the first recorded observers to surmise a long age for Earth was Herodotus, who lived from approximately 480 to 425 B.C. He is best known as the father of history, having written an extensive account of the Persian invasion of Greece culminating around the time of his birth. Herodotus was also a traveler, and visited the Nile River Valley, which was subject to annual cycles of flooding. He came to the important conclusion that the Nile Delta was in fact a series of sediments built up in successive floods. By noting that individual floods deposit only thin layers of sediments, he was able to conclude that the Nile Delta had taken many thousands of years to build up. (In fact, Herodotus coined the term "delta" for the accumulation of sediments at the mouth of a river; the shape of the Nile Delta reminded him of the Greek letter by that name, Δ.)

More important than the amount of time Herodotus computed, which turns out to be trivial compared to the age of Earth, was the notion that one could estimate ages of geologic features by determining rates of the processes responsible for such features, and then assuming the rates to be roughly constant over time. Similar applications of the principle of uniformitarianism were to be used again and again in later centuries to estimate the ages of rock formations, and in particular of layers of sediments that had compacted and cemented to form *sedimentary rocks*.

Throughout the middle ages, European studies of the history of Earth relied on the Bible, and it wasn't until the seventeenth century that attempts were made again to understand clues to Earth's history through the rock record. Nicolaus Seno (1638–1686) worked out principles of the progressive laying down of sediments in Tuscany. However, a Scottish doctor and farmer, James Hutton (1729–1797), was the first to have the important insight that geologic processes are cyclic in nature. Forces associated with subterranean heat, which we deal with in chapter 9, cause land to be uplifted into plateaus and mountain ranges. The effects of wind and water then break down the masses of uplifted rock, producing sediments that are transported by water downward to ultimately form layers in lakes, seashores, or even oceans. Over time, the layers *lithify* to become sedimentary rock. These are then uplifted sometime in the future to form new mountain ranges, which exhibit the sedimentary layers (and the remains of life within those layers) of the earlier episodes of erosion and deposition.

Hutton's concept represented a remarkable insight, because it unified many individual phenomena and observations into a conceptual picture of Earth's history. With the further assumption that these geologic processes were generally no more or less vigorous than they are today, Hutton's examination of sedimentary layers led him to realize that Earth's history must be enormous, that geologic time is an abyss and human history a speck by comparison.

Particularly inspiring to Hutton was Siccar Point, in Scotland, where the sedimentary record was interrupted by an *unconformity*, in which a sequence of layers is missing because of erosion and removal; the layers above and below usually are tilted relative to each other by the tectonic forces that produce uplift and build mountains. Hutton wrote

Here are three distinct successive periods of existence, and each of these is, in our measurement of time, a thing of indefinite duration.... The result, therefore, of this physical inquiry is, that we find no vestige of a beginning, no prospect of an end.

After Hutton, geologists tried to determine rates of sedimentation so as to estimate the age of Earth from the total length of the sedimentary, or *stratigraphic*, record. Typical numbers produced at the turn of the twentieth century were 100 million to 400 million years. These underestimated the actual age by factors of 10 to 50 because much of the sedimentary record is missing in various locations (due to erosion) or compressed (due to metamorphism; see section 8.4) and because there is a long rock sequence prior to half a billion years ago that is far less well defined in terms of fossils (see section 8.6) and less well preserved.

Various other techniques to estimate Earth's age fell short, and particularly noteworthy in this regard were flawed determinations of the Sun's age. It had been recognized by Immanuel Kant (1724–1804) that chemical reactions could not supply the tremendous amount of energy flowing from the Sun for more than about a millennium. Two distinguished physicists, Herman von Helmholtz in the mid-nineteenth century, and William Thompson (Lord Kelvin) at the end of the nineteenth century both came up with lifetimes based on the Sun's energy coming from *gravitational contraction*. Under the force of gravity, the compression resulting from a collapse of the object (whether slowly or quickly) must release potential energy (chapter 3). Ages in the tens-of-millions-of-years range were derived, much less than the geologic estimates of the time.

Although, at present, the Sun's energy is not derived from gravitational contraction, this is a primary source of energy during star formation, when interstellar gas and dust collapse from a more diffuse state (chapter 10), and during some stages of the end of a star's life. This holds true for the planets as well, whose late stages of formation were characterized by material falling at high speeds under the influence of the gravitational fields of the growing planets. For Earth, temperatures during this *accretion* process were likely sufficient to melt rock. Some of the heat that we measure today coming from interiors of Earth and other planets is in fact a remnant of this initial energy of collapse.

It was the discovery of radioactivity at the end of the nineteenth century, independently by French scientists Henri Becquerel and Marie Curie and German physicist Wilhelm Röntgen, that opened the door to a solution of both the Sun's energy source and the age of Earth. From the initial work came a suite of discoveries leading to radioisotopic dating, which quickly led to the realization that Earth must be billions of years old, and to the discovery of nuclear fusion as an energy source capable of sustaining the Sun's luminosity for that amount of time. By the 1960s, analysis of meteorites and refinements of solar evolution models both converged on an age for the solar system, and hence for the Earth, of 4.5 billion years (chapter 5). Figure 8.1 summarizes the steps in the increasing estimates of the age of Earth.

8.4 GEOLOGIC PROCESSES AND THEIR CYCLICAL NATURE

Rocks that comprise Earth's surface originate in the interior as molten or semimolten material (figure 8.2). Some rocks find their way to the surface in volcanic eruptions,

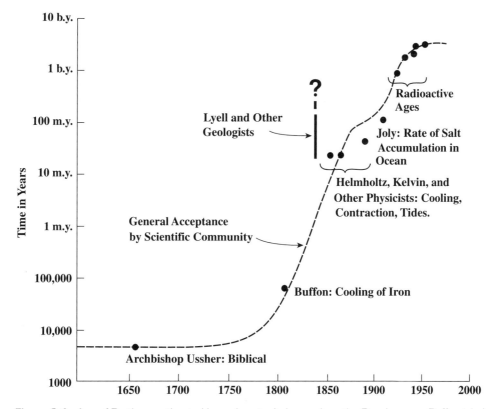

Figure 8.1. Age of Earth as estimated by various techniques since the Renaissance. Buffon tried to use rate of cooling of iron from a molten state to estimate the age of Earth, but the omission of as-yet undiscovered radioactive heating of Earth's interior (chapters 9 and 11) seriously shortened his cooling time. Joly worked out how long it would take to bring the oceans up to their current salinity based on the rate at which rivers carry salt to the sea; he ignored the precipitation of salt out of the ocean water into seafloor sediments. From Press and Siever (1978) by permission of W.H. Freeman and Company.

in which case the solidified rock is a *volcanic igneous* rock. Most continental igneous rocks follow a different route: They may cool and solidify beneath the surface, or in the hidden cores of mountain ranges, as *plutonic igneous* rocks. These may be exposed eventually as elevated terrains such as plateaus or mountains are eroded by wind and water. Often, such igneous rocks will cut through sedimentary or metamorphic (defined later in this section) layers in the final stages of their elevation and solidification. Note that oceanic rocks originate almost entirely from undersea volcanism and have a distinctly different composition from continental rocks; we discuss these differences and their origin in chapters 11 and 16.

Once rocks are exposed at the surface they are subjected to *weathering* processes, which include abrasion and erosion by wind, and, much more important, erosion by water. Water molecules in raindrops will combine chemically with atoms comprising the minerals in the rock, weakening the material and accelerating its crumbling. Water works its way into cracks in the rock and expands when it freezes; this leads to mechanical breaking of rock. Liquid water carves out river valleys, often along pre-existing fractures or *faults* in the rock. Ice sheets, or glaciers, can do the same in colder regions. Some minerals such as limestone are dissolved directly by the action of water, forming caves and exotic-looking structures known as *Karst* topography. The removal of material eroded from the uplands is by streams and rivers, moving debris downward toward lakes and oceans. In regions bordering oceans and other saltwater bodies, corrosive salt injected into the atmosphere via evaporating sea spray enhances erosion.

The long journey of sedimentary debris from highlands to the sea, sketched schematically in figure 8.3, is where geology most commonly and pervasively binds itself to the history of humankind. We rely on streams, rivers, lakes, and groundwater to sustain our existence. Agriculture began and continues to flourish in river valleys. The edge of the sea has always held a special place in human imagination both as a source of sustenance and a gateway to faraway lands.

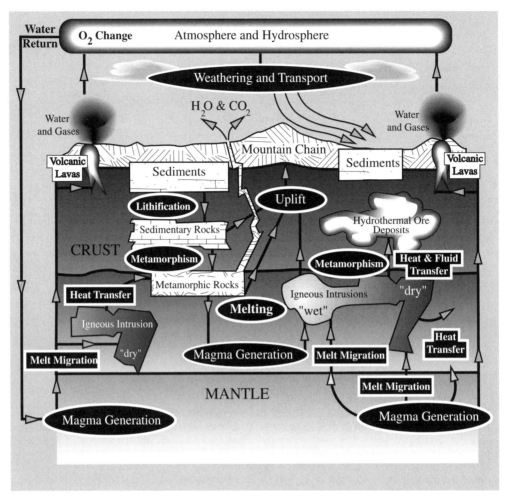

Figure 8.2. Rock cycle of Earth shown schematically. Adapted from Wyllie (1971, p. 48) by permission of John Wiley and Sons, Inc.

The geologic cycle does not end with the deposition of sediments in alluvial plains or on the ocean floor. As sediments accumulate, the underlying material slowly cements to form sedimentary rocks in which the layering and pebble sizes and shapes are preserved. These rocks may be exposed by uplift combined with erosion and usually are tilted in the process. Erosion will remove some layers, but eventually, the scene changes enough that new sediments are deposited atop the old; missing sedimentary layers removed by erosion as well as the tilt of the older layers disappearing suddenly at the interface with the newer deposits are easily recognizable to geologists as unconformities.

Often, sedimentary layers are so deeply buried that they are subjected to high pressures and temperatures. The sedimentary layers may be softened by the temperature, distorted by the pressures, and chemically altered to a greater or lesser degree. These rocks, uplifted once again in a cycle of mountain building, are known as *metamorphic rocks*; those only slightly altered from sedimentary rocks are known as *low-grade* metamorphics, whereas those heavily altered are *high grade*.

Some sedimentary layers, particularly on the seafloor, eventually return to the interior of Earth. These rocks are melted, mixed with other molten and solid rocky material, and ascend as new lavas (via volcanoes) or magmas (via plutonic igneous processes) to start the cycle over. The new igneous rocks, having been cycled through the crust and surface of the earth, generally have a slightly different composition than their progenitors; this is part of the process by which continental rock with its distinct composition is created from other rock types.

8.5 PRINCIPLES OF GEOLOGIC SUCCESSION

Many of the above-mentioned geologic processes deposit layers of rock one atop the other. This is particularly true for sedimentary rocks, but one may see layering in igneous rock associated with successive episodes of volcanic eruptions. Furthermore, metamorphic rocks often

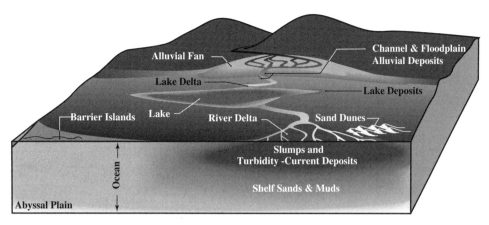

Figure 8.3. A sedimentary journey: sketch of the process of sedimentation from highlands to deposition in lakes or seas. Adapted and redrawn from Press and Siever (1978).

are sufficiently well preserved that the original layering of the precursor sedimentary rocks is retained. Distinct rock layers are called *formations*, and sets of formations constitute *strata* or a *stratigraphic section*.

These strata are, in essence, a relative chronology of geologic events on Earth, a point that was realized by Hutton and others in the nineteenth century. It was at that time that a very basic principle, that of *geologic succession*, was developed: *The chronological sequence of any stratigraphic section depends upon the original order in which the formations are laid down.*

The application of this principle, of course, is complicated. Tectonic processes will tilt or entirely invert a set of layers. Erosion will remove whole sequences of strata, leaving an enormous gap of time between the remaining layers and those subsequently deposited above them. Igneous rock may intrude into sedimentary layers, and the layers themselves could be buried and metamorphosed.

Figure 8.4 shows an idealized example of the inference of relative ages in a stratigraphic sequence. Consideration of the geologic processes described earlier in this chapter allows one to sort out what has happened. Age decreases, generally, from the bottom to the top. The lowest layer, once a sediment, was buried and metamorphosed at some time in the past, but not enough to completely obscure the pattern of the original sedimentary layers. Also seen in the figure is an unconformity, where the lower layers were tilted, and erosion then formed a level surface on which newer sedimentary layers were deposited. Erosion could have removed a number of layers between the tilted and the level sequences, but this cannot be discerned from the figure; in section 8.6, we describe how fossils provide information on such gaps. Intrusions of igneous rocks along faults (*dikes*) are sketched on the figure. Clearly, they must postdate those sedimentary or metamorphic layers through which they cross, and must predate those layers that lie above them and truncate them. With these guidelines, the reader should ponder figure 8.4 and work out the temporal sequence.

8.6 FOSSILS

The determination of relative time for events – that is, the order in which rock sequences were laid down or intruded into pre-existing layers – provides only limited information on the history of Earth. It is difficult or impossible from the rocks themselves to reliably correlate a sedimentary layer or sequence from one part of the world to another, or even from one part of a continent to another. Because of this, filling in the gaps associated with unconformities is often impossible. What is required, other than the absolute timescale afforded by radioisotopic dating, which has been available only in the last half-century, is some sort of indexing system for rock layers. Such an indexing system exists, in the form of fossils.

Defined broadly, a fossil is evidence in Earth's rocks of an organism (plant or animal) that once was alive. Fossil remains fall broadly into one of six categories:

(i) actual remains, including mummified bodies, shark's teeth, etc.;
(ii) petrifications, in which the original organic matter has been completely or partially replaced by mineral matter such as calcite, quartz, chert, pyrite, or others;
(iii) molds or casts of the inside or outside of bodies;
(iv) prints of leaves or of soft-bodied creatures such as jellyfish, and including detectable amounts of unidentified remnant organic matter;
(v) fossilized excrement;
(vi) tracks, trails, and burrows of animals.

Examples of fossils are shown in figure 8.5. The processes that result in fossilization are varied, depending

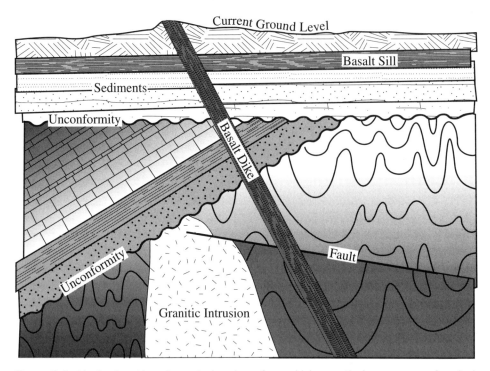

Figure 8.4. Idealized section of a geologic column from which a particular sequence of geologic processes may be inferred as described in the text. The section might be exposed, for example, as a roadcut, or as a cliff face on the side of a mountain. Adapted from Press and Siever (1978) by permission of W.H. Freeman and Company.

on the type of fossil being considered. For tracks or imprints, the sedimentary template must not be modified greatly as the loose grains of sand, mud, or pebbles slowly cement together under modest compression to form a sedimentary rock. Metamorphism, except for the lowest-grade variety, will distort such prints out of existence.

Petrification of biological structures is the sort of fossil with which most people are familiar, and those of large animals make up the most spectacular museum displays. The formation of such fossils depends on the chemical similarity between the elements carbon, the primary element in the dead organism after desiccation, and silicon, a predominant element in both continental and oceanic rocks. Examine the periodic table in figure 2.6, and note that carbon and silicon are in the same column of the table, occupying adjacent rows. As discussed in chapter 2, the significance is that carbon and silicon require the same number of electrons to complete or close a valence "shell"; the particular ways in which these elements combine with others therefore are similar. Because carbon and silicon occupy a central column of the table, many possible bonding combinations are possible; this complexity leads to some differences in their chemical behavior.

A dead organism resting in sediments, and eventually covered by them, exists in a universe that is predominantly rock. Hence there is far more silicon than carbon in the environment of the remains. In any physical system, atoms and molecules are continually exchanging places with other compatible atoms and molecules. Usually (considering now only atoms), atoms will exchange with other atoms of the same element. However, if a chemically similar element is present in far greater numbers, it may substitute for the atoms of the original element. This is the case with organic remains in sediments: Gradually, the carbon is replaced by silicon from the surrounding material. If the sediments are undisturbed, so that the location of the substituted carbon is preserved, the silicon atoms reproduce the structure of the organism. The outlines of the fossil are discernible because its texture and overall chemical composition are distinct from that of the rocky grains around it, even though both contain silicon.

The process of natural fossilization is a chancy and rare event. The vast majority of individual organisms that lived on this planet show no evidence of their existence, their atoms being consumed by other organisms or otherwise dispersed into various natural chemical systems and not replaced point by point with silicon. However, those creatures that have been fossilized, from the microscopic to the gigantic, provide a record of what life existed at the time a particular sedimentary layer formed. Because life has

Figure 8.5. Examples of fossils: (a) sycamore leaf from the Eocene epoch, found in Utah; (b) an Eocene epoch fish in sedimentary rock near Tuba City, Arizona; (c) dinosaur tracks made during the Triassic period, found in Lincoln County, Wyoming; (d) trilobites from the mid-Devonian period, fossilized in sediments from Erie County, New York. Consult the geologic timescale in figure 8.6 for dates corresponding to the epochs and periods given here. Photographs courtesy of Peter Kresan, University of Arizona.

changed over the history of Earth – new forms replacing older forms and whole ecosystems changing in a clearly recognizable way – fossils provide an index of the particular time when sediments were laid down. This timescale is still relative because no absolute measure of dates is available, but it can be correlated from one place to another on Earth. Unconformities can be identified when sedimentary layers with a particular fossil sequence are present in a geologic column in one part of the world but not in another.

Index fossils identify sedimentary rock formations as having occurred at a certain time in the history of Earth. Because fossils are not found in igneous rocks, and only rarely in metamorphic rocks, such rocks must be fitted into the time sequence according to their relationship to sedimentary layers that they may overlie, underlie, or intrude.

8.7 RADIOISOTOPIC DATING OF EARTH ROCKS

The fitting of igneous rocks into the sequence defined by indexed sedimentary layers provides a mechanism for absolute dating of such layers, by radioisotopic dating of the igneous rock. The radioactive clock in an igneous rock begins ticking when the rock solidifies from a melt. Before that, the atoms in the liquid magma are so mobile that daughter elements of a radioactive decay will migrate away from their site of formation, making dating impossible. Thus the age of a terrestrial rock is the time since it last solidified. Metamorphic rocks may be dated *radiometrically*, that is, by radioisotopes, but the age derived is ambiguous. It may reflect the time of metamorphism, if it was of the high-grade variety, but more often the atoms have only partially rearranged themselves, leading to an apparent age that is meaningless. Sedimentary rocks themselves cannot be dated radiometrically; the age thus derived reflects when a particular grain was originally solidified in an igneous rock, or may be uninterpretable if the grain came from a metamorphic assemblage. The time at which the grains were eroded from the rock outcrop and transported by water is not reflected at all in the isotopic age.

8.8 GEOLOGIC TIMESCALE

The relative timetable assembled from sedimentary sequences was divided by geologists according to major changes in the types of rocks present, and the appearance or disappearance of groups of fossils from layer to layer. Some boundaries in the timetable correspond to the apparently sudden extinction of a significant fraction of Earth's species in existence at the time. We discuss in chapter 18 very recent insights into the origins of such extinctions.

There is little regularity to the nomenclature associated with the geologic timetable (figure 8.6). Layers often are named after particular regions in which the first, or most famous, of a particular sequence of rocks is found. For example, the Jurassic period is named after an exposure of limestone rocks in the Jura mountains of Europe. A hierarchy of divisions is employed in naming rock layers: *eon, era, period, epoch*, and *age*, from largest to smallest divisions. Eons are divided according to major transitions in the rock record. The transition between Priscoan and Archean at 4.0 billion years marks the appearance of the oldest rocks on Earth; Archean to Proterozoic at 2.5 billion years was set originally at the age of the oldest discernible fossil *stromatolites*, remains of certain types of bacterial colonies, earlier fossil evidence of life now has been found. The Proterozoic to Phanerozoic transition at 570 million years roughly marks a rapid diversification and complexification in the types of life-forms on Earth.

Eras are finer divisions more closely associated with the appearance of certain fossil groups, and came into use before the division of time into eons. The Paleozoic is the first era of the Phanerozoic eon. The Paleozoic-Mesozoic boundary is set at the beginning of the dominance of dinosaurs in the fossil record; the Mesozoic to Cenozoic similarly marks the sudden disappearance of dinosaurs and proliferation of mammals. Periods are even smaller time divisions, and date back to attempts in the nineteenth century to divide the history of Earth, like a play, into three acts: Primary, Secondary, and Tertiary, with only the last (youngest) name surviving in current usage. Finally, epochs and ages represent small groups of, and individual, stratigraphic units for which variations in nomenclature from continent to continent are allowed.

8.9 A GRAND SEQUENCE

The classical example of a sedimentary sequence is that of the *Grand Canyon of the Colorado*, which cuts through the high plateau country of northern Arizona. From the South Rim to the Colorado River is a vertical drop of about a mile, and more from the North Rim. The canyon was formed as the Colorado River cut through the landscape, which was uplifted into a plateau sometime over the past few tens of millions of years. However, the sequence of rocks through which the river cuts covers a time period from 230 million to 1.7 billion years ago. The sequence consists primarily of sedimentary layers, with metamorphic rocks and then igneous rocks toward the bottom. The sequence reflects not just deposition of material, but periods of erosion as well; unconformities are present, the largest of which represents a missing 600 million years (or 13%) of Earth history. That time is not lost: Elsewhere in the region, including parts of the North Rim, and farther afield various portions of the missing sequence are present.

Figure 8.7 sketches the Grand Canyon sequence. Most of the various rock layers are named after aspects of the history of the Grand Canyon and the region. A given rock layer will correspond in time to many others around the world; those other layers will differ depending on the particular environment present in each locale at the time. A beach environment will look different in the rock record than a lake bottom, ocean bottom, or mountain region. Names of rock formations identify the location and hence the particular kind of layer present; in each locale, these can be tied to a particular time in the geologic timescale by relative or absolute dating techniques.

8.10 THE GEOLOGIC TIMESCALE AS A MAP

The geologic timescale of figure 8.6 is a kind of road map into the past, based on hundreds of years of work by geologists in the field mapping out layers of rock, determining the rock types and kinds of fossils if sedimentary, correlating with radiogenic dates on associated igneous rocks, and tying the sequences together worldwide. Our planet is not the only one with a geologic history – all of the planets and their moons have one. Characteristic of Earth, however, is the predominance of sedimentary processes based on transport of material by water. Only Mars is likely to have a similar (but much less extensive) form of depositional history – one can see evidence for layering in the sides of the major Martian canyon, Valles Marineris, and elsewhere. Whether fossils are present on Mars, though, is an open issue that we address in chapter 15. And we cannot rule out sedimentary processes on other worlds where

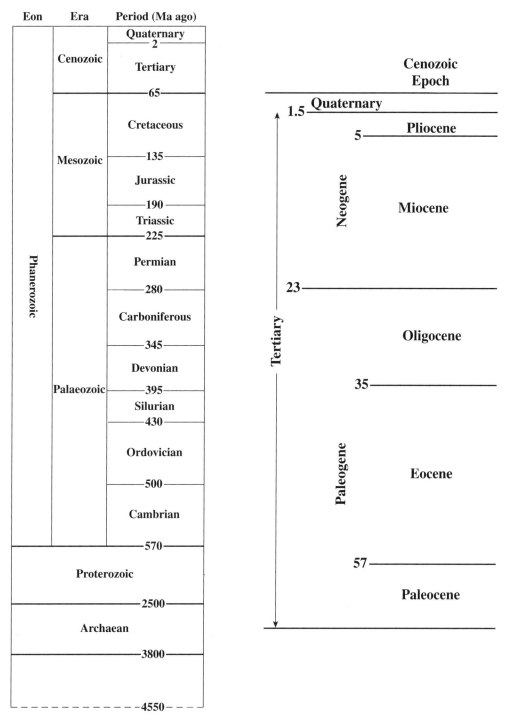

Figure 8.6. Geologic timescale: The left-hand table extends across the whole history of Earth, with times before present indicated in millions of years. In this book the eon of time from 4.55 billion to 3.8 billion years is referred to as the Hadean. The right-hand column is an expansion of the final 65 million years of Earth history, corresponding to the time after the extinction of the dinosaurs. The Tertiary period is further subdivided into epochs as indicated, and the epochs into stages, which are not shown. The final period, Quaternary, is split into the Pleistocene stage (from 1.5 million to 0.01 million years ago) and the Holocene stage (from 0.01 million years ago to the present); these are not given on the chart. The Tertiary is also sometimes broken into two subperiods called the Paleogene and the Neogene, but these divisions are not used further in the book. Adapted from Hawkesworth and van Calsteren (1992) and Rogers (1993).

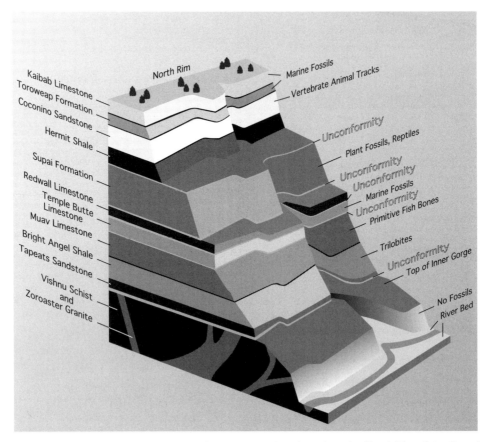

Figure 8.7. Sketch of the sedimentary layers exposed to view along the South Rim of the Grand Canyon of the Colorado River in Arizona. Unconformities and some key fossil types are indicated.

the liquid medium is not water: Deposits of water ice laid down by running liquid methane is an exotica that might exist on Saturn's largest moon, Titan.

8.11 QUESTIONS

a. Speculate on the political and cultural reasons why the enormous age of Earth was not recognized until the eighteenth century.
b. Consult a geologic map for the area in which you live to infer some of its geologic history. What was it like 100 million years ago?

8.12 READINGS

8.12.1 General Reading

Cloud, P. 1988. *Oasis in Space: Earth History from the Beginning*. W.W. Norton, New York.

8.12.2 References

Breed, W.J., Stefanic, V., and Billingsly, G.H. 1986. *Geologic Guide to the Bright Angel Trail: Grand Canyon, Arizona*. American Association of Petroleumn Geologists, Tulsa, OK.

Hawkesworth, C.J., and van Calsteren, P. 1992. Geological time. In *Understanding the Earth: A New Synthesis* (G.C. Brown, C.J. Hawkesworth, and R.C.L. Wilson, eds.). Cambridge University Press, Cambridge, UK, pp. 132–144.

Mason, S.F. 1991. *Chemical Evolution*. Clarendon Press, Oxford.

Murray, O. 1986. Greek historians. In *The Oxford History of the Classical World* (J. Boardman, J. Griffin, and O. Murray, eds.). Oxford University Press, Oxford, pp. 186–203.

Press, F., and Siever, R. 1978. *Earth 2/E*. W.H. Freeman and Company, San Francisco.

Rogers, J.J.W. 1993. *A History of the Earth*. Cambridge University Press, Cambridge, UK.

Wyllie, P. 1971. *The Dynamic Earth*. John Wiley and Sons, Inc., New York.

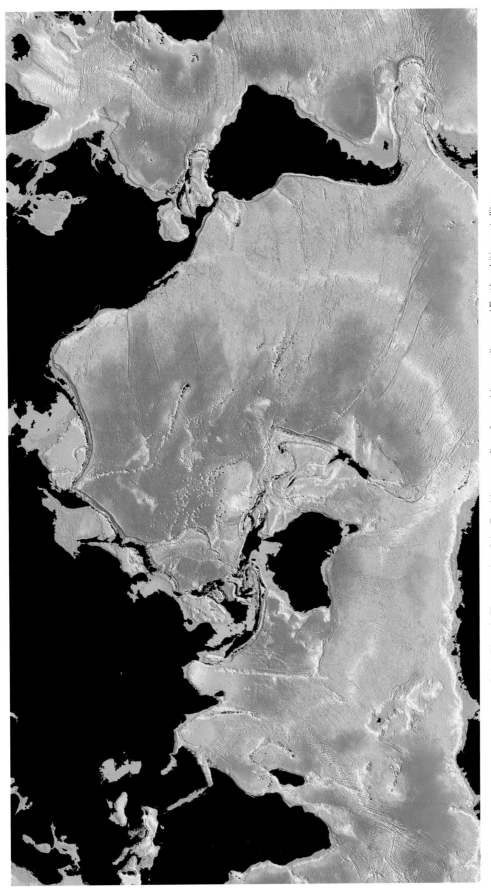

Color Plate I. Topography of the Earth's ocean floor from ship soundings and Earth-orbiting satellite measurements (Smith and Sandwell, 1997). Image courtesy of Dr. David T. Sandwell, Scripps Institute of Oceanography.

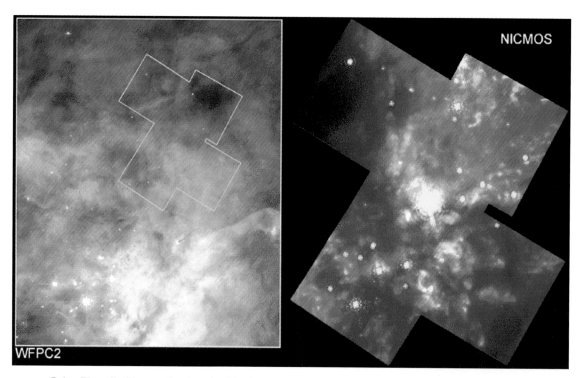

Color Plate II. Giant molecular cloud in the constellation Orion seen by Hubble Space Telescope in (left) visible light and (right) infrared light. The infrared image covers the area outlined on the visible image, stripping away the gas and dust to reveal very young stars. The visible image was produced by C. Robert O'Dell (Rice University) and colleagues, the infrared image by Rodger Thompson (University of Arizona) and colleagues. Courtesy of NASA and Space Telescope Science Institute.

Color Plate III. Hubble Space Telescope image of the Eagle Nebula, showing clearing of the gas and dust by ultraviolet light from newly formed stars which are illuminating the scene. The image was taken by Jeff Hester and Paul Scowen of Arizona State University using the Wide Field and Planetary Camera 2 at visible wavelengths. Courtesy of NASA and the Space Telescope Science Institute.

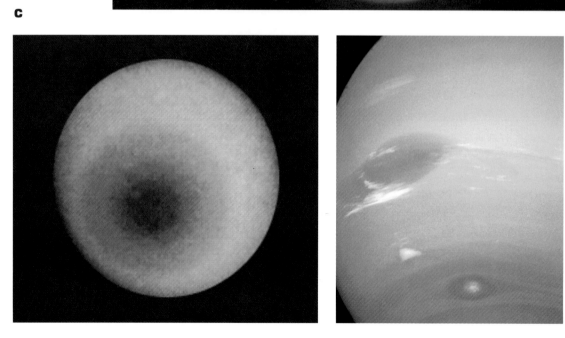

Color Plate IV. The giant planets of our solar system: (a) Jupiter from Hubble Space Telescope; (b) Saturn from Hubble, with colors and contrast exaggerated to show atmospheric patterns; (c) Uranus from Voyager 2, also with enhanced contrast to show very faint banding; (d) Neptune from Voyager 2. Photos (a) and (b) courtesy of NASA and the Space Telescope Science Institute; (c) and (d) courtesy of NASA/Jet Propulsion Laboratory.

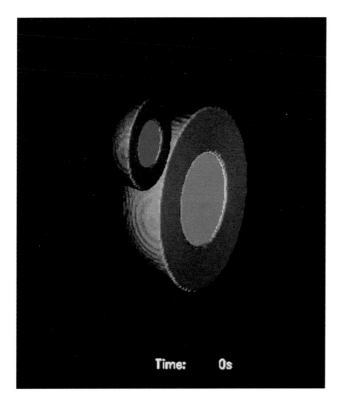
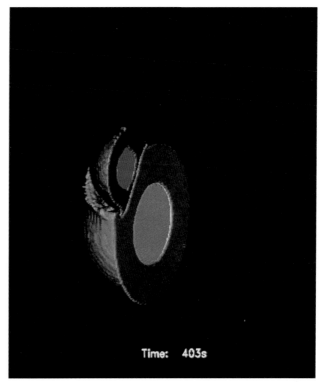
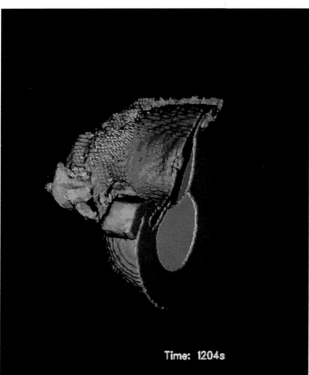
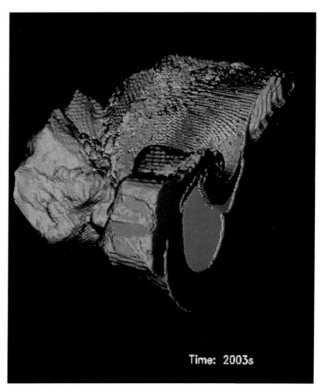

Color Plate V. Computer calculations by M.E. Kipp (Sandia National Laboratories) and H.J. Melosh (The University of Arizona), showing formation of the Moon as a Mars-sized planet strikes Earth. Both Earth and the impacting planet are shown sliced in half so as to reveal what is happening in the interiors. The iron-rich core can be seen as an inner circle in each planet prior to impact. Compared to the mantle of Earth, the core is hardly disrupted. Elapsed time is shown on each panel. Images courtesy of H.J. Melosh.

Color Plate VI. The Mars of today, captured by the Mars *Pathfinder* lander after its successful landing on July 4, 1997, shows hints of earlier times when water rushed across parts of the surface. It also shows the tracks left by the Sojourner rover in its travels across the landing site. The *Pathfinder* imager was developed for NASA by Peter Smith (University of Arizona) and colleagues. Image courtesy NASA/Jet Propulsion Laboratory and Mark Lemmon.

Color Plate VII. Clouds of dark material from multiple impacts, each of which is larger in area than the Earth. Courtesy NASA and Space Telescope Science Institute.

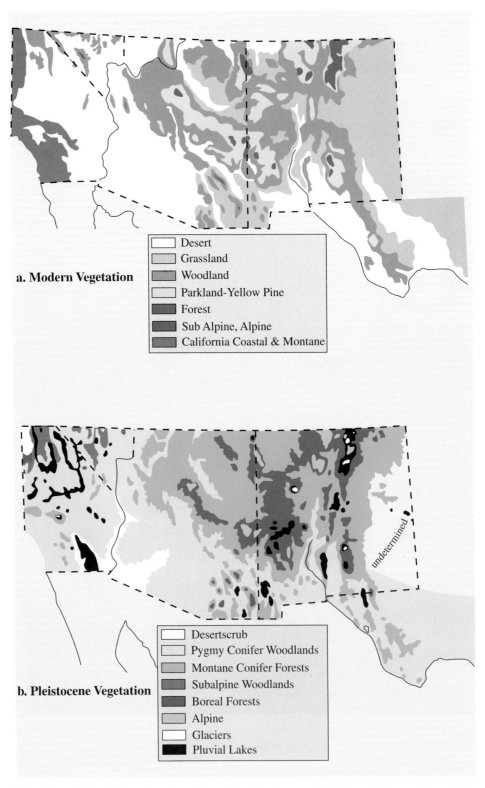

Color Plate VIII. (a) Map of southwestern United States vegetation from today and (b) a similar map from 18,000 years ago, based on packrat midden and pollen studies. Note that keys are slightly different, in part because vegetative associations and species types shifted as well as the altitude of occurrence. Roughly, "Parkland" and "Forest" in (a) correspond to "Montane Conifer Forests" in (b). Panels (a) and (b) are based on Betancourt et al. (1990).

9 PLATE TECTONICS: AN INTRODUCTION TO THE PROCESS

9.1 INTRODUCTION

We close the part of the book on techniques for discerning Earth's history with a conceptual tool. The concept of *plate tectonics*, whereby the outer layer of Earth is divided into a small number of distinct segments called *plates* which move relative to each other, represents a breakthrough in explaining a diverse range of geologic phenomena across our planet. Although the basic ideas are now 30 years old or more, this picture or concept of how Earth's geology works, in a unified way, continues to provide fresh insights into evolution of Earth, the stability of the gross climate of our planet, and the distinctions between Earth and the other planets. Because of its importance, we introduce the concept early to allow the reader to gain an understanding of the basic ideas. We come back to plate tectonics again and again as a fundamental process on Earth driving climate change, erosional processes, atmospheric chemistry, and even the nature of life.

9.2 EARLY EVIDENCE FOR AND HISTORICAL DEVELOPMENT OF PLATE TECTONICS

Revolutions in scientific thinking often take place when increasing numbers of observations challenge existing theories, which in many cases have become dogmatic over time in the face of conflicting data. Particularly satisfying is the synthesis of widely diverse data into a single framework that explains well all of the data.

As early as Sir Francis Bacon over 350 years ago, but mostly since the early nineteenth century when maps of the world became good enough to reveal the true shapes of the continents, the significance of the curious matching of the edges of distant continents has been pondered. Africa and South America seem to fit fairly well into each other, and those people inclined toward jigsaw puzzles found they could put North American, Europe, Africa, and Asia into a plausible single landmass (figure 9.1). What this implied for the origin of today's continents, that somehow they broke apart from one or several bigger landmasses, seemed ludicrous to most geologists used to working on a rather solid Earth.

Other lines of evidence caused trouble for static models of Earth's continents. In the nineteenth century, it was noticed that fossil organisms in a given region, particularly on southern continents, reflected ecosystems with widely varying climates, ranging from cold desiccated regimes to lush tropical jungles. Some scientists argued that Earth has tumbled through its history, shifting the positions of the high latitudes (i.e., poles) through the various continents over time. However, also noticed was that, prior to the Cretaceous period of Earth history, similar or identical fossil plants and animals were found on continents separated today by deep oceans. During and after the Cretaceous, these biological linkages largely disappeared. The idea of temporary land bridges was advanced to allow nonmarine organisms to transfer between continents. However, it was evident to some that allowing continents to split apart and drift away from some common original supercontinent could explain both types of observations. Further support came from the observation that North and South American mountain ranges that terminate at the Atlantic Ocean line up very well with ranges in Europe and Africa, as if they had once been continuous belts within the inlands of a single larger continent.

In the late nineteenth century, catastrophic models were proposed to provide mechanisms for breaking up and moving continents. The idea was advanced that Earth's Moon was originally material blown out of Earth by asteroidal collision from what is now the Pacific Ocean, and the resulting stresses on a single continent on the other side of the world caused fracturing which separated the

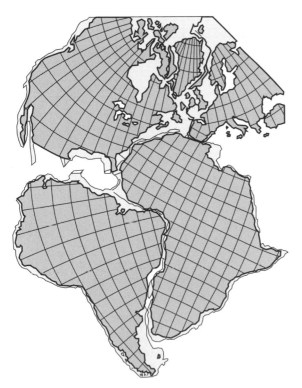

Figure 9.1. Modern version of a jigsaw puzzle worked since the 1600s: the fitting together of the continents around the Atlantic Ocean. The solution shown was done by computer to yield the least amount of overlap. The gridded portion of continents shows land currently exposed; the outlined but ungridded area is *continental shelf*, which is the part of the continental landmass that is currently underwater. (For this fitting exercise, only that part of the shelf extending down about a kilometer below sea level was taken.) The fit is not perfect, but many of the misfits arise from portions of continental material that turn out to be younger, that is, added later than the time when the continents are thought to have split apart. Modified from Wyllie (1971) by permission of John Wiley and Sons, Inc.

continents and opened the Atlantic Ocean. (As we see in chapter 11, current models of lunar origin are not too dissimilar from this basic concept, though very different in the details.) The most famous proponent of continental drift, German meteorologist Alfred Wegener, suggested in 1915 that, because Earth is slightly oblate, with a larger radius at the equator, continents might pull apart from a single continent originally centered at the poles. He also argued that a westward drift might be caused by the gravitational pull of the Sun and the Moon. He saw the continents, made mostly of granite, as being able to plow through the basaltic crust of the oceans.

Wegener's enthusiasm for continental drift did not play well in conservative scientific circles, and in fact, the gravitational tidal forces he invoked to induce drift are much too weak to do so. As a result, the concept gained the flavor of a crackpot model in mainstream scientific circles, and helped delay the realization and acceptance of plate tectonics by four decades.

9.3 GENESIS OF PLATE TECTONICS AFTER WORLD WAR II

9.3.1 Seafloor Topography

The decades after World War II saw a burst of activity in many sciences, including geophysics. Techniques were developed – in large part from military *sonar* technology, which uses echoes of sound waves to locate underwater structures – to map the ocean floor's topography, that is, the distribution of elevation caused by undersea mountains and trenches, among other features. The results were striking, as shown in figure 9.2 (see also color plate I). The ocean floor is subdivided by long ridges of mountains, stretching over thousands of kilometers. These ridges are cut transversely by a series of fractures, or *faults*. Near the edges of some continents or major island chains, long trenches are present, the deepest of which is the Marianas trench extending down 11 km from the surface of the ocean, some 7 km below the average seafloor depth.

Even more striking is the distribution of height over the surface of Earth, shown in figure 9.3. Essentially, Earth has two different kinds of surfaces, characterized by their depth: continents and oceans. The range of depths within each kind of surface is generally less than the difference in depth between them, leading to a *bimodal distribution*. As we see in chapter 15, this is dramatically different from the situation on Mars and Venus, suggesting that Earth's geology has been shaped by a global set of processes unique to our planet. Whether plate tectonics is in fact unique to Earth is explored critically in chapter 16.

The mid-ocean ridges and trenches are distinguished not merely by their topography, but also by the fact that earthquakes and volcanoes are concentrated along their lengths (figure 9.4). Earthquakes also are concentrated along certain fault systems, such as the famous San Andreas fault of California. Because earthquakes are the result of stresses built up by movements in the outer layers of Earth, the presence of narrow belts of activity suggest that the crust is characterized by organized motions rather than by random distortions or deformations.

9.3.2 Magnetic Imprints on Rocks

Further evidence for moving continents came from an entirely different field of study, called *paleomagnetism*. The magnetic force is one expression of the general

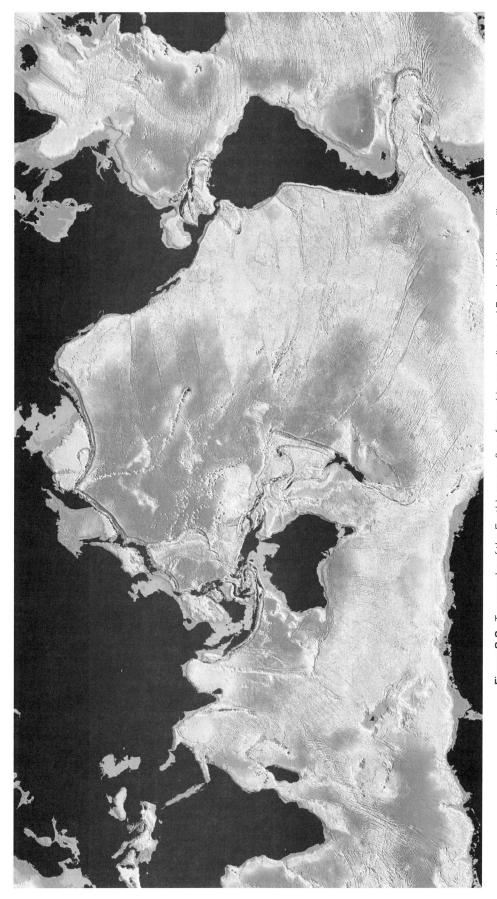

Figure 9.2. Topography of the Earth's ocean floor from ship soundings and Earth-orbiting satellite measurements (Smith and Sandwell, 1997). Image courtesy of Dr. David T. Sandwell, Scripps Institute of Oceanography.

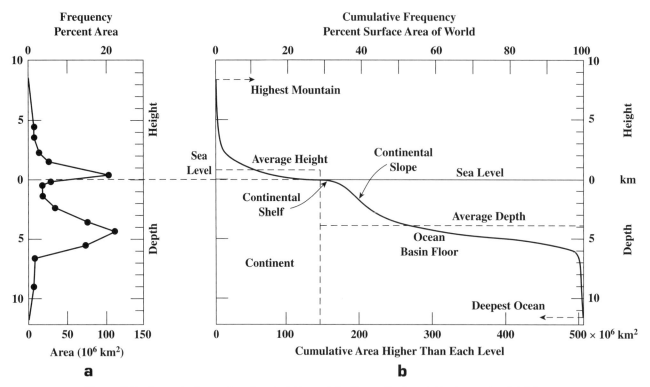

Figure 9.3. Amount of area at various heights and depths on Earth, relative to sea level: (a) Amount is expressed in terms of total area occupied by surfaces with a given height (or depth) in kilometers. The bimodal nature of Earth's topography is evident. (b) The same data are expressed in terms of cumulative area higher than a given altitude; this gives a sense of the gross profile of continents and ocean floors. From Wyllie (1971) by permission of John Wiley and Sons, Inc.

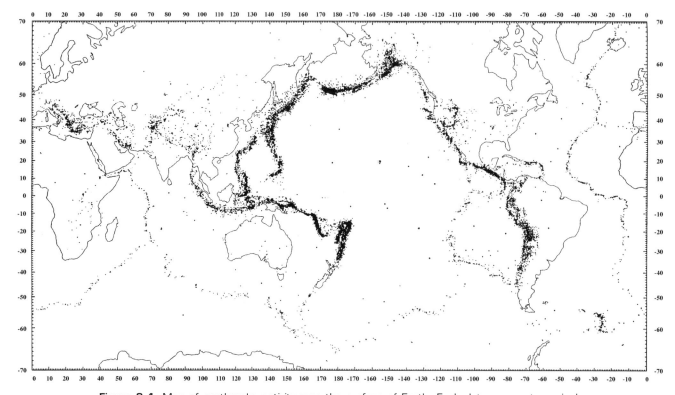

Figure 9.4. Map of earthquake activity over the surface of Earth. Each dot represents a single epicenter. The geographic pattern of volcanic eruptions is very similar. From Isacks et al. (1968).

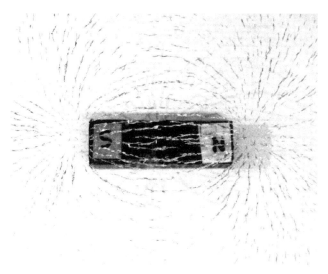

Figure 9.5. Appearance of iron filings when a permanent magnet, in this case bar-shaped, is placed underneath the glass upon which they rest. Lab setup courtesy of Larry Hoffman, University of Arizona Physics Department.

electromagnetic force discussed in chapter 3, which is manifested by charged particles that are in motion. Certain materials, such as the mineral *magnetite*, exhibit the ability to retain a permanent magnetization associated with the alignment of the spins of the electrons in the atoms of which they are comprised. Such a magnetic force will cause tiny scraps of iron (iron filings) to line up in a particular manner, which defines the direction of the *magnetic field* of the grains in the mineral (figure 9.5).

Where does the magnetic field of magnetic or, more precisely, magnetizable, minerals come from? Anyone who has played with a compass is familiar with Earth's own magnetic field, which, as with a bar-shaped magnet, has a definite directionality to it. A magnetized iron needle suspended so as to rotate freely (i.e., a compass) will point in the general north-south direction – not quite due north-south, because Earth's magnetic field is not precisely aligned with the geographic poles. The origin of Earth's magnetic field remains an outstanding puzzle in planetary sciences and is discussed further in chapter 11.

Important to us here is that, as minerals crystallize from molten rock, they do not immediately become capable of holding a magnetic field. This is because at high temperatures the electrons are too mobile to acquire a permanent fixed direction to their spins. Instead, the mineral grains must pass below a temperature known as the *Curie point* to retain a permanent magnetization. This temperature is roughly 800 K and is well below the crystallization point at which the minerals solidify from the melt. Because of this, the direction of the magnetic field inside a permanently magnetic mineral depends on the orientation of the mineral, relative to Earth's magnetic field, at the time it cools through its Curie point.

Hence, if a mineral cools and magnetizes, it becomes a kind of compass. If the rock formation in which the mineral is embedded is rotated by 90 degrees, the "north pole" of the magnetic field in the mineral will point roughly east or west, not north. This would seem to provide an excellent means for determining the movement of continents relative to Earth's magnetic field, provided that one can date the rocks by radioisotopic techniques. However, interpretation of this information up through the 1950s often focused on the idea that the rotational poles of Earth had shifted with time – either that Earth slowly tumbles through space, or that the entire outer layer of Earth slips over the interior in one piece. In this interpretation, there is no relative drift of the continents but only synchronous changes in latitude around the world.

Studies of continental rocks indicated that rotations in the apparent field directions could not be accounted for solely, or even primarily, by a coherent shifting of all of the continents. Instead, continental drift had to be invoked to reconcile the directions of remnant magnetization in rocks of various ages and locations. But how were the continents moving relative to each other? The key observation came from the direction of magnetization of rocks on the seafloor.

Study of continental rocks revealed that the Earth's magnetic field has reversed direction many times in the past – a compass held fixed during a reversal would swing from south to north, or vice versa. These reversals could not be accounted for by a 180-degree rotation of the rock itself, because rocks of the same age from a variety of locations and orientations show the same reversal. The origin of the reversals lies in the way Earth's magnetic field is generated, deep in its interior, but is as yet very poorly understood. Regardless of their origins, magnetic reversals provide yet another way of delineating the progression of geologic time in the rock record, provided the ages of the reversals are determined by independent dating of rocks, for example, by radioisotopes.

In the 1950s, technologies developed to detect submarines by their magnetic signatures began to be employed to map the magnetic orientations of seafloor rocks from surface ships. Very quickly, a remarkable regularity emerged at the mid-ocean ridges, illustrated in figure 9.6. Stripes of differing magnetic field intensity are laid out parallel to the ridge itself. The variations in intensity are most straightforwardly interpreted to be caused by the direction of Earth's magnetic field at the time the rock cooled from a magma. The pattern mimics that of the

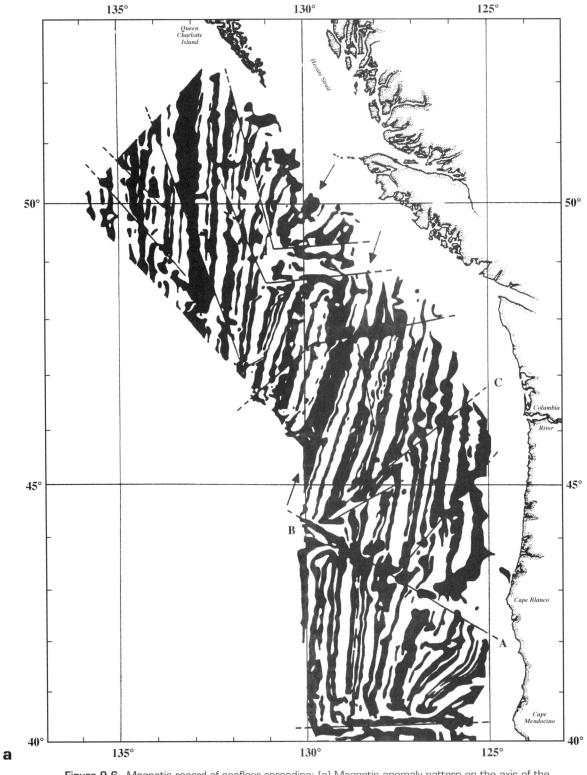

Figure 9.6. Magnetic record of seafloor spreading: (a) Magnetic anomaly pattern on the axis of the Juan de Fuca ridge, near Vancouver island. Black indicates current magnetic field direction, white the reversed field. From Morgan (1968).

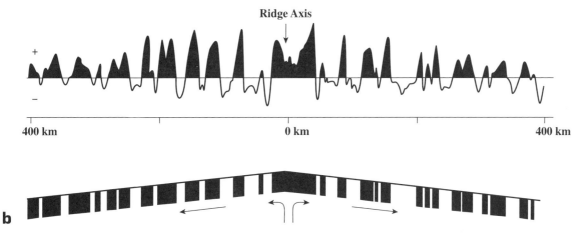

Figure 9.6. (Continued) Magnetic record of seafloor spreading: (b) Interpretation of the cause of magnetic reversals, shown as a cross-section through the top of the Earth's oceanic crust. Arrows indicate the sense of spreading. On the upper chart, the axis of the mid-ocean ridge is marked, and the relative strength of the magnetic field as a function of distance from the axis is given. Positive (black) is the present direction of the field. From van Andel (1992).

field reversal history recorded in rocks on the continents, indicating that the youngest rocks are closest to the mid-ocean ridge, increasing in age farther out. The simplest and most straightforward interpretation of the pattern was advanced in the 1960s: The ridges are sites where new ocean crust is being created, moving like a conveyor belt away from the ridges. As the new crust cools after extrusion, the field direction is recorded in magnetic minerals as they cool below their Curie point. This interpretation quickly led to another question: If new seafloor is being created at ridges, where is it being destroyed?

9.3.3 Geologic Record on Land

By the 1960s it was becoming increasingly difficult to refute the notion that continents moved about Earth, and that seafloor spreading was somehow involved. Geologic patterns on continents were now being re-evaluated in light of the apparent mobility of Earth's surface. Radioisotopic dating of similar types of igneous rocks on the eastern end of South America and western end of Africa showed a remarkable correspondence – a well-delineated boundary separating rock of 2-billion-year and 600-million-year ages in western Africa was present in eastern South America as well, and in just the right place for a good jigsaw puzzle fit.

Rocks in mountain ranges in northern California, on the west side of the very active San Andreas fault, matched up well in type and age with rocks a couple of hundred kilometers to the south, in southern California, on the *east* side of the fault line. It was more than tempting to simply slide the east and west portions of California, along the fault, so that these rock types matched. Measurements spanning many decades of ground slippage resulting from earthquakes showed a northward movement on the western side of the fault of roughly a centimeter per year. Using that figure, the now-separated rock formations would have been together some tens of millions of years ago.

Drift timescales of centimeters per year seemed to fit in other parts of the world. The magnetic anomaly pattern on the mid-Atlantic ridge displayed ocean floor ages consistent with spreading of a few centimeters per year. Studies of the Hawaiian island chain showed that the age of the islands increases to the northwest, with active volcanism confined to the Big Island (Hawaii) on the southeast end and a still-growing submerged island (*seamount*) just to the southeast of it. If interpreted in terms of the ocean crust drifting northwestward over a source of volcanism, drift rates of some centimeters per year again are obtained.

With mounting evidence in the 1960s for continental drifting and seafloor spreading, the question of how the growing crust was accommodated was answered through the study of earthquakes.

9.3.4 Earthquakes and Subduction

Networks of *seismometers*, which measure the local shaking of the ground due to near and distant earthquakes, are capable of inferring both the geographic location (epicenter) and the approximate depth at which the sudden shifting of rock occurred which caused the quake. (More on this is given in chapter 11.) Most earthquakes occur at shallow depths beneath Earth's surface. Careful comparison of figures 9.2 (color plate I) and 9.4 shows that quakes are common at mid-ocean ridges, trenches and at sites of lateral movement such as the San Andreas fault in California. However, earthquakes that originate at depths greater than 100 km are confined to the trenches, where some quakes originate as deep as 600 km below Earth's surface (figure 9.7).

Earthquakes generally occur where stresses build up in rocks and that stress is relieved suddenly by a failure or fracturing of the rock. Stress buildup is generally the result of some movement that forces rocks up against each other; in zones such as the San Andreas, such motion is a sliding one in which rocks lock up against each other and then break free. In the trenches, a map of earthquake locations showed an interesting pattern: The deeper quakes were actually displaced, usually toward the continental side of the trenches (figure 9.7). The most natural interpretation is that the trenches are regions where ocean floor is being forced under continents; as the movement occurs, lockup of rocks causes stress buildup and eventual release through fracturing.

It was clear from this and other evidence that the trenches are sites at which ocean crust is destroyed: the other end of the conveyor belt that begins at the mid-ocean ridges. Ocean crust is forced under continental crust

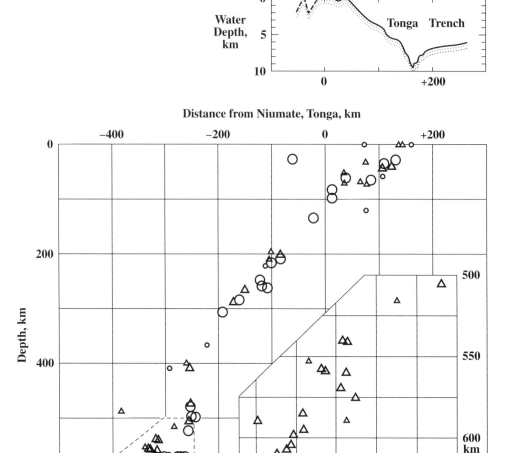

Figure 9.7. Earthquake location and depth near the Tonga subduction region in the Pacific. A profile of the topography (exaggerated 13 times) is shown in the upper small panel, positioned so that the zero point is aligned with that on the big figure. Note that the position of the quakes with depth seems to outline quite nicely a subducting slab of crust moving diagonally under Earth's surface. The symbols distinguish between earthquakes that occurred north of the station at Niumate, Tonga, (circles) and those occurring south of Niumate (triangles). The inset shows the deepest quake locations in more detail. From Isacks et al. (1968).

because the former is denser than the latter. Much the same effect occurs when cold air is forced under hot air: On a small scale, this causes heated air in a room to cling to the ceiling; on a larger scale, it generates uplift of the warm air, which produces violent thunderstorms. Returning to Earth's crust, the timescales are much longer, but the basic physics is the same: The continents force oceanic crust to dive underneath them, eventually reaching temperatures at which the crust melts and is destroyed. The density difference is chemical in nature, ocean crust being basalt which is a relatively dense rock compared to continental granites. The origin of this difference, one of the key enablers of plate tectonics and a distinguishing feature of Earth compared with the other planets, is considered in chapter 16.

With understanding of the nature of the great seafloor trenches, the basic picture of floating continents was completed in the 1960s and became the theory of *plate tectonics*. Tectonics refers to the study of movement and deformation of the outermost layer of Earth, its crust; plate refers to the emerging concept that Earth's crust was divided into discrete plates. Some plates carry continental masses like passengers on rafts; seafloor is created at plate boundaries called mid-ocean ridges and destroyed at *subduction zones* which are the ocean trenches. This basic realization tied together a wealth of geologic data accumulated over the first half of the twentieth century and has shaped our thinking about the geology of Earth and other planets since then.

9.4 THE BASIC MODEL OF PLATE TECTONICS

The Earth's crust is broken up into a small number of relatively rigid plates that move slowly across the surface in response to forces generated beneath the crust, in the *mantle*. Strictly speaking, the terms crust and mantle refer to chemical differences between the layers. It is more accurate to refer to the plates as comprising the Earth's *lithosphere* (rock-sphere), a rigid outer shell that rides on a hotter and plastic (not molten) layer called the *asthenosphere* (weak-sphere). The crust-mantle boundary is not the same as the lithosphere-asthenosphere boundary, but there is significant overlap between the mantle and the asthenosphere. As discussed in chapter 11, knowledge of the presence of a plastic layer beneath a rigid outer shell comes from observing the slowing of earthquake waves which must pass through this layer; the chemical differences are inferred from material erupted to the surface in certain volcanoes.

Figure 9.8 illustrates the nature of the basic plate boundaries. Ridges and trenches where crust is brought up and forced back down, respectively, are not sufficient to

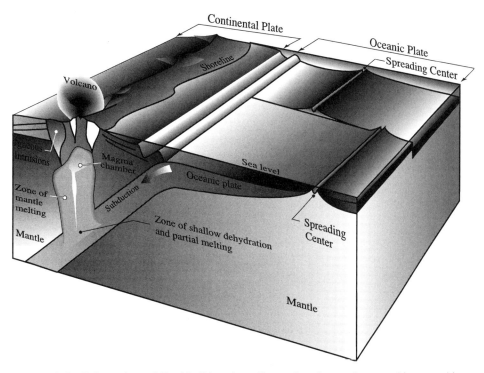

Figure 9.8. Oblique view of Earth's lithosphere illustrating the motion at mid-ocean ridges (spreading centers) and subduction zones. The spreading center is shown split by a transform fault.

accommodate plate motion on a spherical Earth with irregular continents. Instead, the ridges are sliced through by *transform* faults, where lateral motion occurs. The San Andreas fault is part of a transform system, which connects the eastward-moving subduction of the Pacific plate on the west coast of South America with the northerly and northwesterly subduction occurring along western Canada across to Asia. A simplified map of ridges, trenches, and transform faults around the world is given in figure 9.9.

The speeds at which plates move range from 1 to 10 centimeters per year, corresponding to a thousand kilometers in 10 million to 100 million years. Aside from the geologic evidence that tells us, among other things, how long ago certain well-separated rock formations were together, and magnetic reversal stripes on the seafloor that give us the rate from the calibrated history of reversals, the space age provided direct measurement capability. Astronauts placed reflectors on the surface of the Moon, which geologically appears to lack plates and to be a rigid surface. Bouncing laser beams off the Moon from various continents on Earth allows the relative movements of the stations to be determined directly. More recently, Earth-orbiting satellites that calibrate their position by making accurate observations of deep-sky objects have been able to make similar direct determinations of relative movements of ground stations. Finally, large radio telescopes on Earth that also determine their position by staring at the sky, and are linked together by computers, can determine continental drift as well. The measured speeds are similar to inferred speeds at which the very plastic rock in the asthenosphere of Earth slowly turns over, removing heat to the surface.

Current plate motions and associated geology provide a good tour of the phenomena of plate tectonics (figure 9.9). The concentration of earthquake and volcanic activity along the Pacific rim is the result of the northwestward movement of the Pacific plate, where it is taken up by subduction along the Aleutian islands and the east coast of Asia. Southward from Indonesia, the Pacific plate adjoins the Australian plate, which generally is moving northward. Although the plates are approximately rigid

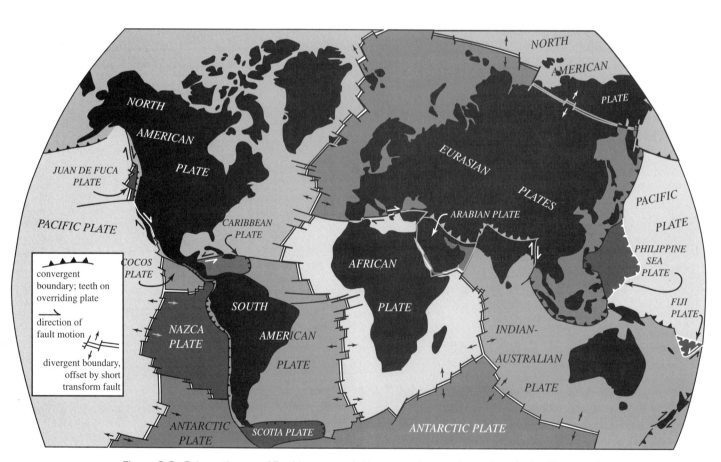

Figure 9.9. Schematic map of Earth's system of ridges, trenches, and transform faults. Plates and the general directions of their motions are labeled. Adapted from Cloud (1988) by permission of W.W. Norton Company.

bodies, collision between continents on separate plates leads to compression and uplift of mountains or plateaus. The collision of India on the Australian plate with Asia is the most dramatic example of this, pushing the Himalayan plateau up to the greatest altitudes above sea level anywhere on the planet.

On the east side of the Pacific, the plate's northward motion is expressed by transverse sliding of the plate along the North American plate. The San Andreas fault system is where this motion is accommodated on the continent, leading to the prodigious earthquake activity in California. Farther north, the Pacific plate undergoes subduction, the expression of which includes volcanoes such as Mount St. Helens. The material brought upward from the heating of subducting plates includes basalts, rhyolites (a volcanic form of granitic rocks), and a kind of hybrid between basaltic and granitic compositions, called *andesite*. These volcanic products reflect a complex sequence of mixing of oceanic crust with scraped-off detritus of the underlying portions of the adjacent continent, followed by melting and eventual ascent to the surface.

Farther south, subduction of the Cocos and Nazca plates under South America is expressed in extensive mountain building and volcanic activity along the Andes. Spreading between the Cocos and Nazca plates, and between the Pacific plate and both of these smaller plates, creates a complex *triple junction*, the motion of which is accommodated by the subduction and transverse motions along the western end of North America.

The situation in the Atlantic is simpler and quieter. Spreading of that ocean along the mid-Atlantic ridge, dividing the African and Eurasian plates to the east and the South American and North American plates to the west, continues. The African plate is also rotating in such a way that the African continent is moving northward, pushing the floor of the Mediterranean Sea into southern Europe and building the Alps. The Arabian landmass is moving into Asia, and a zone of spreading between Africa and Asia includes the Arabian Sea and very substantial rift valleys on the east coast of Africa itself.

9.5 PAST MOTIONS OF THE PLATES AND SUPERCONTINENTS

Reconstructing the positions of the continents back in the past is straightforward for recent times, because one simply has to take today's plate motions and run them in reverse. Eventually, however, one reaches a point at which the current continents have amalgamated into two continents, and then even earlier, into one supercontinent. At that point, roughly 200 million years ago, plate motions cease, and the earlier history of plate tectonics must be deduced from similarities between ancient rock formations on different continents, as well as the magnetic orientations of rocks.

Figure 9.10 shows the history of spreading from 200 million years ago (Jurassic times) to the present. *Pangaea*, a name derived from the Greek pan (all) and gaia (earth),

Today

65 Million Years Ago

135 Million Years Ago

200 Million Years Ago

Figure 9.10. Reconstruction of the drifting of continents from Jurassic times to the present. Adapted and redrawn from Cloud (1988).

was a single Jurassic supercontinent comprising all current landmasses. It began to break up with North America and Eurasia forming *Laurasia*, from the Greek laura, meaning passage or channel, and the remaining continents comprising *Gondwana*, named by a contemporary of Wegener, for unknown reasons, for an ancient tribe in India. The *Tethys* seaway between the two is the ancestral Mediterranean, meaning mid-land, Sea. By 135 million years ago, Gondwana had begun to break up into Africa/South America, Antarctica/Australia, and India. At the close of the Cretaceous, some 65 million years ago, Africa had reconnected with Eurasia, but the major new event was the opening of the Atlantic Ocean, the growth of which continues at present. Extrapolating 50 million years into the future, the Atlantic will widen, a new sea will open in East Africa, Australia will cross the equator, the present southern California coast will shift north of San Francisco and the Gulf of California will merge into the Pacific.

While this scenario generally has been considered sound for about a quarter of a century, new work has illuminated the history of the continents prior to 200 million years ago, again from detailed detective work on rocks and their magnetic orientations. Rocks from prior to the Mesozoic era in North America and Africa have different magnetic directions, such that they could not be pieced back together into a single Pangaea prior to about 260 million years ago. Evidence showing that ancient ocean basins actually closed to form Pangaea at this time comes from volcanic rocks of ocean crust composition appearing in mountain ranges of Argentina, Africa, and parts of the Appalachians, these rocks being forced up onto the continents by the closure.

Reconstructing the continental configurations prior to Pangaea raises an important point: Because of the continuous process of subduction of ocean crust, there are few seafloor rocks older than roughly 250 million years before present. On the continents, which do not subduct, rocks dating back billions of years are not uncommon. Here and there on the continents, very ancient oceanic crust can be found, where it was pushed upward and sutured onto continents during collisions. These preserved remains of ocean floor, or *ophiolite suites*, stand in stark contrast to the vast bulk of the oceanic crust that is part of a youthful and dynamic conveyor belt of material moving from ridge to trench in a matter of some hundred million years or so.

Expeditions to Antarctica by teams of scientists, including I.W.D. Dalziel of the University of Texas at Austin, have found sandstones in the mountains of that continent from about 700 million years ago that contain fossil remains of worms identical to those in rocks of similar age in Wisconsin and other locations around the world. This and other resemblances in widely separated rocks of the era push one toward the existence of another supercontinent, named Rodinia, in existence about 750 million years ago. The breakup of that continent and eventual reassembly into Pangaea some half-billion years later has been pieced together by Dalziel and numerous other geologists, using computers to aid in reconstructing continental configurations based on rock relationships on different continents. Figure 9.11 illustrates the possible sequence.

Prior to Rodinia the record of continental motions is extremely limited. The fossil record is very poor before 750 million years ago, in part because the variety and abundance of preservable organisms was impoverished prior to that time. The explosion of new and complex biological types is a remarkable watershed about 600 million years before present that we consider in chapter 18. The rock record itself is spottier, both fossils and magnetic signatures being more poorly preserved because of subsequent modification of the rocks. Furthermore, the decreasing abundance of rocks at earlier and earlier ages is not simply a matter of preservation: The continents themselves have grown over time, and prior to several billion years ago, much of the landmass we see today did not exist (chapter 16). Thus the history of plate tectonics prior to a billion years before present, including assembly and disruption of supercontinents, is not well understood.

What is now understood from the efforts to track the motions of continents over hundreds of millions of years is that creation and destruction of supercontinents must have had important effects on the climate and biota of Earth. This is not simply due to the changing latitudes of individual localities but more profoundly to the global effects on ocean currents and landmass ice sheets of having continents and oceans in different configurations. We discuss these possible effects in chapter 19.

9.6 DRIVING FORCES OF PLATE MOTIONS

What drives the motion of the plates? To understand fully the origin of plate tectonics requires thinking about how Earth gets rid of its heat, a subject covered more fully in chapter 11. However, some important aspects of the forces behind plate motion can be considered here.

The mechanics of plate motion suggest two mechanisms for moving plates. At mid-ocean ridges, hot material is welling up from the asthenosphere/mantle to form new crust. This hot material must be less dense than its surroundings in the mantle, in order for it to rise. As it reaches the surface it must move laterally away from the ridge, creating a pushing force as more material rises to take its place. This *ridge push* has been invoked by some as a means to drive spreading motions. At subduction zones, or trenches, oceanic crust is forced under continents back

a. NEOPROTEROZOIC
(750 million years ago)

b. MIDDLE CAMBRIAN
(530 million years ago)

c. MID-SILURIAN
(422 million years ago)

d. EARLY PERMIAN
(260 million years ago)

Figure 9.11. Continental drift from 750 million to 250 million years ago, from Rodinia to Pangaea. The initial figure shows Rodinia 750 million years ago with mountain ranges as they exist today (white lines) to illustrate their continuous nature in the ancient supercontinent. The remaining drawings show the continents at various times. Based on Dalziel (1995).

into the mantle. For this process to take place, oceanic crust must be denser than continental, and indeed, the oceanic basalts are significantly denser than the continental granites and other rocks. However, for subduction of the slab to continue downward some 600 km or more (based on earthquake data), the slab also must be denser than the surrounding mantle. How can ocean crust be buoyant relative to mantle at ridges while not being so at the trenches? Cooling of the ocean crust apparently leads to a sufficient increase in density of the rock that it sinks into the mantle. This process is aided by the thinness of the ocean crust, which allows it to be easily bent and redirected into the mantle at the trenches.

A very crude analog of this process is seen at volcanic lava lakes. Hot material rides to the surface, cools, solidifies, and gets dense enough to sink. In the case of ocean crust and mantle, the basalt within a ridge is molten, but it solidifies quickly as it rises out of the ridge. The density increase that allows sinking occurs in the solid basalt as it moves away from the ridge and cools further.

The effect of slabs of ocean crust sinking is to produce a pulling motion on the plates. This force also has been invoked as that which sets the plates in motion. Detailed modeling of both slab-pull and ridge-push suggests to some geophysicists that the pulling action is most important. However, this cannot be the case when the continents are all combined into a single supercontinent, which has happened at least twice in Earth's history. The initiation of continental breakup must be caused by another phenomenon, one not yet well understood. What has been suggested is that the supercontinent, having stalled plate motion, tends to prevent heat from escaping from Earth's interior. Hot spots in the mantle are formed, which move buoyant material rapidly upward to the base of the supercontinent, causing rifting and eventual breakup. Another hypothesis is that the hot spots do not originate because of the presence of the supercontinent, but arise very deep in Earth for unknown reasons, and those appearing under supercontinents cause their breakup.

What is the origin of the chemical differences between mantle, oceanic crust, and continental crust? Armed with an understanding of the basic plate tectonic mechanisms, we discuss this issue in chapter 16. It is a critical one, because the subtle density differences between these different types of rock enable plate tectonics to occur, and these differences in part depend on the presence of liquid water. The continued stability of liquid water on Earth over time is dependent on a climate regime that may have been moderated by plate tectonics, leading to a tightly coupled and complex relationship between water and plate motions.

It is valid and worthwhile to ask whether other planets and moons in the solar system possess plate tectonics. The answer seems to be no for the other terrestrial planets. Mars has a thick crust and cool interior which long ago stalled plate tectonics. Venus, so similar in size to Earth, has a surface characterized by massive volcanic flows and little or no evidence for plate tectonics. It seems the lack of water and very high surface temperatures have played a role in preventing Earth-type tectonics from starting there. We delve deeper into the reasons for Earth being unique in regard to plate tectonics in chapter 16.

For those who were educated in geology prior to the 1950s, the processes shaping Earth seemed to be vertical ones: mountains get built upward, and subsidence creates ocean basins. Plate tectonics turned geologic thinking sideways: The primary motions are horizontal, but in the compression and breakup of plates lie the origin of most mountain ranges and new seas. Throughout the book we see profound effects of cycling crust over time, as important atmospheric gases get captured in ocean sediment and cycled beneath continental crust, only to be released again. The interplay between plate tectonics and Earth's atmosphere and biosphere is one of the surprises of late twentieth century geology and planetary sciences, one with a sobering impact on the possibilities for stable climates equable for life elsewhere in the cosmos.

9.7 AN END TO TECHNIQUES AND THE START OF HISTORY

In this part of the book we have considered tools and concepts essential for our examination of the history of Earth and other planets. We turn next to that history, starting at the beginning of our solar system and its central star, the Sun.

9.8 QUESTIONS

a. Why do you suppose geologists of the early twentieth century were so reluctant to consider continents moving across the globe, in view of the fact that they accepted as plausible large *vertical* movements?
b. The technologies available to geologists after World War II provide an excellent example of how military technology can create scientific revolutions. What other areas of science or medicine were revolutionized as a result of military developments in World War II?

9.9 READINGS

9.9.1 General Reading

Cloud, P. 1988. *Oasis in Space: Earth History from the Beginning*. W.W. Norton, New York.
Press, F., and Siever, R. 1978. *Earth*. W.H. Freeman and Company, San Francisco.

9.9.2 References

Browne, M.W. 1995. Experts ponder causes of breakup of ancient supercontinent. *New York Times*, Oct. 3, p. B5.
Cloud, P. 1988. *Oasis in Space: Earth History from the Beginning*. W.W. Norton, New York.
Dalziel, I.W.D. 1995. Earth before Pangea. *Scientific American* (Jan.), 272(1), 58–63.
Isacks, B., Oliver, J., and Sykes, L.R. 1968. Seismology and the new global tectonics. *Journal of Geophysical Research* 73, 5,855–5,899.
Morgan, W.J. 1968. Rises, trenches, great faults and crustal blocks. *Journal of Geophysical Research* 73, 1,959–1,982.
Smith, W.H.F. and Sandwell, D.T. 1997. Global sea floor topography from satellite altimetry and ship depth soundings. *Science* 277, 1956–1962.
van Andel, T.H. 1992. Seafloor spreading and plate tectonics. In *Understanding the Earth: A New Synthesis* (G.C. Brown, C.J. Hawkesworth, and R.C.L. Wilson, eds.). Cambridge University Press, Cambridge, UK, pp. 167–186.
Wyllie, P. 1971. *The Dynamic Earth*. John Wiley and Sons, Inc., New York.

III. THE HISTORICAL PLANET: EARTH AND SOLAR SYSTEM THROUGH TIME

10 FORMATION OF THE SOLAR SYSTEM

10.1 INTRODUCTION

Having dealt with some of the tools and key concepts to which we will return as we develop the history of Earth and the other planets, we are ready now to consider that history. Five centuries after the beginning of the European Renaissance, humanity's explorations of Earth and the cosmos have exposed an intriguing, perhaps profound, paradox. Earth and the other planets of the solar system seem to be explainable as manifestations of common physical processes that have operated over very small and very large scales, to produce a range of cosmic phenomena. In this sense we are neither special nor particularly important in the grand scheme of things.

On the other hand, in our own solar system, we now understand Earth as the one planet with a uniquely stable climate, equable for liquid water over the billions of years required to bring forth intelligent life. Although Mars may have come close to this state at one time, no other planet seems to have the combination of characteristics needed to sustain life. Other solar systems may be common and life may flourish elsewhere, but it is also possible, with what we know today, that we are a rare or even unique speck in the cosmos. We will know more over the decades to come, but for now we seek to understand how this planet came to be, and how physical processes have operated to make it habitable for billions of years.

10.2 TIMESCALE OF COSMOLOGICAL EVENTS LEADING UP TO SOLAR SYSTEM FORMATION

Based on the observed red shifts of galaxies described in chapter 2, the age of the universe, that is, the time since the Big Bang explosion, is somewhere in the range of 8 billion to 20 billion years. Recent Hubble Space Telescope observations that improve galactic distance determinations argue for the lower end of the age range. The long-lived radioactive isotopes bound in primitive meteorites, when analyzed as described in chapter 5, suggest that elements heavier than helium began to be produced about 10 billion to 15 billion years ago. This gives a lower limit to the time that stars began to form in the universe. Although there is current controversy as to whether these two age determinations are in fact consistent or contradictory (because of the need, based on Hubble data, to choose a lower age for the Beginning), they at least provide a rough estimate for the scale of cosmic ages.

The beginning of element formation defines the earliest epoch of the evolution of our galaxy (the Milky Way) and others in the universe. Two generations of stars likely were formed before the gas of our galaxy, now enriched in heavier elements and dust, brought into being our Sun and the planets of our solar system. Isotopic dating of meteorites using the rubidium-strontium system and others indicates that the most primitive rocks in the solar system are 4.56 billion years old, determined rather precisely (chapter 5). This is the time when solid matter was first assembled into what would become the solar system. At that epoch in galactic history, roughly 0.1% by number of the atoms in the galaxy were in the form of elements heavier than helium, and these played a key role in processes that led to planet formation.

Figure 10.1 sketches the cosmological timescale. What we seek to understand is how the material available in galactic clouds of gas and dust some 4.6 billion years ago evolved into a system of the Sun and the planets, and whether such a process is likely to be a common one. Twenty years ago, much of the discussion attached to this subject was speculation because of a lack of observations. Today, ample evidence in nearby star-forming regions exists of structures that appear to be the precursors

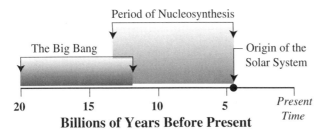

Figure 10.1. Timeline of cosmic events from the Big Bang through formation of the solar system.

of planetary systems. With new technologies applied to astronomical observations, we are able to identify and study the prenatal structures in stellar nurseries out of which planetary systems likely form.

10.3 FORMATION OF STARS AND PLANETS

10.3.1 Molecular Clouds and Star Formation

Observations reveal regions of gas and dust dispersed among the hundreds of billions of stars that make up the spiral arms of our galaxy. The largest of these gaseous and dusty regions, *giant molecular clouds*, contain enough gas to make 100,000 stars each the mass of the Sun. The closest major molecular cloud, the Orion molecular cloud, is some 1500 light-years from Earth. It is so large that it spans an area of the sky equal to 15 full Moons, but is not visible to the eye because it is so dim (figure 10.2 and color plate II).

Most of the gas in molecular clouds is hydrogen and helium but, as noted above, one out of every thousand atoms in the molecules making up the gas and dust is an element heavier than helium. Virtually all of the atoms are combined into molecules, because of the low temperatures in the cloud and the relatively large number of atoms packed into every cubic centimeter (i.e., high gas density) compared to other cosmic environments. In addition to molecular hydrogen (H_2), many different kinds of molecules occur with abundances that vary in complex ways from cloud to cloud and even within the same cloud. In the colder parts of molecular clouds many or most of the molecules are bound up in rocky and icy grains.

Determining the abundances of molecules in neighboring molecular clouds, some hundreds to thousands of light-years from the solar system, depends on the technique of spectroscopy (chapter 3). Because temperatures in molecular clouds are low, most of the material radiates photons at very long wavelengths, in the microwave part of the spectrum. Where stars are forming, dust and gas falling into the nascent star may be heated to high temperatures, and

light in the infrared and optical parts of the spectrum can be observed as well. Very precise microwave spectroscopy, such that the light is split into very fine wavelength bins so that spectral lines can be measured precisely, allows not only composition but also velocities of the gas to be determined. This in turn allows astronomers to map out regions of infall or collapse of gas and dust into nascent stars.

To get a sense of what a dense molecular cloud corresponds to in terms of terrestrial conditions, consider that the air in the room that you are occupying holds over 10^{19}, or ten million trillion, molecules of air, mostly nitrogen and oxygen, in every cubic centimeter. The average space between the stars, *interstellar* space, holds about 1 atom of hydrogen in each cubic centimeter of space; under these conditions, hydrogen is in the form of individual atoms rather than molecules of H_2. The densest clumps of dust and gas in a typical molecular cloud have 10^7 (10 million) atoms per cubic centimeter. Again, the density in clouds is determined from observing spectral lines of common molecules, such as CO (carbon monoxide), and tracing changes in the strength and shape of the line in different regions of a molecular cloud. Conditions – temperature, density, abundance of different molecules – vary widely between different parts of a given molecular cloud, ranging from cold tenuous portions grading into interstellar conditions all the way to dark, dense, localized clumps.

Most molecular clouds contain very bright but small areas of elevated temperature and strong energy emission. These glow in the infrared and their energy distribution (number of photons as a function of wavelength) is well simulated by computer models of stars surrounded by gas and dust. Such stars, called *T-Tauri* stars after the first one discovered, are very bright and are best explained as newly formed stars, or stars still forming by the processes described below.

10.3.2 The Start of Star Formation

It appears, then, that molecular clouds are sites where stars form, and that the Orion molecular cloud might be akin to a molecular cloud out of which the Sun and the planets formed 4.5 billion years ago. Why do stars form in such clouds? There is plenty of hydrogen, helium, and heavier trace elements available to form the stars. The key to the stars' formation lies in the high-density regions, which are gravitationally unstable. The dense dark parts of molecular clouds are cold, and calculations show that the density of dust and gas is high enough that the mutual gravitational pull of the gas and dust should cause the material to come closer together, that is, to become denser. As the stuff becomes denser, the mutual gravitational pull

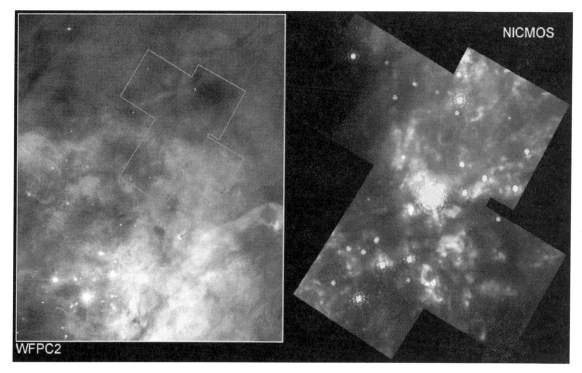

Figure 10.2. Giant molecular cloud in the constellation Orion seen by Hubble Space Telescope in (left) visible light and (right) infrared light. The infrared image covers the area outlined on the visible image, stripping away the gas and dust to reveal very young stars. The visible image was produced by C. Robert O'Dell (Rice University) and colleagues, the infrared image by Rodger Thompson (University of Arizona) and colleagues. Courtesy of NASA and Space Telescope Science Institute.

becomes stronger and stronger, further increasing the density. This instability continues ad infinitum, the material falling in to a common center and attracting more and more gas and dust.

How do such unstable clumps arise? The cloud of gas is threaded with magnetic fields, which are known to exist by remote measurements from Earth. These fields attract charged particles in the gas and force the particles to move along the magnetic lines of force. The charged particles are a small but important fraction of the gas and, as they collide randomly with the neutral (uncharged) particles, they impart a pressure to the whole gas, a pressure caused by the force of the magnetic field on the charged minority in the gas.

This process ought to prevent collapse forever, but there is a curious effect here: Charged particles form in the gas when neutral particles absorb high-energy light – ultraviolet (uv) photons, defined in chapter 3 – from stars embedded in the cloud. These charged atoms, or ions, last only a certain time before they capture free electrons and become neutral again. Thick clumps of gas prevent the uv photons from traveling far, and so, the thicker clumps of gas in the cloud have fewer charged atoms. The fewer the charged particles, the less pressure that is exerted on the gas by the magnetic field. Thus, the magnetic field is least effective at inhibiting the collapse of the densest cloud fragments. The cloud therefore is in a state of unstable equilibrium where, if a clump of gas forms randomly, it will lose its ion population, lose its magnetic support, and begin collapsing to form a denser and denser core. This collapsing core is the beginning of the formation of a star or group of stars.

10.3.3 A Star Is Born

As a core collapses in the molecular cloud, material falls deeper and deeper into the core's gravitational well, deepening the well. The molecules making up the gas and dust collide with increasing vigor toward the center of the core, converting uniform motion of collapse into heat. Temperatures at the center of the core become enormous – tens of millions of degrees – and pressures rise to billions of atmospheres according to computer simulations. (Astronomers can measure the brightness of such cores in the Orion molecular cloud, which helps constrain these calculations.) Recall from chapter 4 that these conditions are enough to initiate the fusion of hydrogen into helium, with release of energy. The energy generated creates a tremendous outward pressure in the core – the implosion of the gas has created an explosion at the center. A balance is achieved between the outward and inward pressures: Too much expansion shuts off the fusion, reinitiating collapse, whereas too little expansion causes further implosion, a faster fusion rate, and higher outward pressure. This newly balanced core of fusing hydrogen, surrounded by infalling gas and dust, is the picture that astronomers and physicists have developed of a newly formed star.

10.3.4 Figure Skaters and Astrophysicists: The Formation of Planets

Is there room in this picture for planets? Indeed, the formation of planets may be a natural consequence of the intrinsic spin or angular momentum of the gas. The entire Milky Way galaxy consists of stars and gas moving in orbits about a common center. This circular motion is not completely uniform and, in particular, the gas in molecular clouds has eddies and turbulence that provide an intrinsic spin to the gas. A fundamental law of physics is that *momentum*, the product of velocity and mass of an object, is conserved; that is, it will not change unless a force acts upon it. This holds true for momentum associated with spinning motion, called *angular momentum*.

As a clump of gas collapses to form a core and then a newborn, or *proto-*, star, the gentle spin intrinsic to the extended tenuous gas becomes faster and faster as the clump becomes more compact. Why? To conserve angular momentum, the gas spins faster as it becomes more compact. The effect is just that of a figure skater: As the skater's arms contract she will spin faster even if she imparts no further force with her skates. (For this to work, her contact with the floor must involve little friction, hence the desirability of ice.) The collapsing core of a molecular cloud must shrink by a factor of 10^8 to become the size of a typical star like the Sun. Long before this size is reached, the spin rate of the gas becomes too large to allow continued infall to the center: the angular momentum forces the gas into an orbit around the protostar, along the spin direction. Thus a disk is formed within the collapsing gas, but if the angular momentum of the original clump is too high, it actually splits into two cores to form a binary star. This process is complicated: Some of the gas, with little spin or angular momentum, falls right to the center. The rest is arranged according to angular momentum, with the gas having the highest angular momentum on the outer edge of the disk.

It is remarkable that most of the mass of our solar system is in the Sun and most of the angular momentum is in the planets. The disk out of which our solar system formed had to have possessed efficient mechanisms for moving mass to its center while retaining angular momentum in the dwindling disk material. Much of the extensive computer simulation work to understand the nature of disks from which planets form has focused on how enough angular momentum and mass could be transported in opposite directions (outward versus inward) during the limited lifetime of the disk. The lifetime itself is set by astronomical observations which show that stars that are older than a few million years (based on spectral appearance and models) generally do not possess massive gas and dust disks.

Conceptually, it is possible to divide the evolution of a protoplanetary disk, or (for our Sun), "solar nebula," into four stages, as has been done by the Harvard astrophysicist A.G.W. Cameron. The rationale for such a division lies as much in conceptual convenience as it does in observations. It is likely that, if one could watch the evolution of such a disk, one would see the stages merge into each other and vary in their distinctiveness from one disk to another.

The four stages are:

(i) *Formation of the nebula.* The parent molecular cloud collapses to form a disk, perhaps because of the loss of magnetic support, as described above. The amount of material per square meter (the *surface density*) in the disk is increasing. This stage lasts perhaps a few hundred thousand years, very short compared to other astrophysical timescales.

(ii) *Dissipation in the nebula.* As material is added to the disk, some of it falls into the very center, forming the core of what will become the central star. The gas and dust in the disk begin to interact in three important ways. The heating of the disk sets up circulations of gas and dust, causing eddies that convert motion into heat and transfer angular momentum outward through the disk. Also, the gravitational force of material in the disk sets up waves in the gas, creating a pattern very similar to that seen in spiral galaxies (figure 10.3). These waves act to create a force on the disk that causes further outward transport of angular momentum and heating. Finally, a small fraction of the gas is in the form of charged particles that are forced to move in a direction different from the bulk gas, because of the remaining presence of a magnetic field. All three of these processes – eddies, spiral waves, and magnetic effects – cause energy of rotation to be lost as heat, forcing more material to fall inward while shedding angular momentum to the outer extremities of the disk. The stage of most vigorous dissipation lasts perhaps 50,000 to 100,000 years. Evidence for it comes from disk systems, located in other star-forming regions, which suddenly

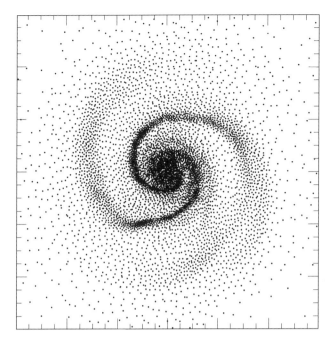

Figure 10.3. Spiral density wave pattern in a computer model of the protoplanetary disk, that is, the solar nebula, from which the solar system formed. The view is looking down on the face of the disk with the growing Sun (too small to be shown in this simulation) at the center of the figure. The disk is represented in the model by 8,000 discrete points; in reality the solar nebula was made up of countless more gas molecules and grains. The spiral pattern seen in the disk is reminiscent of the much larger-scale structure seen in spiral galaxies. Simulation by A. Nelson, D. Arnett, W. Benz (University of Arizona), and F. Adams (University of Michigan).

brighten as seen from Earth; the best-studied example is a disk around the star FU Orionis in the Orion star-forming region.

(iii) *Terminal accumulation of the star*. Accumulation of more gas and dust has slowed dramatically. A wind of charged particles emanating from the star acts to erode the disk from the inside out; the present-day *solar wind* is the pale shadow of this primordial gale. Material also is ejected along the poles of the newly formed star in spectacular jets (figure 10.4). Within the disk, the building blocks of planets – grains of rock and ice – are agglomerating together to form comet-sized bodies called *planetesimals*. In our own protoplanetary disk that became the solar system, the giant planets may have formed during this time, before the gas of the disk was blown away by the wind. This stage lasts perhaps one million to a few million years. Stars in such a phase are readily visible in molecular clouds because of the action of their winds; they are called T-Tauri stars after the best-studied example of their class.

(iv) *Residual static nebula*. The central star has finished growing and is shining stably by virtue of hydrogen fusion. The vigorous wind that eroded away the nebula in stage (iii) has largely but not completely abated and continues to drive off residual gas. Rocky planetesimals near the star agglomerate to form planets such as (in our solar system) the terrestrial planets. Observations of residual disks of dust around other stars, such as the star Beta Pictoris, whose disk was first imaged in 1984, suggest that this stage lasts from a few million to 30 million years.

Are the disks themselves, vastly smaller than the grand lanes and billows of the molecular cloud, observed? Until a decade ago, the answer was no. But observations with radio telescopes linked to form a giant radio eye can penetrate the densest parts of molecular clouds such as Orion, and disks of gas and dust have been observed around newly forming stars (figure 10.5).

10.3.5 Disks Around Protostars – the Source of Planets?

The stages of star and disk formation outlined above can be observed indirectly or directly in the Orion and other neighboring molecular clouds. However, the act of planet formation in disks has never been observed. We have roughly 10 definitive examples of planets around normal stars, including our own; the others were detected beginning in 1995 using a variety of techniques described in section 10.5. The idea that Earth and the other planets formed from a disk of gas and dust is centuries old. The co-planarity and common orbital direction of the planets of our solar system led seventeenth century scientists to propose such a hypothesis. Beginning in the 1960s, study of the putative properties of the disk, called the solar nebula (meaning gas around the early Sun) has been based on analysis of planetary atmospheres and primitive meteorites. Observations of disks in other star-forming regions, beginning in the 1980s, lent additional support to the notion that this is how planets form.

The source of planet formation in a disk is the turbulent motion of dust and gas. As material of different angular momentum sorts itself out according to distance, it collides with other material and generates heat. The collisions tend to cause material to fall ever inward until most of it ends up in the protosun. However, some of the dust sticks upon collision. The process of sticking, or accretion, can continue to ever larger sizes, from dust to pea gravel to golf balls to boulders. The process of continued agglomeration in the disk can be simulated by a computer, and indicates that on a timescale of millions to tens of millions of years, bodies the size of Earth can be formed.

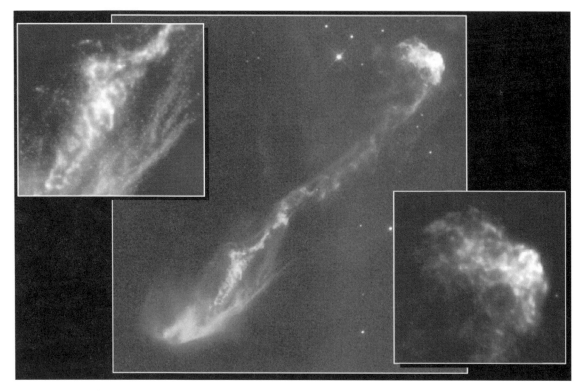

Figure 10.4. Hubble Space Telescope image of a jet of material ejected from a newly formed star. The star is hidden in the lower left portion of the image behind a dark cloud of gas and dust. The insets show details of the jet, which stretches outward trillions of kilometers from the star. This Wide Field and Planetary Camera-2 image was taken by J. Morse (Space Telescope Science Institute) and colleagues. Courtesy of NASA and Space Telescope Science Institute.

The nature of the dust depends on position in the disk, and evolves with time. As accretion of material into the disk slows, the disk cools. In the inner disk, collisions of gas and dust are vigorous and heat the gas to hundreds or even thousands of degrees throughout the lifetime of the disk. Where the temperature is below 1,500 K, abundant rocky and metallic grains can survive: This region from 0.5 to 5 AU from the Sun is today the realm of the inner planets. Beyond about 5 AU, the gas was cold enough during much of the history of the disk to allow water ice to condense out and survive, and so, grains of both ice and rock were stable. It is in the outer solar system that we see bodies made of rock and ice – the moons of the giant planets.

Although this process of collision and sticking explains the rocky terrestrial planets – Mercury, Venus, Earth, and Mars – it does not directly account for the giant planets. These gas giants might have formed directly by collapse of the gas in the outer disk, but information on their composition, derived as explained in chapter 11, suggests against this. The composition of Jupiter and Saturn differ from the Sun's in being enriched in elements heavier than hydrogen and helium; some of this heavier material appears to be concentrated in cores at the centers of these planets. Uranus and Neptune are smaller objects that are a bit like Jupiter and Saturn but with most of the hydrogen and helium envelopes absent.

One explanation for the internal structures of the giant planets is that their formation started with the accretion of rock and ice which produced a body large enough to gravitationally attract the gas of the solar nebula. As the gas concentrated near the growing planet, the gravitational field increased, drawing yet more gas, ice, and rock into the planet. Based on computer simulations, Jupiter and Saturn could have formed this way in a few million years. Uranus and Neptune may have taken longer to form, perhaps up to 10 million years longer based on recent computer simulations, and literally ran out of gas to make the envelopes as the solar nebula dissipated.

The terrestrial planets may not have acquired gaseous envelopes because temperatures in the gas were too high to maintain stable envelopes, or perhaps because vigorous solar wind and uv radiation from the parent star stripped the planets of any hydrogen envelopes during the dissipation phase of the nebula. Perhaps the lack of water ice in the terrestrial planet zone slowed accretion because there

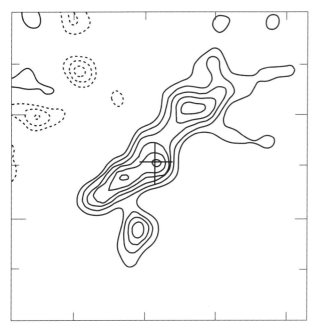

Figure 10.5. Map of a disk of gas and dust, a possible precursor of a system of planets, around a newly forming star HL Tauri. Photons at radio wavelengths emitted by carbon monoxide gas near the star are plotted as contour lines on the map; they roughly define a disk oriented diagonally on the figure. To see this amount of detail some 160 parsecs from Earth required linking together radio telescopes in a technique called interferometry. The total length of the disk is 2,000 AU. From Beckwith and Sargent (1993) by permission of University of Arizona Press.

was less material to collide and stick together, so that by the time the planets grew big enough to attract gas, the nebula was largely dissipated.

As the giant planets formed, they produced disks of gas and dust out of which their satellites, or moons, formed. The formation of Earth's Moon, which is not too much smaller than Earth, must have occurred a different way, and this is discussed later. The formation of Pluto is also an enigma, but it may represent the largest of the *Kuiper Belt* class of objects that were left over from planet formation: Pluto and its moon Charon, a smaller object in the outer solar system called Chiron, and the innumerable asteroids and comets. Comets, 10-km agglomerations of ice and rock, may be incredibly numerous in orbits beyond Pluto, perhaps totaling to tens or hundreds of earth masses – the leftover detritus of planetary accretion.

10.3.6 The End of Planet Formation

As the newly formed Sun reached a steady state between collapse and outward pressure from hydrogen fusion, its tremendous energy generation produced not just a high luminosity of photons but a wind as well: a tenuous atmosphere of charged particles pushing outward from the solar atmosphere. We detect such a wind from the Sun today by spacecraft. Early in the Sun's history, the wind may have been enormous, profoundly affecting the entire solar system. Astronomers have observed in newly formed T-Tauri stars in other molecular clouds powerful winds that are driving the gas and dust away from the new stars. They also observe slightly older stars that have lost much or all of the surrounding gas and dust, and are no longer true, classical T-Tauri stars.

In the case of the Sun, this strong wind would have dispersed the solar nebula, stripped the early terrestrial planets of much or all of their original atmospheres, and driven very small grains of dust out of the solar system. With the gas and dust gone, accretion of new planets would have stopped because they were robbed of the raw material required.

The blowing-away of gas by wind and uv radiation from the newborn star extends beyond the circumstellar disk itself to the surrounding clumps of gas and dust in the molecular cloud. As star formation reaches a crescendo, and there is some evidence that it may be episodic, winds and uv radiation from multiple star systems erode significant portions of the gas and dust. A graphic example of this process comes from a Hubble Space Telescope image of a portion of a molecular cloud called the Eagle Nebula (figure 10.6 and color plate III). Lanes of gas and dust are interspersed with clear areas, the whole illuminated by the light of newly formed stars that are driving the removal of the placental gas and dust.

Based on observations of other newly formed stars and theoretical models of their age, the entire solar system formation epoch took of order 10 million years. By this time, the Sun had settled into a steady state and the planets of the solar system, including Earth, were in place. Some models suggest the formation of the terrestrial planets, including Earth, took up to 100 million years, but the giant planets must have formed more quickly, so that the gas of the solar nebula was still present.

10.4 PRIMITIVE MATERIAL PRESENT IN THE SOLAR SYSTEM TODAY

The planets have evolved through time after formation, their composition being altered by *outgassing* and *chemical differentiation*. Details of these processes are presented for Earth and the other planets in chapter 11. Because no planetary body, even the larger moons, has escaped this evolution, what record do we have of the original

Figure 10.6. Hubble Space Telescope image of the Eagle Nebula, showing clearing of the gas and dust by ultraviolet light from newly formed stars which are illuminating the scene. The image was taken by Jeff Hester and Paul Scowen of Arizona State University using the Wide Field and Planetary Camera 2 at visible wavelengths. Courtesy of NASA and the Space Telescope Science Institute.

composition of solid material throughout the solar system? Some meteorites have been known for many years to have a composition that mimics that of the Sun over certain classes of elements, and these are almost certainly unevolved remains from the solar system's formation. Tiny bits of dust that make their way into Earth's atmosphere – *interplanetary dust particles*, or IDPs – also appear to be unaltered samples of rock-forming materials. If we want to examine primitive *ices* (water, ammonia, carbon dioxide, methane, nitrogen, carbon monoxide, among others) we must go farther out in the solar system to Kuiper Belt objects and comets, discussed further in section 10.4.2.

10.4.1 Remnants of the Beginning: Meteorites

An important record of the earliest time in the history of the planets, when large solid bodies first were present in the solar nebula, is in meteorites – rocks that fall to Earth from the sky. The origin of meteorites lies in larger *parent* bodies from which fragments have been blown off by impacts. The wide range of meteorite types, from nearly pure iron and nickel to those with largely silicate (rocky) and organic (carbon-bearing) composition, require that they come from at least several dozen distinct parent bodies. Most of these are likely to be asteroids either in the main asteroid belt or on other orbits in the solar system. About a dozen meteorites each are recognized to be from the Moon and Mars.

Some of the meteorites have abundances of elements over a certain range of volatility that are very similar to those in the Sun. (Recall that volatility is simply a measure of the tendency of a material to vaporize into a gas; water is more volatile than rock.) Elements that have an affinity for incorporating in rocky matter are generally all present. Elements such as carbon generally are depleted, but do exist in *organic* (carbon-bearing) phases of the meteorite. The more volatile elements apparently did not condense into the grains from which meteorites formed, but are present in significantly depleted amounts trapped in the rocky or organic phases of the primitive carbonacous chondrite meteorites (chapter 5).

The general consensus is that the chondrites could have come from parent bodies that formed relatively close to the Sun – the asteroid belt, for example. Nonetheless, they might bear a strong resemblance to the rocky component of primitive bodies from farther out in the solar nebula, that is, comets. Differences between the chondrites and rocky components of comets would provide very strong constraints on the distinct processes that operated in the hot inner part of the solar nebula versus the cold outer part. These exciting comparisons await detailed examination of a comet nucleus.

The carbonaceous chondrites may have survived relatively unchanged from the period of planet accretion. Their texture, differences in elemental abundances from the Sun, and patterns of isotopic abundances tell us much about the solar nebula. In particular, the gas in the region where the terrestrial planets formed must have been hot enough, very early in the solar nebula's history, for most elements to have been in the gaseous phase, followed by condensation to form the primitive meteorites as temperatures in the nebula dropped. This may be a very different situation from the region of the outer planets, where the rocky component of solid material was never vaporized, but instead fell into the nebula as largely intact grains. (The water ice, on the other hand, likely evaporated and then recondensed, at least in the region corresponding to the orbits of Jupiter and Saturn.)

In addition to the primitive matrix material, chondritic meteorites contain tiny, nearly spherical, nodules called *chondrules*, which appear to be bits of the rock that melted, resolidified, but did not change significantly in chemical composition. The origin of the chondrules is a mystery, but seems to imply that there were sudden episodes of heating in the solar nebula, perhaps caused by electrical discharges akin to terrestrial lightning, or by bursts of energy associated with a strong magnetic field in the nebula. They illustrate the complex nature of the processes from which the planets formed.

10.4.2 Comets and Kuiper Belt Objects

Beyond Neptune's orbit lie other remains of solar system formation, bodies ranging in size from tiny grains to minor planets hundreds of kilometers across (figure 10.7). Objects currently residing in the Kuiper Belt very likely have been there since the beginning of the solar system, because those orbits are not readily accessible from closer in toward the Sun. The Kuiper Belt objects are likely to be a mixture of rock and ice, with very volatile species such as methane, nitrogen, and carbon monoxide still present as ices themselves, or trapped in void spaces in the water ice. The very low temperatures (30 to 50 K) in the Belt ensure that evolution of these bodies is slow. So, modified perhaps slightly by collisions and by internal heat sources for the larger bodies, Kuiper Belt objects should preserve a record of the ice and rock composition of the original grains of the outer solar nebula, those that formed the moons and cores of the giant planets. Meteorites were part of parent bodies that formed too close to the Sun to have retained the volatile components as ices, and so, Kuiper Belt objects and comets provide a distinct record.

A visit to a Kuiper Belt object is many years away, but could be achieved as part of a robotic mission to Pluto which might be undertaken early in the twenty-first century. Pluto and its moon Charon are the largest Kuiper Belt objects; Pluto's orbit is very similar to a number of recently discovered smaller objects in the region and occupies a part of the Kuiper Belt that is not disturbed by Neptune. Until the first spacecraft flyby, these objects are so faint that chemical examination of their surfaces, using spectroscopy to break the light into component wavelengths, requires the largest Earth-based telescopes. Recent successes in identifying surface ices on Kuiper Belt objects, besides Pluto and Charon, represent remarkable progress in astronomical spectroscopy.

Comets are more readily examined, and they too are icy bodies from the farthest reaches of the solar system. However, their detectability stems from their very noncircular

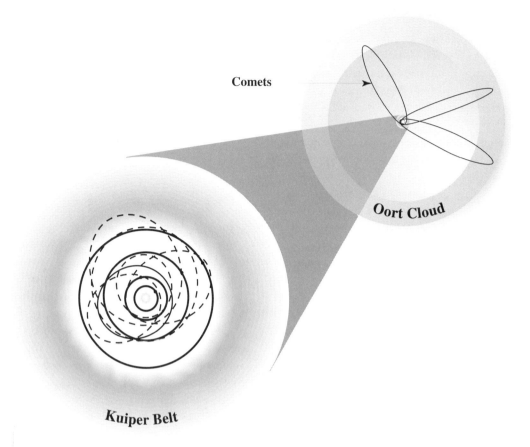

Figure 10.7. Sketch of the outer solar system, showing the location of the Kuiper Belt as the shaded region. The solid orbits are (from inside outward) Jupiter, Saturn, Uranus, and Neptune. Pluto's orbit, not shown, places that planet in a stable part of the Kuiper Belt. The dashed lines show orbits of Centaur objects, which probably have been perturbed inward from the Kuiper Belt by Neptune's gravity as well as, perhaps, through collisions in the Belt itself. On the upper right is a diagram showing the much larger Oort Cloud and a few sample long-period comet orbits. The Oort cloud is so much larger than the Kuiper Belt that the latter can hardly be seen on the scale of the former.

orbits that bring them close to the Sun, vaporizing ices from the outer layers and lofting dust, that allows examination remotely from Earth. Material from only one comet, Halley, has been analyzed directly by spacecraft. Although comets contain a valuable record of the primitive composition of ices from the outer solar system, their very noncircular orbits that bring them close to Earth also make it difficult to understand where they originally formed. Although these orbits seem to have their origin in a region called the *Oort cloud*, extending to 100,000 AU from the Sun (3,000 times the Pluto-Sun distance), it may be that comets did not form at such distances. More likely is that the Oort cloud comets were formed closer in, where the giant planets reside today, and then were flung outward in gravitational close encounters with the giant planets (figure 10.8). Also, certain classes of comet orbits, such as those possessed by the *Jupiter family short-period* comets, could not have evolved from the Oort cloud, but instead these objects are likely to be Kuiper Belt bodies perturbed out of their original orbits by collisions or repeated gravitational perturbations by the giant planets.

Thus, the record of composition contained in comets is difficult to relate directly to a particular region of the early solar system, except perhaps in the case of the Jupiter family short-period comets. Nonetheless, the accessibility of comets makes them interesting. The European Space Agency plans a mission, Rosetta, to rendezvous with and land on a comet sometime in the second decade of the twenty-first century. An American robotic probe, *Stardust*, is planned for the first decade of the 21st century to fly through the coma (dust and gas cloud) of a comet and collect dusty material for return to Earth.

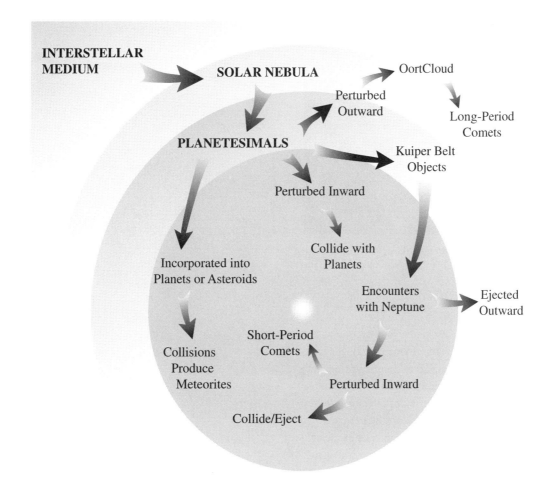

Figure 10.8. Possible history of small bodies in the solar system, from the original molecular cloud of the interstellar medium in which the solar system was born, through the solar nebula phase to the accumulation of planetesimals into larger bodies. Some planetesimals became part of the growing planets and asteroids; later collisions among asteroids produced fragments, some of which reach Earth as meteorites. As giant planets formed and their gravity increased, orbits of remnant planetesimals were increasingly perturbed; some planetesimals were ejected to the Oort cloud, others inward to collide with the terrestrial planets. The Oort cloud became the source of the long-period comets. Remnant planetesimals just beyond Neptune constitute the Kuiper Belt, and some of these have survived to the present day. Others, perturbed principally by Neptune's gravity, were either ejected outward or shunted inward to form Centaur objects. These either collide with the giant planets or have their orbits further altered to become short-period comets. Based on a scheme by Cruikshank (1997).

10.4.3 Interplanetary Dust Particles

IDPs are microscopic bits of cosmic dust that enter Earth's atmosphere and fall slowly toward the ground, or are collected from Earth orbit. Their fluffy nature (figure 10.9) and very small size ensure a gentle descent through Earth's atmosphere, and they can be collected by airplanes with appropriate sampling tools. The origin of IDPs is not certain, but their composition resembles the primitive carbonaceous chondrites. It has been suggested that they are derived from the dust (non-ice) component of passing comets. Unfortunately, IDPs are too close to the Sun for ice to be stable, and hence an important clue linking them to comets is missing. More detailed information on the silicate and organic components of comets is required to make this linkage.

10.5 THE SEARCH FOR OTHER PLANETARY SYSTEMS

The problem of identifying planets around other stars is one of enormous challenge. Jupiter's brightness, at optical wavelengths to which our eyes are sensitive, is one-billionth (10^{-9}) that of the Sun. At infrared wavelengths, where cooler objects tend to radiate (as described in chapter 3), Jupiter is still a mere 10^{-5} as bright as the Sun.

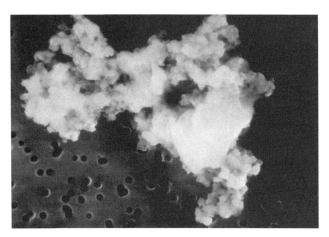

Figure 10.9. Photograph, using an electron microscope, of an interplanetary dust particle, roughly 10 microns across. The dark holes in the background (used to help mount the particle) are 0.4 microns across. Image courtesy of Professor Don Brownlee, Washington University.

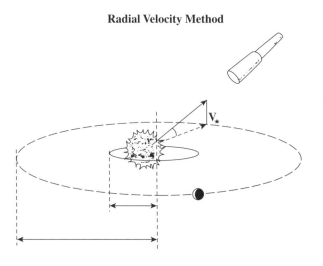

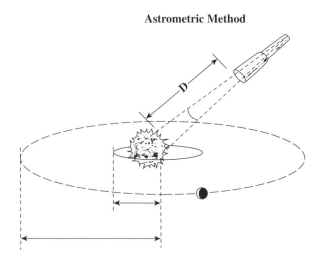

Figure 10.10. Examples of indirect techniques for detecting planets. Shown in both sketches is a star and its companion planet. The planet forces the star to be nonstationary, that is to orbit the common center of mass of the system. The observer on Earth is symbolized by the telescope, which is extremely far from the star and its companion planet. In the radial velocity technique, the distant observer on Earth sees the component of the star's motion (actually its velocity, V) directly toward or away from the Earth via Doppler shift. In the astrometric method, the star's slight shift side to side in the sky, due to its orbital motion, is detected on Earth. In both cases, the planet itself is lost in the glare of the parent star, and is detected only by its gravitational influence on the star. Adapted from a drawing by NASA.

Imagine looking at our solar system from a great distance, many light-years away. Any normal telescope system will see the Sun, the central star, but even Jupiter is lost in the glare caused by light scattered across imperfections in the telescope mirrors and across the telescope structure.

To overcome these problems, astronomers pursue several types of approaches to detect planets around other stars. *Indirect techniques* rely on observing the effect of the planet on the motion or appearance of the parent star. *Adaptive optics* and *interferometric imaging* are both *direct* techniques to overcome telescope glare and the smearing effects of our own atmosphere (through which Earth-bound astronomers must look) to achieve images of planets orbiting another star.

10.5.1 Indirect Techniques

Indirect techniques illustrated in figure 10.10 involve watching the position of the star oscillate in the sky, caused by the gravitational effect of a planetary companion. Motion back and forth can be seen using precision position measurement determinations referred to as *astrometry*. Star wobble toward or away from the observer can be identified by looking at one or more spectral lines from the star that reveal the *radial velocity* through the Doppler effect described in chapter 2. As the star moves toward the observer, the line is *blue-shifted*; as the star moves away, the line is *red-shifted*. The mass of the planetary companion can be determined from the magnitude of these effects.

Another indirect technique works if the orbit of the planet around the star is close to the line of sight to Earth. Then, as the planet passes in front of its parent star, it blocks some of the light and causes a partial eclipsing of the starlight. To accurately see such dimming, this *eclipse-photometric* technique requires a small telescope above the atmosphere of Earth observing a large number of stars.

Microlensing detects planets through an entirely different effect. As seen from Earth, if a star passes in front of a more distant, background star, the light from the background star is temporarily enhanced by the bending of light rays around the nearer star, in accordance with the general theory of relativity which predicts that gravitational fields bend light rays. If a planet is present in the right position around the nearer star, it produces a further brightening which is distinguishable from that of its parent star. Though such microlensing events are rare, a modest-size telescope can automatically scan many hundreds of thousands of stars to catch those rare signatures of the focusing of light by a passing interloper and its planet.

Indirect techniques have already yielded some important results. A Swiss team of astronomers in 1995 discovered the radial velocity signature of an object orbiting very close – 0.05 AU or one-tenth the orbit of Mercury around the sun – to the star 51 Pegasi. The mass of this object is between half and twice the mass of Jupiter. This planet takes only four days to orbit its parent star, and hence many orbits have been tracked by detecting the cyclical red and blue shifts of the star's spectral lines. Based on the discussion of planet formation earlier in this chapter, we would not expect a gas giant to have formed so close to its parent star. Perhaps it spiralled in from a more distant orbit. It is also possible that the planet began as a companion star which then lost much of its mass to its parent star through tidal (gravitational) forces. Since 1995, over a dozen planets, all roughly the mass of Jupiter or larger, have been discovered around other stars by indirect techniques. Some of these detections are so difficult to make that they may turn out to be false alarms. For example the planet around 51 Pegasi has been challenged as an artifact in the observations caused by small oscillations in the star itself, though this challenge was recently withdrawn in the face of further observations.

In 1992, radial velocity techniques were employed in a very different fashion to detect planets around a pulsar. A pulsar is the ultradense neutron star core of an exploded star, one that has finished the chain of fusion reactions described in chapter 3. Most such neutron stars have very strong magnetic fields, which result in charged particles streaming along the magnetic poles of the star, creating a beacon that can be detected at radio wavelengths. Using the Arecibo radio telescope in Puerto Rico to measure the Doppler shift to progressively shorter (bluer) and then longer (redder) wavelengths of radio energy, National Radio Astronomy Observatories (NRAO) astronomer D. Frail and team were able to infer the presence of at least two and, from 1994 observations, possibly as many as four, planets orbiting the pulsar PSR1257 + 12. These planets range in mass from several times that of Earth to a mass as small as that of Earth's Moon.

How planets could have survived the pulsar-creating explosion of the original star is a mystery. One idea holds that the planets did not exist prior to the supernova explosion but were instead created from debris of the explosion in a process mimicking planet formation around very young stars. The presence of these planets suggests that such formation processes can occur in many different kinds of environments around stars.

10.5.2 Direct Techniques

Direct techniques include adaptive optics, in which the primary telescope mirror, or a second mirror, is made flexible and its shape is adjusted by actuators to compensate for atmospheric turbulence. An artificial reference star may be created by a laser exciting molecules in our atmosphere, and a computer is used to guide the shaping of the mirror to make the reference star as point-like as possible. This technique, originally military technology, has in the past year or two demonstrated its feasibility.

A different approach is to use *imaging interferometry*, in which two mirrors are linked together on a long beam placed in space and the images combined as if the mirrors were part of one huge telescope. This provides exceptional resolution, or sharpness, of the image. If the whole structure is spun in space like a celestial pinwheel, as suggested originally by Stanford physicist Ronald Bracewell, the light from the mirrors can be combined in such a way as to suppress the light from the parent star and enhance the light coming from the region adjacent to the star where planets might be present.

In an improvement of this technique, conceived by University of Paris scientist Alan Legèr and University of Arizona astronomer Roger P. Angel, four mirrors on a space-borne tower allow *Earth-sized* planets, if they are present, to be imaged around the hundred stars nearest to the Sun (figure 10.11). The system is then capable of taking the light from each discovered planet, breaking it into a spectrum, and looking for the signature of gases such as carbon dioxide (indicative of an atmosphere), water (a sign of life-sustaining surface liquid water), and ozone (indicator of an oxygen-rich atmosphere capable of sustaining complex life). Such a space-borne interferometer might cost 1 billion to 2 billion dollars (US) and would require some technology developments before it could be put into space. Its sensitivity is so great that it would need to be moved to an orbit beyond the asteroid belt to get it out from under the glare of dust in the inner part of our own solar system. Such a device will not likely be launched until the end of the first decade of the twenty-first century, but would be the culmination of a search that already has some positive results.

Direct techniques from the ground have yielded evidence of massive planets. Caltech astronomer S. Kulkarni and colleagues used the 5-meter telescope on Mt. Palomar, California, to image an object orbiting about 40 AU from the star Gliese 229. No special adaptive optics

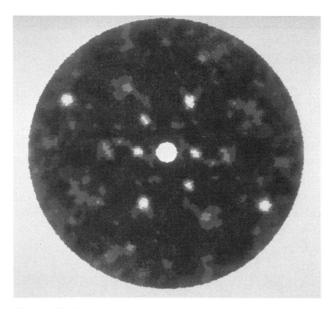

Figure 10.11. A solar system with four Earth-like planets, 10 parsecs away, might look like this simulated image from a rotating infrared interferometer. The image of each planet appears twice, because of the way the interferometer detects planets, but further processing would reveal which of the two dots is the actual planet. Courtesy of Roger Angel and Neville Woolf, University of Arizona.

technique was necessary because the companion is bright and orbits relatively far from its parent star. Spectra taken of the object reveal the presence of methane, which is not chemically stable in an atmosphere above a temperature of about 1200 K. This and other aspects of the observation constrain the size of the companion to about 20 times the mass of Jupiter. This places the object on a threshold between massive planets and *brown dwarfs*, objects that form like stars but are not massive enough to undergo hydrogen fusion. The observations performed by Kulkarni and colleagues provide the confidence to try much more difficult imaging and spectroscopy by a space-borne interferometer in searching our galactic neighborhood for other earths.

10.6 SUMMARY OF PLANET FORMATION

Twenty years ago our solar system was known as the singular example of the end state of the process of planet formation, with no observable examples of such systems actually in formation. Today, evidence exists that star-forming regions have abundant disks of gas and dust, and at least some of these likely evolve into planetary systems. We have evidence of planetary-mass objects around over a dozen nearby stars similar to the Sun. Although none of these planetary systems resembles our solar system in terms of the arrangement or nature of the planets themselves, they represent the first inroads into the very challenging problem of detecting other planetary systems. Terrestrial planets like Earth will be much more difficult to detect, but the technology anticipated to be available within a decade will be able to do so. By the end of the first decade of the new millennium, we could well have the answer to the question: Is our solar system a unique or rare peculiarity of the evolution of stars and galaxies, or are we one of many such systems in the heavens?

10.7 QUESTIONS

a. Given that giant planets have been discovered very close to a handful of stars (much closer than Jupiter is to our Sun), does the picture of planet formation presented here require revision? How would you revise it?
b. What kinds of measurements would you make to determine whether IDPs come from comets or asteroids, and if the latter, what particular type of asteroid?

10.8 READINGS

10.8.1 General Reading

Goldsmith, D. 1997. *Worlds Unnumbered: The Search for Extrasolar Planets*. University Science Books, Sausalito, CA.

10.8.2 References

Anders, E., and Grevesse, N. 1989. Abundances of the elements: Meteoritic and solar. *Geochimica et Cosmochimica Acta* 53, 197–214.

Angel, J.R.P., and Woolf, N.J. 1996. Searching for life on other planets. *Scientific American* 274(4), 60–66

Beckwith, S.V.W., and Sargent, A.I. 1993. The occurrence and properties of disks around young stars. In *Protostars and Planets III* (E.H. Levy and J.I. Lunine, eds.). University of Arizona Press, Tucson, pp. 521–541.

Cameron, A.G.W. 1995. The first ten million years in the solar nebula. *Meteoritics* 30, 133–161.

Cruikshank, D.P. 1997. Organic matter in the outer solar system: From the meteorites to the Kuiper Belt. In *From Stardust to Planetesimals*, ASP Conference Series Vol. 122 (Y.J. Pendleton and A.G.G.M. Tielens, eds.). Astronomical Society of the Pacific, San Francisco, pp. 315–333.

Levy, E.H., and Lunine, J.I. (eds.) 1993. *Protostars and Planets III*. University of Arizona Press, Tucson.

Lunine, J.I. 1992. Planetary clues from a collapsed star. In *Science Year 1993*. World Book, Inc., Chicago, pp. 244–245.

Lunine, J.I., and Snow, T.P. 1994. Astronomy. In *Science Year 1995*. World Book, Inc., Chicago, pp. 221–229.

Mason, S.F. 1991. *Chemical Evolution*. Clarendon Press, Oxford.

Taylor, S.R. 1995. Potassium tells a tale. *Nature* 376, 20–21.

11 THE HADEAN EARTH

11.1 INTRODUCTION

The period from the formation of Earth, some 4.56 billion years ago, to the time when the oldest rocks still in existence today were formed, roughly 3.8 billion to 4.0 billion years ago, is called both the *Hadean* era and *Priscoan* eon of Earth. The term Hadean, referring to the classical Greek version of hell, is well chosen, because all evidence that we have is that the Hadean Earth was very hot and extremely active, with widespread volcanism and frequent impacts of debris left over from planetary formation. This time encompasses the assemblage of Earth from the smaller *planetesimals*, dramatic internal rearrangements such as core formation, the creation of the ocean and earliest atmosphere, and the origin of Earth's Moon. Forces that acted on Earth were essentially the same as those acting on Mars and Venus, and a traveler visiting Earth would have seen little to distinguish it from the two neighboring terrestrial planets.

Each planet initially had a molten, or nearly molten, silicate surface, followed by cooling and establishment of a solid crust. Each had an atmosphere dominated by carbon dioxide (CO_2), with little free molecular oxygen (O_2). Evidence exists that each planet had liquid water on its surface during a portion of the Hadean era. Most important, no sign of life could be seen on any of these three planets – conditions were too severe and variable to allow life-forms to survive except near the end of the Hadean on Earth, and perhaps at about the same time on Mars.

The close of the Hadean saw a fall-off in impact rates as the solar system's planets swept up much of the remaining debris, and on Earth the stabilization of a liquid water ocean and the buildup of chemically processed, buoyant crustal fragments that stood above the level of the sea: protocontinents. It was during this time, perhaps 4 billion years ago, that life began to survive continuously on Earth. Life also might have gained a foothold on Mars, and even briefly Venus, as well, but beyond the Hadean, Earth's evolution diverged from that of its planetary neighbors in terms of its atmosphere, surface, and growing abundance of life.

In this chapter we focus on key events and processes on the Hadean Earth, and also take a close look at the chemistry of materials that build terrestrial planets. This *geochemistry* will be an important part of the story of interaction between Earth's crust, atmosphere, and life in evolving a habitable world.

11.2 BULK COMPOSITION OF THE PLANETS

We can measure the masses of the planets using the periods and distances of their natural satellites (Kepler's laws, described in chapter 3) or (for moonless Mercury and Venus) via observing their small gravitational effects on the slightly elliptical orbits of neighboring planets or on flyby spacecraft. Knowing their distances from us, we can measure the size of each of the planets using powerful telescopes and resolving the disk of the planet. Given the mass and the size, it is possible to determine the density of each planet, that is, the mass divided by the volume. The density for the terrestrial planets is given in figure 11.1.

The bulk density of a planet is determined both by the material out of which it is made and the amount of compression of that material caused by the planet's own gravitational field. Because different materials compress to differing extents (for example, a soft pillow versus a slab of rock), the composition and compression are coupled. The more massive the planet, furthermore, the more compression. Also listed in figure 11.1 are the uncompressed densities, that is, the densities of the planetary materials in the absence of self-compression. We can display these

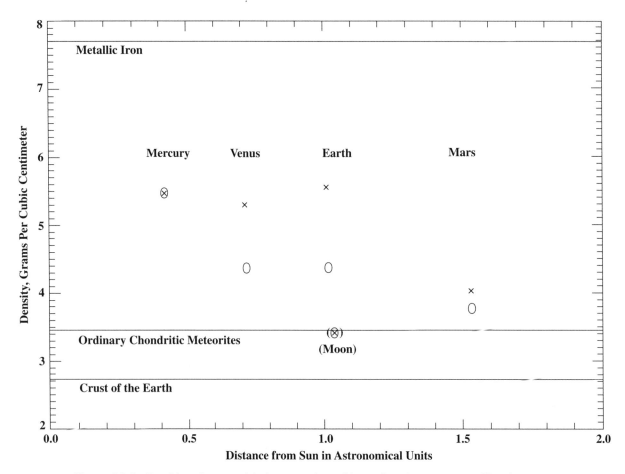

Figure 11.1. Densities of terrestrial planets and candidate mineral components. The planets are plotted as a function of their distance from the Sun. The symbol X indicates the measured density; the symbol O refers to the uncompressed density of the planet when the effect of gravitational compression of the material is removed. The compressed and uncompressed densities for the Moon are shown in parentheses, to distinguish them from Earth's. The density of Earth's crust also is shown; for comparison, the density of liquid water under low pressure is 1 gram per cubic centimeter.

as single numbers only for the solid planets; for the giant planets, most or much of the uncompressed material would be an ideal gas for which the density depends on the pressure and temperature under which the gas is contained.

Using the uncompressed densities, the abundances of the elements shown in figure 3.8 of chapter 3, and some chemical knowledge of how these elements tend to combine in the interiors of planets, we can infer the most abundant constituents in each of the planets. Obviously, this is not a simple deductive exercise because the uncompressed densities given in figure 11.1 were computed on the basis of some assumed composition. Instead, it is an *iterative* exercise, wherein one eliminates certain materials for certain planets because they do not produce the right compressed density. We will not go through the details of the exercise, but instead summarize the results for the solid planets, and then for the giant planets.

11.2.1 Solid Planets

The designation solid planets includes the terrestrial planets (Mercury, Venus, Earth, Mars, and the Moon), Pluto, and the larger moons of the outer solar system. The terrestrial planets all have uncompressed densities in the range of 3 to 6 grams per cubic centimeter (g/cm^3); by comparison, the density of liquid water at normal conditions is $1 g/cm^3$. Chondritic meteorites, composed to a large extent of minerals containing silicon, magnesium, and oxygen, have a density in the 3- to 4-g/cm^3 range. The meteorites provide us with clues as to the nature of planet-building materials in the right density range. In terms of cosmic abundance, the rock-forming elements silicon and magnesium are less abundant than elemental oxygen. The silicon and magnesium combine with abundant oxygen to form minerals such as enstatite ($MgSiO_3$), forsterite (Mg_2SiO_4), and other "rocky" compounds.

11.2 BULK COMPOSITION OF THE PLANETS 117

Figure 11.2. The giant planets of our solar system: (a) Jupiter from Hubble Space Telescope.

Densities of the silicate minerals are too low to account fully for the uncompressed density of all but Earth's Moon and perhaps Mars. A clue to the identity of a denser material lies in the high abundance of iron in chondrites, as well as in the existence of the *iron meteorites*, which have densities of 7.5 g/cm³, approaching that of metallic iron. Nearly as abundant as silicon and magnesium, iron is an excellent candidate for the material that raises the terrestrial planet densities beyond those of silicates. Mercury has by far the largest abundance of iron (most of its mass); the Moon has the least (close to zero). This is interesting in view of the fact that these two heavily cratered planets have similar sizes, the smallest of the terrestrial planets; clearly, their histories and probably their origins were quite different, given the distinct compositions. Nickel also is present in meteorites at roughly 6% of the abundance of iron, and in planets, it is expected to be present in similar amounts.

About one-third the mass of Earth is in the form of iron. Much of this may be chemically combined with sulfur or oxygen, and is known to be mostly segregated in a core, for reasons we discuss in Section 11.3. Hence Earth is stratified with the densest material toward the center; this is likely to be the case for all of the planets and most of the moons. The outermost chemical layer, or crust, of Earth is composed of minerals containing largely silicon and magnesium with an admixture of lower-density minerals. Aluminum, for example, underabundant relative to silicon and magnesium but with similar mineral-forming properties, is more abundant in the crust of Earth than throughout the rest of its interior. Venus' strong resemblance to Earth in density and size leads us to conclude that, in bulk composition, it is similar to Earth. Limited chemical measurements of the surface from landed, Russian space probes suggest this to be the case, even though the geologic processes shaping the face of Venus appear to be different from those on Earth (see chapter 15).

Figure 11.2. (Continued) The giant planets of our solar system: (b) Saturn from Hubble, with contrast exaggerated to show atmospheric patterns.

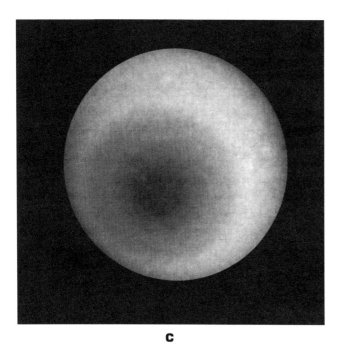

c

Figure 11.2. (Continued) The giant planets of our solar system: (c) Uranus from Voyager 2, also with enhanced contrast to show very faint banding.

Most of the major moons of the outer solar system, and Pluto, have densities around 2 g/cm^3. This is too low to be accounted for by common silicate minerals, but too high for pure ices. A sensible inference is that these bodies are roughly one-half ice and one-half silicate by mass. Because oxygen is significantly more abundant than carbon, nitrogen, or other ice-forming elements, it is a logical assumption that water dominates the ice component of these moons, with admixtures of ammonia, methane, carbon dioxide, and other ices. Spectroscopic identification of water ice, and of other ices in the cases of Triton and Pluto, seem to confirm these general ideas. Two major moons of Jupiter – Io and Europa – exceed 3 g/cm^3 in density. Io is almost entirely silicate; Europa is lower in density and appears to have a water-ice veneer. They both may have lost water early on, or never acquired significant quantities in the first place.

Interpreting planetary compositions in terms of the internal arrangement, or *structure*, of the various major components is a challenge that requires additional observational tools. We describe the monitoring of earthquakes to infer Earth's internal structure in Section 11.3.

11.2.2 The Giant Planets

Determining the detailed composition of the giant planets, pictured in figure 11.2 (see also color plate IV), is difficult

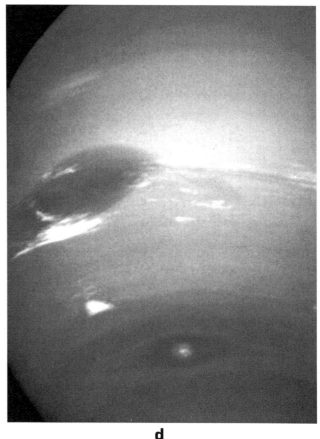

d

Figure 11.2. (Continued) The giant planets of our solar system: (d) Neptune from Voyager 2. Photos (a) and (b) courtesy of NASA and the Space Telescope Science Institute; (c) and (d) courtesy NASA and the Jet Propulsion Laboratory.

because of their distance from Earth, and the inaccessibility of their vast interiors. Density can be measured from size and mass, and the values are 1.33 for Jupiter, 0.69 for Saturn, 1.27 for Uranus, and 1.64 for Neptune, all in units of grams per cubic centimeter. These are much lower than the densities of the terrestrial planets plotted in figure 11.1. Equally important to understanding composition is the determination of the shape of the giant planet and hence its gravitational field. Such information provides constraints on whether especially dense layers are located near the planet's center, and to what extent the outer gaseous layers are pure hydrogen and helium. Measuring the gravitational field requires precise tracking of the orbits of a planet's moons, particularly its closest ones, and this must be done using flyby robotic spacecraft such as Voyager. Tracking the paths of the spacecraft themselves also yields gravity data. The spin rate of the planet also must be measured, because a faster spin tends to flatten gaseous planets significantly, affecting the distribution of mass in their interiors and hence their gravitational fields.

Both Jupiter and Saturn have such low densities that they must be made up mostly of the light, primordial elements hydrogen and helium; spacecraft spectroscopic measurements show that helium is 10 to 15% the abundance of hydrogen in the outer layers of Jupiter, Uranus, and Neptune, but much lower in Saturn. The giant planets much more closely resemble the Sun in composition than do the terrestrial planets. However, there must be much more of the elements heavier than hydrogen and helium in the giant planets than in the Sun, based on their densities and observed gravitational fields. The composition of the deep interiors cannot be sampled directly but is likely to be largely the abundant rock- and ice-forming elements such as oxygen, carbon, nitrogen, silicon, magnesium, and iron. The tremendous pressures, 70 million times Earth's sea level pressure at the center of Jupiter, force these materials to exist in chemical configurations different from those we are used to seeing on Earth. A schematic slice of Jupiter's interior is shown in figure 11.3.

Spacecraft and Earth-based identification of hydrogen and helium by spectroscopy and other techniques confirm their presence, at least in the outermost layers. However, other molecules are found to be present, such as methane and ammonia, which likely contain most of the carbon and nitrogen atoms in the outer layers; water is probably present but temperatures in the atmospheres are low enough that it is condensed out below the measurable outer layer of these planets. If Jupiter and Saturn had an overall composition equal to that of the Sun, we would expect that no more than 1% of the mass of each planet would be in the form of heavy elements. However, 10% or more by mass of each planet must be elements heavier than hydrogen or helium, based again on gravity tracking, and this important fact drives scientists to the model described in chapter 10 in which solid cores form first and then gravitationally attract nebular gas to form the giant planets. In effect, the genesis of Jupiter and Saturn begins with the formation of terrestrial- or ice-moon-type bodies, a process that does not stop until these protoplanets are drowned in hundreds of Earth masses (in the case of Jupiter) of hydrogen-helium gas. Continued infall of icy planetesimals during and after this process "salts" the gas envelopes of these planets with more ice- and rock-forming material.

Uranus and Neptune hold far less hydrogen-helium gas than do Jupiter and Saturn but vastly more than the terrestrial planets. Though not as rich in hydrogen and helium as are Jupiter and Saturn they may contain large amounts of water. Because of their great distances from the Sun, their atmospheres are extremely cold; the water is frozen out of the upper, observable, atmospheres, where instead clouds of methane are seen to form.

Careful measurement of the absorption of sunlight and emission of heat from each of the giant planets reveals that, with the exception of Uranus, each releases more energy in the form of heat than it receives as sunlight. There must be an internal source of energy in each planet, but it cannot be hydrogen fusion, because temperatures and pressures computed for the center of each body are too small to overcome the repulsive forces that prevent protons from fusing together. Even deuterium fusion, which is easier to initiate, cannot be achieved; computations show

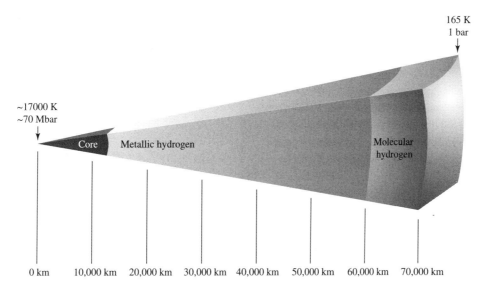

Figure 11.3. A slice through the interior of Jupiter, with distances to the center marked, as well as the pressure and temperature at the top and in the center.

that a body must be 13 times the mass of Jupiter for such reactions to take place.

The most plausible source of heat comes from the formation of the giant planets themselves. In compressing gas into a self-gravitating, bound sphere, from a state in which the gas originally is spread over a large region of space, potential energy is lost and converted into random energy of motion of the atoms and molecules, that is, into heat. It is the same process that heats the air that you pump into your bicycle tire, but the source of compression is gravitational energy rather than the stored chemical energy in your muscles that you use to move the pump piston. The initial energy of formation cannot be lost all at once; processes such as conduction, radiation, and convection, which transport heat from the inside of a large body such as a planet, do so in a finite amount of time. Hence heat is still being lost today. University of Arizona scientist W.B. Hubbard and colleagues showed, back in the 1970s, that the excess heat emitted by Jupiter today is consistent with residual heat of formation if Jupiter formed some 4 billion to 5 billion years ago – consistent with the age of the solar system. Neptune's heat yields a similar result.

Saturn and Uranus, however, are a mystery. Saturn emits almost twice as much heat as it should if the energy is that of its initial collapse some 4.5 billion years ago. Is Saturn younger than the other giant planets? A more elegant and sensible explanation comes from the helium abundance, which is depleted in the upper layers as measured from spacecraft. David Stevenson of the California Institute of Technology and Edwin Salpeter of Cornell showed, over two decades ago, that Saturn's interior is cool enough (in contrast to Jupiter's) that helium is chemically insoluble in hydrogen. The helium has been separating out as droplets that fall toward Saturn's center because their density is larger than that of the surrounding hydrogen. This slow rainout of helium contributes additional gravitational energy which is detected as excess heat emission.

The process of *planetary differentiation*, in which the interior materials are sorted out according to density, is important in Saturn because the ringed planet is less massive than Jupiter, and hence its interior cooled quickly to the point where separation could begin. Subsequent calculations by Stevenson suggested that helium rainout also is occurring in Jupiter, but began more recently than in Saturn. *Voyager* and *Galileo* data on the helium abundances of these giant planets, showing a strong helium depletion in Saturn's atmosphere, but not Jupiter's, appear to support the model.

Measurements showing that Uranus emits essentially no heat, other than what it derives from sunlight, do not yet yield to a tidy explanation. Nearly Neptune's twin in terms of size, mass, and density, we expect it to have a similar source of internal heat. However, Uranus spins on an axis that lies parallel to the plane of its orbit around the Sun, rather than close to perpendicular as with Earth and most of the other planets. Over the past decade, Uranus has had one pole tipped toward the Sun, and hence is receiving solar energy in a very different distribution of latitudes than is Neptune. It is possible that this has bottled up or redirected the internal heat of Uranus so that it is not observable. As Uranus moves in its orbit so that the equator, rather than the poles, points toward the Sun, interior energy may be released. That this is happening is suggested by Hubble Space Telescope images of Uranus in the late 1990s showing clouds becoming more abundant on its surface, a surface that *Voyager 2* found to be bland in 1989 (figure 11.4). Perhaps the "cork" has been popped from the planetary bottle as Uranus moves from a solstice orientation to one of equinox, that is, from summer/winter to spring/fall.

11.3 INTERNAL STRUCTURE OF EARTH

We cannot drill more than a few kilometers into the solid Earth with current technologies, yet the center of Earth lies over 6,000 km from the surface. *Apollo* astronauts have drilled only a meter into the Moon, and the *Viking* spacecraft have dug just a few centimeters into Mars. How, then, can we possibly know anything of the internal structure of these planets if drilling significant distances is not possible? Mapping the gravitational fields of the terrestrial planets reveals important information akin to that for the giant planets. Such mapping has revealed the root structure of continents on Earth, information on the patterns of convection (bulk motions removing heat) in Earth's mantle, the nature of the structure beneath the highlands on Venus, and the presence of very dense rock under portions of the Moon's surface.

Earthquakes provide the key to inferring the structure of Earth, the only planet upon which a dense network of seismometers, or earthquake detectors, exists. (A much smaller network, but lacking global coverage, was placed on the Moon during the *Apollo* landings.) An earthquake is an oscillatory movement of the ground, generated by the sudden slippage of rock along a fault (fracture) in Earth's crust. Earthquake waves travel literally around and through Earth, because solid rock is an excellent medium for conducting the wave motion. Rock motions include compression-rarefaction of the rock, or *P-waves*, and shearing motions, or *S-waves*. P-waves move easily through solid or liquid media; S-waves cannot move through liquid

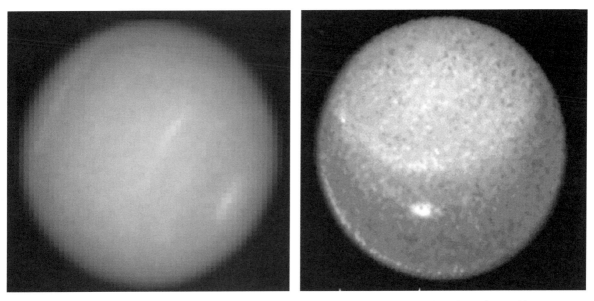

Figure 11.4. Uranus and Neptune from Hubble Space Telescope, showing continued weather on Neptune (left) and a surge of storms on Uranus (right). Uranus image by Kenneth Seidelman (U.S. Naval Observatory); Neptune images by David Crisp at NASA's Jet Propulsion Laboratory and Heidi Hammel at Massachusetts Institute of Technology Courtesy of NASA and Space Telescope Science Institute.

but instead are damped out, because shear forces cannot be maintained in a fluid.

Networks of seismometers recording ground motions were placed around Earth early in the twentieth century, allowing measurement, at many places around the globe of the time for P- and S-waves to reach various stations from a given earthquake. These times yield the velocity of the seismic waves through the interior and, because P- and S-waves travel at different speeds, it is possible to precisely locate the geographic point at which an earthquake occurred by measuring arrival times at different stations.

Figure 11.5 illustrates the most important result of such measurements, the inference of the presence of a chemically distinct *core* to Earth. P-waves are observed to be

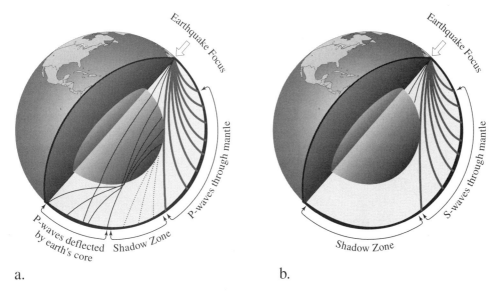

Figure 11.5. (a) P-waves moving through an earth with a central structure different from the bulk outer shell are bent in such a way as to create a shadow zone. (b) S-waves do not propagate at all through a portion of the deep interior, though near the very center, some S-waves do form and propagate a finite distance. Based on Press and Siever (1978) and other sources.

absent from seismometers for certain distances around Earth from the *epicenter* or *focus* of an earthquake, and more highly concentrated elsewhere. This is a strong indication of a structure inside Earth that is bending the paths of, or refracting, the P-waves. The structure could be liquid or solid, but it must have a sharply different density and/or composition from the material above it. S-waves do not accompany those P-waves whose path takes them through this inner structure of Earth, indicating that they cannot propagate through the structure. Hence, there must be a sphere of liquid in the deep interior of Earth – the *core*.

However, the core turns out to be not entirely liquid. The velocities and paths of P-waves are altered in such a way that there must be another structure even deeper in Earth's interior – an *inner core* that is solid and surrounded entirely by the liquid *outer core*.

A promising candidate for a material making up the core is iron, given that the overall density of Earth is significantly higher than the silicates that comprise our planet's outer layers. Iron is a cosmically abundant element, and meteorites with both a stony and an iron phase suggest that, in larger bodies, the iron separated from the less-dense silicates and formed a discrete core under the influence of the gravitational field of the body – a gravitational separation akin to that of helium from hydrogen in Saturn. By estimating the size of the core from the earthquake data and assuming that it is mostly iron, Earth's overall density is reproduced adequately.

However, the core cannot be made of pure iron. Chemically compatible and abundant nickel is likely to be present as well, in small quantities (less than 10%). But more important, the behavior of the core in having a liquid outer region and a solid inner region cannot be reproduced by iron for any plausible temperatures and pressures that are computed for the core. Instead, other elements must be chemically combined with iron in such a way as to allow iron to melt at a lower temperature than it otherwise would, and to provide a means for iron to segregate into a solid inner and a liquid outer core. The velocities of seismic waves in the outer core also argue for the presence of elements in addition to iron.

It is a common chemical fact that combining certain elements or molecules that are compatible in terms of atomic size or valence can lead to the production of liquids at temperatures lower than those for which the pure substances would melt. These are called *eutectic* or *peritectic* solutions. Iron combined with oxygen or sulfur forms such solutions and yields the right melting properties under the tremendous interior pressures of Earth (the central pressure is 4 million atmospheres – 4 million times sea-level pressure on Earth) to account for the dual liquid and solid cores. These chemical, and density, considerations indicate that roughly 10 to 15% of the core is nonmetallic elements: oxygen, sulfur, and perhaps even silicon.

There is much structure near the outer edge of Earth as well. A clear increase in the velocity of seismic waves was established by seismologists at the turn of the twentieth century. The discontinuity occurs some 6 km beneath the ocean floor, but 30 to 40 km below the surface of the continents. This change, known as the *Mohorovičič* or Moho discontinuity (after a Balkan seismologist), defines the boundary between Earth's *crust* and its *mantle*. The boundary is both chemical, in terms of a change in composition, and physical, in that the density changes discontinuously. The chemical change is from crustal rocks that are rich in silicon and aluminum to mantle rocks rich in silicon and magnesium.

Other sharp changes in seismic velocity are seen in the upper 700 km of Earth's mantle. Understanding their origin rests on laboratory simulation of the relevant pressures and temperatures, as well as some theoretical analyses of the chemistry involved. Most notable of these is a sudden drop in velocity at about 70 km indicating that the mantle has transitioned from a stiff upper layer (the lithosphere, which includes the crust of Earth), to a soft, even plastic, behavior (the asthenosphere). Still farther down (400 km) is a sharp increase in velocity where the mineral olivine (a magnesium silicate bearing some of the iron that is not sequestered in the core: Mg_2SiO_4 and Fe_2SiO_4) is forced by pressure to assume a more compact mineral form, called spinel (figure 11.6). (The iron in the mantle is a residue left behind by the original core formation process; estimates from volcanic lavas generated in the mantle yield only about 10% of the Earth's iron as residing in the mantle and crust.)

At 700 km, pressure forces another phase transition as some of the silicon, magnesium, iron, and aluminum are forced into simpler *oxide* forms: SiO_2, MgO, FeO (*wüstite*), and Al_2O_3. The bulk of the magnesium and iron assume mineral structures called *magnesium silicate* and *iron silicate* (*perovskite*), with the chemical formulas $MgSiO_3$ and $FeSiO_3$. Although the chemical formulas are similar to those of some minerals found in meteorites, the configurations of the atoms are very compact. Below this transition, the mantle is remarkably simple: There is no variation in the depth of layers as there is closer to the surface (that is, no "interior" topography, by analogy with variation of height on Earth's surface), and no phase changes until within a few hundred kilometers of the boundary of the outer core, which resides some 3,000 km below the surface.

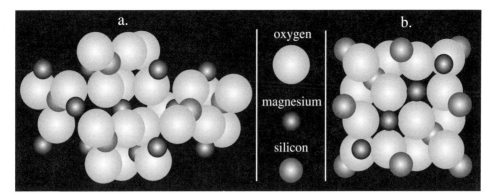

Figure 11.6. Crystal structures of (a) olivine and (b) spinel. After Press and Siever (1978).

Within this transition zone between core and mantle, some remarkable chemistry might be taking place. It has only been possible in the past decade or so to determine the structure of the boundary, using sensitive seismometers on the surface of Earth that determine not only arrival times of P- and S-waves from earthquakes, but the shapes of the waves as well. Work by geophysicist Thorne Lay of the University of California and colleagues has revealed a complex topography lying on top of the core-mantle boundary. The layer in which this topography resides (called the D'' layer for historical reasons) ranges in thickness from less than 10 km (thinner than this cannot be detected) to several hundreds of kilometers.

What is happening within this layer? One very controversial possibility is suggested by very recent laboratory experiments by Elise Knittle and Raymond Jeanloz of the University of California, and those of other groups. These experiments attempt to reproduce in the laboratory the conditions within the D'' layer. Interestingly, liquid iron (representing the outer core) placed in contact with magnesium perovskite (the primary constituent of the lower mantle) reacts chemically to produce the oxides of magnesium, iron, and silicon mentioned above, along with lesser amounts of iron silicide (FeSi). Iron can combine with silicon in this way only because the paired oxygen atom – in wüstite – acquires the electron valence of sulfur, which is just below oxygen in the periodic table of figure 2.6. The intense pressure at the core-mantle boundary – over one million atmospheres – is enough to distort the electronic structure of the elements.

The technology to simulate these pressures rests on very precisely machined diamonds, whose faceted ends are pushed together by small mechanical presses, with the sample mounted between these ends. A laser heats the sample, and the color of the glowing sample material determines its temperature (see chapter 3). Such *diamond anvil cells* are in use at laboratories around the world.

Why does the D'' layer exist? The picture offered by Lay, Jeanloz, and colleagues is that the liquid iron (with nickel, sulfur, oxygen) of the outer core seeps into cracks in the rock of the lower mantle. Capillary action draws the liquid upward perhaps hundreds of meters into the rock, where it reacts. The oxides and other products of the reaction are much closer in density to the mantle material than they are to the denser liquid metal of the outer core. Hence, any circulation patterns in the mantle will sweep the products of the reaction upward. *Solid-state convection*, in which warm rocks move upward and cooler rocks sink, provides the source of the circulation patterns. Computer models suggest that the upwelling can pile the reaction products into "mountains" hundreds of kilometers high; reactions in sinking regions correspond to "valleys" in the D'' layer.

The overall model for Earth's interior is shown in figure 11.7. What is striking about the model is that most of the chemical and structural complexity of Earth is confined to the crust, where the buoyant distillates of the mantle reside, and the D'' layer. The controversial picture for the D'' layer discussed here is offered to give the reader a flavor for current thinking on the poorly understood core-mantle interface. The model has been criticized on the grounds that it is very difficult to maintain chemical instability in the zone. The delivery of additional chemically active material from the upper mantle and the removal of fully reacted products may require invoking large-scale bulk motions beyond thermally-driven convection such as sinking of some subducted slabs all the way to the core-mantle interface.

11.4 ACCRETION: THE BUILDING UP OF PLANETS

As material is added to the forming planets by collisions (little chunks of rock agglomerating to make bigger rocks),

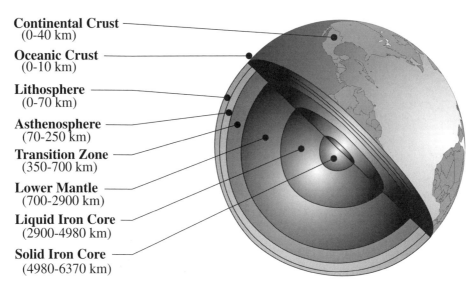

Figure 11.7. Cutaway sketch of Earth's interior from seismic data, laboratory experiments, and theoretical models. The crust is too thin to show at its true scale, and the thickness increases by a factor of four from ocean to continent. The lower mantle is defined as starting around 700-km depth where magnesium and iron perovskites become stable. After Press and Siever (1978).

the kinetic energy of impact is converted to heat. An alternative way to look at this is that the potential energy of the dust and small rocks, distributed over a large region of the primordial nebula, is larger than the potential energy of the material gravitationally bound to the growing planet. The lost potential energy shows up as heat, just as it does for the nebular gas going into the giant planets.

The heat added per unit of material (conveniently measured in mass, for example, per kilogram) in the outer layers of the growing planet is equal to the gravitational potential energy lost in each kilogram. The potential energy, in turn, is proportional to the density and radius of the growing planet, which defines the gravitational well into which the material falls. Hence, among the terrestrial planets, Earth and Venus were heated the most, Mercury and the Moon, the least. Two important complications are the average size of the impacting planetesimals at the end of accretion (bigger bodies deposit their energy deeper into the growing planet) and assembly time (longer times allow more heat from the impacts to leak out at the planetary surface, creating lower temperatures overall). These are suggested schematically in figure 11.8.

11.5 EARLY DIFFERENTIATION AFTER ACCRETION

Earth, Venus, and perhaps Mars achieved temperatures throughout parts of their interiors, by virtue of accretion, above the melting point of most silicate minerals. Hence, the earliest Earth had a massive molten region, extending from the surface down partway through the interior. Heat flowed both toward the colder center region, and outward toward space. Elements that previously were bound in the solid crystalline structures of the planetesimals were free in the liquid to rearrange themselves, associating with other compatible elements.

On the basis of their valence structure and the effective size of the atoms, elements can be divided into lithophiles, siderophiles, and chalcophiles. A *lithophile* (from the Greek, "rock-loving") element tends to associate with silicate phases, a *siderophile* ("iron-loving") element with the metal phases, and a *chalcophile* ("ore-loving") in the sulfur-bearing, or sulfide, phases. Chalcophile elements also can be distinguished by their tendency to be volatile and hence to escape from the solid phase. Table 11.1 lists the size of the ions of various elements, where the particular ions chosen are those common in Earth's mantle or crust. Ions significantly larger than the host magnesium or silicon ions have difficulty fitting into the solid crystal structure, and hence tend to stay in the molten rock.

In the early melting of the outer layers of Earth, large ions such as sodium (Na) and potassium (K) tended to reside in the liquid and float to the top of this massively deep *magma ocean*. How much differentiation occurred during the time after Earth reached its present size is controversial, because the precise temperature increase and hence extent of the magma ocean due to accretion cannot be pinned down. Furthermore, Earth's crust has been

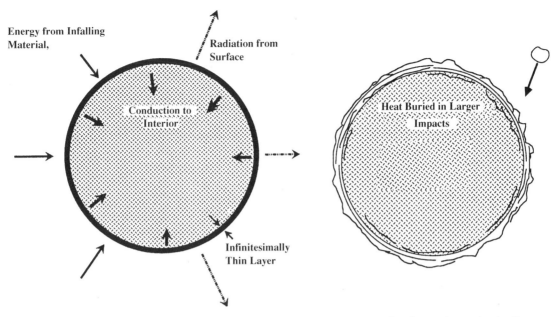

Figure 11.8. Two extreme ways that solid planets can accrete, by small or large planetesimals. On the left, a planet grows by accumulating small grains or boulders which, as they hit, deposit their heat on the surface. Some of the heat is radiated away by photons; the temperature increase depends on the rate of impacts compared to the rate at which the heat is radiated away. On the right, a planet grows by giant impacts, which gouge out the surface and bury the heat of impact in the planetary interior. The amount of heat that each impact provides depends in this case on the complex details of the impact process. Actual accretion involves both large and small impactors. Adapted from Melosh et al. (1993) by permission of University of Arizona Press.

Table 11.1 Ionic Radii for Selected Elements

Element	Size[a] (Angstroms)
Be	0.34
P	0.35 (lithophilic)
Si	0.39 (lithophilic)
Al	0.57 (lithophilic)
Li	0.78 (lithophilic)
Mg	0.78 (lithophilic)
Fe	0.82 (siderophilic)
Na	0.98 (lithophilic)
Ca	1.06 (lithophilic)
Sr	1.27 (lithophilic)
O	1.32
F	1.33 (lithophilic)
K	1.33 (lithophilic)
Rb	1.49 (lithophilic)
S	1.74 (chalcophilic)
Cl	1.81 (lithophilic)
Br	1.96 (lithophilic)
I	2.20 (lithophilic)

[a]Ionic radii are given for the element's usual form in Earth's crust. Data from Broecker (1985) and Mason (1991).

geochemically cycled and processed extensively in the 4.5 billion years after formation, erasing evidence for an early episode of differentiation. We expect the degree of early geochemical evolution on Venus to be the same as that of Earth, and less on Mars, Mercury, and the Moon commensurate with their smaller sizes. Only Mercury and the Moon are small enough to have undergone little crustal evolution after the initial accretional episode, and may preserve the chemical evidence of that earliest part of their history. In the case of the Moon, as we discuss in section 11.8, the material out of which it formed may have been derived in large part from an already partly processed Earth. An eventual sampling of the crust of Mercury may be the best way to learn how planetary interiors were altered during the last stages of accretion.

11.6 RADIOACTIVE HEATING

The building blocks of the terrestrial planets were likely similar to the chondritic meteorites, with more or less of the volatile elements included depending on the distance from the Sun at which particular grains condensed out. Among the elements present were uranium, thorium, and potassium, each of which has isotopes (^{235}U, ^{238}U, ^{232}Th, and ^{40}K) that are radioactive. The half-lives of these isotopes are given in table 5.1. Interestingly, both uranium and thorium have large ionic radii like potassium, and hence over time have become concentrated in the rocks of the crust, particularly in granite, where the radioactive isotopes of these species average 20 parts per million (ppm) in abundance. This is enough to produce, each year, 0.03 joules of energy in every kilogram of rock. Although this does not seem like much energy (a billion kilograms of granite is required to put out a watt of power), it is still substantial when the entire mass of granitic crust is considered. Roughly 2×10^{22} kilograms of granite are in the crust, leading to a total annual production of energy of 6×10^{20} joules: 20 trillion watts of power.

In the bulk of Earth, the present radioactivity abundances are much smaller; estimates from mantle-derived volcanic rock suggest about 0.1 ppm and an energy production rate at present of 0.0001 joules per kilogram every year. However, the entire mantle, which is roughly 4×10^{24} kilograms in mass (200 times more massive than the granitic part of the crust), generates over 10 trillion watts of power from the decay of radioactive potassium, uranium, and thorium.

At present, then, over half of the heat coming out of continental rock is generated within that rock, with heat from the deeper mantle being the other source. Oceanic crust, however, is depleted by a factor of six from continental in terms of its store of radioactive elements; most of the heat coming from ocean crust had its ultimate origin in solid-state convection in the mantle.

The effect of radioactive heating depends on a planet's size in two ways. The smaller the body, the less radioactive material that is present to heat the interior, and the larger is the ratio of surface area to volume. As a sphere shrinks, the surface area decreases more slowly than the volume. Reduce a planet to half its original size (while retaining the shape of a sphere), and the surface area drops by a factor of four while the volume (and hence the number of radioactive atoms within the planet) drops by eight. Since more relative surface area allows faster cooling, smaller objects cool more quickly than bigger ones. Based on relative sizes, Venus' thermal history was similar to Earth's, but Mars likely cooled more quickly than Earth, and Mercury even more rapidly. We see the evidence for this in the heavily cratered surface of Mercury and in the bimodal nature of Mars, wherein both massive volcanoes and heavily cratered terrains exist.

Sufficient heating is occurring today in Earth's mantle to soften the rock and allow bulk flow to remove the heat. The core of Earth is releasing heat to the mantle as well, so that the nature of the heat flow is somewhat complicated (figure 11.9). Simple patterns of bulk convective motion

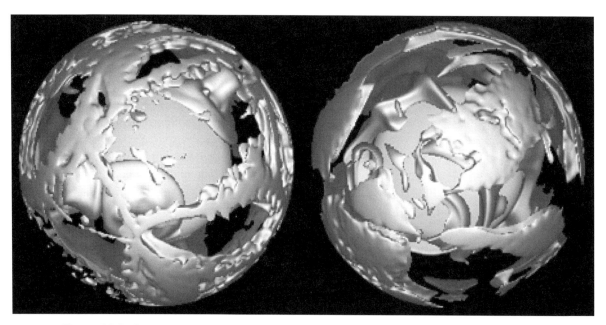

Figure 11.9. Computer model of convection in Earth (Tackley, 1995; Tackley et al., 1994). The model is three-dimensional and includes the presence of the phase transition at the upper-lower mantle interface. The left panel shows hot upwelling currents; the right panel shows cold downwelling currents. The inner sphere, which can be partly seen through the mantle currents, indicates the boundary with the iron core, which convects separately. Figures courtesy of Paul Tackley, University of California at Los Angeles.

of the mantle are interrupted by plumes of hot material driven by heat from the core. These deep-seated plumes may reach the surface in the form of large volcanoes, which are then dragged laterally by plate motion to form island chains such as Hawaii.

11.7 FORMATION OF AN IRON CORE

No more than a hundred million years after accretion brought Earth to essentially its present size (about 4.5 billion years ago, from the meteorite ages), temperatures throughout the deep interior were enough to partially melt the mixed solids of silicate, and iron. Iron melts at a temperature a couple of hundred degrees below the melting point of the major silicate component, magnesium silicate, and would be expected to sink to the Earth's center by virtue of its higher density. However, a plausible mechanism for iron core formation requires that a substantial fraction of the silicates melted as well, to allow the denser iron to separate readily from the surrounding material and sink. Because the iron core formation involves taking denser material from a distributed state and placing it in the very center of the planet, gravitational energy is released. The sinking of helium to the center of Saturn, creating extra heat from gravitational energy, is an entirely analogous process discussed earlier in the chapter. The total iron content of Earth corresponds to 32% of the mass of our planet, and the density of iron is about 50% larger than that of silicates, and so, the differentiation process releases an amount of heat not very much less than the total accretion energy of Earth; this undoubtedly helped to ensure melting of Earth's upper layers at that time.

The iron core is able to generate a magnetic field. As the core convects to remove heat to the cooler mantle layers above it, the motions of the electrically conductive iron have the potential to induce magnetic fields. Schematically, if a "seed" magnetic field is initially present in the core (left over from magnetic fields in the solar nebula that magnetized rocks and iron grains), then the moving fluid generates electric currents, which in turn generate a stronger magnetic field. This self-perpetuating process, energized by the heat slowly leaking from the core, is called a *magnetic dynamo*.

When did core formation occur? Theoretical calculations suggest that temperatures were high enough to initiate mantle melting at or shortly after accretion, but it is important to have an independent constraint on the time of initiation and duration. Isotopes of lead provide that determination. The element lead is chemically compatible with iron and hence followed iron into the core. Uranium, on the other hand, tended to stay in the crust and mantle.

Heavy isotopes of lead (^{206}Pb and ^{207}Pb), however, are daughter products of uranium decay, with long half-lives (4.5 and 0.7 billion years, respectively). Thus, by measuring the abundances of these daughter isotopes we have a potential way of determining when the core separated from the mantle. Ancient lead-bearing rocks on Earth's surface are compared with lead isotopic abundances in meteorites to infer that core formation occurred during the first 100 million years of Earth's history, essentially coincident with the late stages of accretion. A check can be made using xenon isotopes, which corroborate this determination.

11.8 FORMATION OF THE MOON

The origin of the Moon has always been a difficult issue because our natural satellite is unusually large relative to its primary (Earth) and resides in a circular orbit. Capture of the Moon after its formation is possible but extremely unlikely, requiring just the right set of conditions; capture into a tight circular orbit (the Moon's orbit has been slowly evolving outward with time because of the dissipating effects of ocean tides) is even more improbable. Formation in place at the same time as Earth also has difficulties when one tries to model the process by computer. Finally, fission, wherein a rapidly spinning molten Earth split off the Moon, also has some problems with physical plausibility, but neither this nor formation in place could be ruled out completely on theoretical grounds.

The Apollo missions to the Moon returned rock and dust samples that virtually eliminated all three models considered above. In spite of the Moon's small size, and hence limited geologic activity over time, the rocks were more typical of Earth's mantle than of primitive meteorites. However, even more chemical processing beyond that of Earth's mantle was implied: The rocks were strongly depleted in certain elements as volatile or more so than potassium, relative to those of Earth's mantle. In a very crude sense, one could obtain lunar material by taking terrestrial mantle rocks, heating them to temperatures at which they could vaporize, and recondensing only the less volatile constituents. (The term "very crude" must be taken literally, because the described process does not fully explain the lunar composition.)

This geochemical puzzle prompted planetary scientists in the mid-1980s to consider that the Moon might be the product of a huge collision between Earth and another planet-sized body: *a giant impact*. Conditions in the early solar system were right for such an impact. Early on, planetesimals were small and were in roughly circular orbits which resulted in gentle collisions, and hence sticking or accretion. As planets grew from planetesimals, close passes of bigger bodies altered orbits to make them elliptical, and hence increased relative collision speeds. By the time the terrestrial planets were formed, encounter velocities with solar system debris, on highly elliptical orbits, ensured catastrophic collisions in most cases. This was the case both in the inner and the outer solar system: The newly formed giant planets stirred up nearby planetesimals and ejected them into distant orbits which we recognize today as the cometary Oort cloud. The rate of impacts on planets decreased exponentially with time over the first few hundred million years of solar system history, as debris was swept up or ejected (see chapter 7).

Small bodies hitting big ones would vaporize and melt, disseminating their products in the crust of the big bodies. Big bodies hitting other big bodies could have more devastating consequences. A giant impact with Uranus likely tipped that planet on its side and spun out a disk from which its moons formed. Detailed computer simulations show that a planet one to several times the mass of Mars striking the Earth could have spun off a large amount of the Earth's mantle, very little iron core, and a fraction of that debris would have entered circular orbit around Earth while the remainder was lost into orbit around the Sun or reaccreted onto Earth (figure 11.10 and color plate V).

Much of the material that shot into orbit was vaporized, with only the least volatile material remaining solid. Some recondensation occurred, but in the absence of a nebular gas providing the conditions for full recondensation, much of the volatile material (water and the volatile lithophilic elements) was lost. Absence of debris from Earth's core resulted in little iron being present, and the

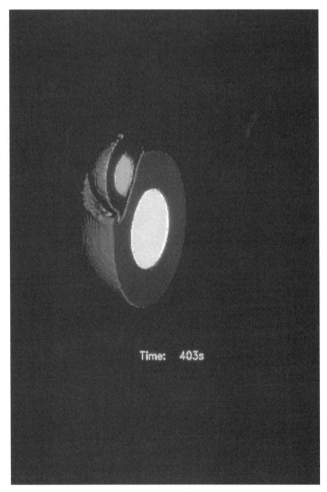

Figure 11.10. Computer calculations by M.E. Kipp (Sandia National Laboratories) and H.J. Melosh (The University of Arizona), showing early stages in the formation of the Moon as a Mars-sized planet strikes Earth. Both Earth and the impacting planet are shown sliced in half so as to reveal what is happening in the interiors. The iron-rich core can be seen as an inner circle in each planet prior to impact. Compared to the mantle of Earth, the core is hardly disrupted. Elapsed time is shown on each panel. Images courtesy of H.J. Melosh.

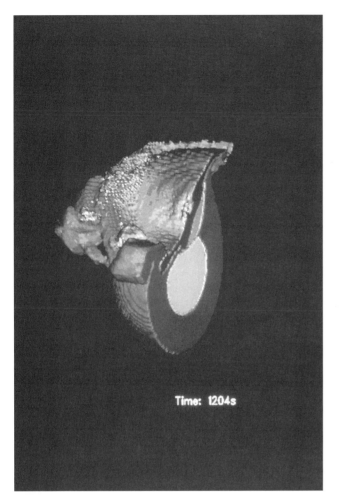 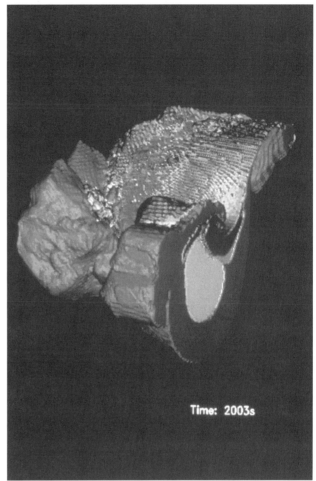

Figure 11.10. (Continued)

Moon's present density is consistent with little or no iron. Accretion of the material in circular orbit to form the Moon was apparently enough to cause melting of the upper 500 km or so of Earth's new satellite, because geochemical analysis indicates that the lunar surface is strongly enriched in lower-density minerals that likely floated to the top during a molten phase. The ancient lunar highlands are especially enriched in these minerals. Higher-density minerals that resemble basalts on Earth have flooded large basins on the Moon, forming the *mare*.

When did the Moon's formation from Earth occur? The oldest lunar rocks found, from the highland provinces, date by radioisotopic techniques (chapter 5) at 4.4 billion to 4.5 billion years ago; certainly the Moon is no younger than this. This also sets a limit on the time when the Earth's core formed: It had to be before the lunar-forming impact because the Moon is so depleted in iron. Most likely is that the lunar impact occurred extremely early in Earth's history, close to or before 4.5 billion years ago. Earth was not a single planet for very long. Venus, on the other hand, does not possess a moon, and hence either never suffered a giant impact or experienced one that left it in retrograde rotation without a companion, in which case, the ejected material was either reaccreted or lost to solar orbit. Pluto has a moon, Charon, that is even closer in mass to its primary than is the Moon to Earth. It may have formed from a large impact on Pluto, probably by another large Kuiper Belt object whose orbit was stirred up by a close pass to Uranus or Neptune.

What was the origin of the impactor that struck Earth? This remains a mystery, but it is clear from the geochemistry of the Moon that the impactor had to have had a composition similar to that of Earth. Because some of its mass went into the debris that formed the Moon, gross compositional differences would show up in the lunar rocks. Because those rocks do so closely resemble a devolatilized Earth's mantle, the impactor could not have been very different from terrestrial mantle composition.

Figure 11.11 summarizes the timescales for the earliest events in Earth's history, up through core formation. The

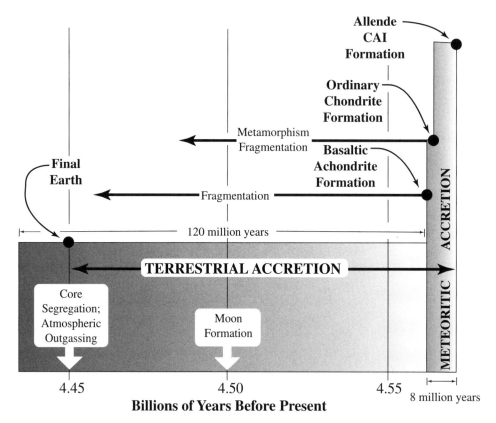

Figure 11.11. Timescales for the formation of Earth and early events in its history, as developed by Claude Allègre and colleagues from radioisotopic analyses of meteorites and lunar rocks. "Allende CAI" refers to particular phases in the Allende meteorite that predate formation of the bulk portion of the chondrites. "Basaltic achondrites" are a class of meteorites that have undergone chemical differentiation and hence are less primitive than the chondrites. Redrawn from Allègre et al. 1995 by permission of Elsevier Science Ltd.

enormous upheavals in the first 2% of Earth's history, in large measure, are a reflection of the crowded solar system environment at the time: The final stages of growth of Earth by sweep-up of smaller debris heated the planet to high temperatures (with a contribution from internal radiogenic elements as well), and the apparent presence of large bodies in eccentric orbits that crossed those of the planets set the stage for the catastrophic collision that led to lunar formation.

11.9 ORIGIN OF EARTH'S ATMOSPHERE, OCEAN, AND ORGANIC RESERVOIR

Earth's earliest atmosphere was a cloud of silicate vapor surrounding it during its accretion and core formation. As accretion stopped and core formation ended, the surface cooled and the silicate vapor condensed to form molten and solid rock. If this process concluded early enough, and this is uncertain, Earth would have been surrounded by a remnant primordial atmosphere of molecular hydrogen and trace amounts of other gases. This primordial atmosphere very quickly was swept away by the strong solar wind and is of little consequence to the rest of Earth history.

From whence came the gases that made up the "permanent" atmosphere? Outgassing from Earth's interior, of trace gases trapped in rocks, could have put hydrogen sulfide, carbon dioxide, and a large amount of water (all originally dissolved in the early magma ocean) in the atmosphere and on the surface. The origin of these volatile materials may not have been the vicinity of the forming Earth – where temperatures were too high to condense water – but instead may have been farther out in the forming solar system. Impactors that came from the outer solar system – comets – were rich in water ice, organics, carbon dioxide, carbon monoxide, and ammonia.

The comets are detritus from the formation of outer solar system bodies. Although hundreds of earth masses of comets now reside in orbits far from the Sun, early in the history of the solar system comets were more commonly in orbits that intersected the orbits of Mars, Earth, and Venus (based on computer studies of solar system formation). Collisions of comets with the planets would have released the cometary ices and gases into the atmospheres of the target planets. Early in Earth's history, the first couple of hundred million years, cometary material including water might have been episodically added to the atmosphere in large amounts. As Earth cooled, after core formation and the moon-forming catastrophe, water could have condensed permanently onto the surface, and water and carbon dioxide trapped in the crust as well.

The entire Earth's ocean is about one billion trillion (10^{21}) kilograms of water. A typical comet has perhaps one million billion (10^{15}) kilograms of water, and so, only a million (10^6) comet impacts are required to supply the present Earth inventory of water. This is not an unreasonable number, based on estimates of the total number of comets now in existence, which must be only a fraction of the number in the early solar system. (A large amount of water is trapped in the crustal rocks of Earth, and so, our estimate for the required number of comets is a lower limit.)

In addition to water, as noted above, comets brought in carbon dioxide, carbon monoxide, methane, ammonia, nitrogen, and other gases. Carbon dioxide also could have been available from rocks in Earth's mantle, and the early atmosphere likely was dominated by this gas after condensation of water. Molecular oxygen is essentially nonexistent in comets, is nearly absent from Mars and Venus, and was absent from the early Earth atmosphere. That this is so is demonstrated in part by minerals in ancient rocks that would have been unstable in an atmosphere composed of oxygen (chapter 17).

Comets undoubtedly were not the only source of terrestrial planet atmospheres. Carbonaceous chondrites contain large amounts of organic (carbon-hydrogen) compounds, as well as nitrogen and other volatile elements. However, these volatiles are depleted relative to the abundances in comets. An indication that comets were likely a more important source of atmosphere than chondrites comes from the noble gas record. Laboratory studies by Akiva Bar-Nun (Tel-Aviv University), Tobias Owen (University of Hawaii), and colleagues of how various elements are trapped in water ice at low temperatures reveal that the noble gas pattern in terrestrial planet atmospheres is better fitted by an icy source (comets) than a rocky source (chondrites) (figure 11.12). Although this interpretation of the evidence is not unique (other processes might have led to the noble gas pattern), it is sufficiently compelling that it deserves further careful study.

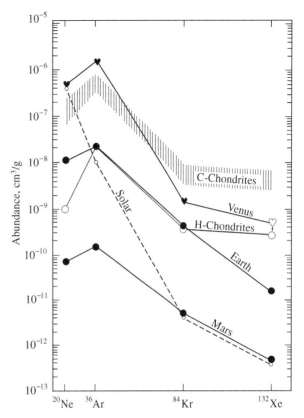

Figure 11.12. Abundances of noble gases in the atmospheres of Venus, Earth, and Mars compared to solar abundances (see chapter 4) and to abundances in two different kinds of chondritic meteorites. The three planets' atmospheres differ in total abundance of the noble gases relative to the total mass of their atmospheres, but the relative pattern of abundances (neon to argon, for example) is similar. The lack of neon relative to solar abundance indicates that these atmospheres were not borrowed directly from the solar nebula. The high argon-to-xenon and krypton-to-xenon ratios relative to abundances in meteorites suggest that comets, which can supply the needed argon and krypton, were an important source of atmosphere, and hence also of Earth's ocean. From Owen and Bar-Nun (1995) by permission of Academic Press, Inc.

How was this cometary material delivered to the terrestrial planets? Computer simulations suggest that the giant planets played a vital role. Their massive gravitational fields had three important effects on the early solar system:

(i) Jupiter pumped up the eccentricities of orbits of planetesimals inward of its orbit, leading to destructive collisions that prevented planet formation in the asteroid belt and stunted the growth of Mars.

(ii) Uranus and Neptune stirred up the orbits of neighboring planetesimals, putting them in highly eccentric orbits the aphelia of which were very distant from the Sun. Passing stars and the influence of nearby molecular clouds helped reshape these orbits into an inner

Oort cloud. Further gravitational effects of distant stars and clouds pumped up the orbits farther into an outer Oort cloud, and also sent comets from the Oort cloud into the inner solar system, where they have collided with Earth and other terrestrial planets over the history of the solar system. Icy planetesimals not perturbed outward by Uranus and Neptune formed the Kuiper Belt; occasional close encounters with Uranus and Neptune have sent some of those objects into the inner solar system as *short-period* comets.

(iii) All of the giant planets, especially Jupiter, very effectively cleared the solar system of planetesimal debris, with much of the material being ejected permanently into distant orbits, or forced into the inner solar system where the icy material collided with the terrestrial planets. Had the giant planets not swept the solar system clear, the impact rate in the inner-planet region might have remained high for billions of years, making for an unstable environment on Earth and frustrating the earliest origin and survival of life.

Although atmosphere-supplying impactors hit Earth with high velocity, much of the material may have fragmented in the protoatmosphere and reached the surface at low speeds. A significant portion of the organic molecules present in the comets and meteorites may have survived intact to the surface. Thus, the early ocean likely was seeded with large amounts of organic compounds, with complexity up to and including amino acids, the building blocks of proteins (see chapter 12), which have been found in meteorites. As the impact rate declined and Earth's surface began to stabilize, the materials necessary to initiate a biosphere were very likely in place.

11.10 FROM THE HADEAN INTO THE ARCHEAN: FORMATION OF THE FIRST STABLE CONTINENTAL ROCKS

Even after the early earth crust stabilized, continuing impacts and the vigorous convective activity of Earth's mantle discouraged the preservation of the crustal material over time. The early crust may have had a composition somewhat similar to present-day oceanic crust, depleted in magnesium relative to the composition of the mantle. Continental-type crust required repeated cycling of crustal basalts, with separation of silicon and other elements from the magnesium; such crust was later in coming (see chapter 16).

The oldest whole rock samples on Earth date back almost 4.0 billion years. These ancient rocks, seen in northern Canada, are composites of *mafic* (magnesium and iron-rich) and *felsic* (less iron- and magnesium-rich, more abundant in silicon) rocks. The former are typical of oceanic basalts, the latter of more continental-type rocks. The samples show evidence of having been metamorphosed (subjected to episodes of modest pressure and high temperature) in a way that suggests processing in and beneath a primitive basaltic crust. Also present in these rocks are rounded pebbles that appear to be sedimentary, that is, laid down in an environment containing liquid water. Belts of these rocks appear to be the remnants of the earliest continents. They indicate that continental-type crust, floating buoyantly atop a denser mantle, began to appear about 500 million years after the formation of Earth; whether continents could have formed much earlier is unknown. The chemistry of oceanic and continental rock formation is explored in more detail in chapter 16.

This Hadean earth, while vastly different from the present planet, set the stage for what was to follow. By 3.8–4.0 billion years ago, the growth of continents, the stabilization of liquid water, and the decreasing impact rate made for an increasingly predictable and benign environment. Increasing environmental stability characterized the transition from the Hadean era to the Archean eon of Earth.

11.11 QUESTIONS

a. Some meteorite properties suggest that rocky bodies were strongly heated by ^{26}Al, a very short-lived radioisotope of aluminum. How might the asteroids help determine whether this heating actually occurred? What would you look for?

b. What might have been different about Earth's Hadean and Archean history had the Moon not been present?

11.12 READINGS

11.12.1 General Reading

Broecker, W.S. 1985. *How to Build a Habitable Planet*. Eldigio Press, Palisades, NY.

Cloud, P. 1988. *Oasis in Space: Earth History from the Beginning*. W.W. Norton, New York.

Press, F., and Siever, R. 1978. *Earth*. W.H. Freeman and Company, San Francisco.

11.12.2 References

Allégre, C., Poirer, J.-P., Humler, E., and Hofmann, A.W. 1995. The chemical composition of the Earth. *Earth and Planetary Science Letters* **134**, 515–526.

Jeanloz, R., and Lay, T. 1993. The core-mantle boundary. *Scientific American* **268**(5), (May), 48–55.

Mason, S.F. 1991. *Chemical Evolution*. Clarendon Press, Oxford.

Melosh, H.J., Vickery, A.M., and Tonks, W.B. 1993. Impacts and the early environment and evolution of the terrestrial planets. In *Protostars and Planets III* (E.H. Levy and J.I. Lunine, eds.). University of Arizona Press, Tucson, pp. 1339–1370.

Owen, T., and Bar-Nun, A. 1995. Comet, impacts and atmospheres. *Icarus* **116**, 215–216.

Press, F., and Siever, R. 1978. *Earth*. W.H. Freeman and Company, San Francisco.

Spudis, P.D. 1992. Moon, geology. In *The Astronomy and Astrophysics Encyclopedia* (S.P. Maran, ed.). Van Nostrand Reinhold, New York, pp. 452–455.

Squyres, S., Reynolds, R.T., Cassen, P.M., and Peale, S.J. 1983. Liquid water and active resurfacing on Europa. *Nature* **301**, 225–226.

Tackley, P.J. 1995. Mantle dynamics: Influence of the transition zone. *Reviews of Geophysics* **33** (Suppl.), 275–282.

Tackley, P.J., Stevenson, D.J., Glatzmaier, G.A., and Schubert, G. 1994. Effects of mantle phase transitions in a 3-D spherical model of convection in the Earth's mantle. *Journal of Geophysical Research* **99**, 15,877–15,901.

Taylor, S.R., and McLennan, S.M. 1995. The geochemical evolution of the continental crust. *Reviews of Geophysics* **33**, 241–265.

Weissman, P. 1992. Comets, Oort cloud. In *The Astronomy and Astrophysics Encyclopedia* (S.P. Maran, ed.). Van Nostrand Reinhold, New York, pp. 120–123.

12 THE ARCHEAN EON AND THE ORIGIN OF LIFE: I. PROPERTIES OF AND SITES FOR LIFE

12.1 INTRODUCTION

The close of the Hadean and opening of the so-called Archean eon is defined and characterized by the oldest whole rock samples found on Earth, 4.0 billion years old. At the opening of the Archean, Earth had an atmosphere, rich in carbon dioxide, with perhaps some nitrogen and methane but little molecular oxygen, and liquid water was stable on its surface. Mantle convection had begun producing oceanic basalts and continental-type granitic rocks. The rate of impacts of asteroidal and cometary fragments had decreased significantly. The Moon, formed from Earth at the end of accretion some half billion years before, could be seen in the terrestrial sky.

By 3.5 billion years ago, rocks were present that recorded definitive evidence for life; less definitive evidence exists back to almost 3.9 billion years. Layered formations in ancient limestones contain concentric spherical shapes, stacked hemispheres and flat sheets of calcium carbonates (calcite), and trapped silts. These *stromatolites* are best understood as the work of bacteria from 3.5 billion years ago, precipitating calcium carbonate in layers as one of the byproducts of primitive photosynthesis. If the interpretation is correct, life on Earth was present then and somewhat earlier as well, because such bacteria constitute already reasonably well-developed organisms.

It therefore appears that, as Earth settled down from the chaos of accretion, core formation, and impacts, life was able to exist on its surface (figure 12.1). The same might be true for Mars, but the evidence discussed later in the chapter is vague and controversial. How did life arise on the Earth? Could it have arisen on the neighboring planets as well? Is there life in other planetary systems? Why was Earth able to sustain life over billions of years of change, and the other terrestrial planets not? How did life alter the Earth environment?

These are questions whose explorations constitute the remainder of the book, including part IV, where humankind's role is examined. In the present chapter, we outline the definition of life and the essential structures that make it possible.

12.2 DEFINITION OF LIFE AND ESSENTIAL WORKINGS

12.2.1 What Is Life?

No completely satisfactory definition of life – or of "living things" – has yet been devised. Most simple definitions of life – something that grows spontaneously, or something that replicates itself – fail because they either include demonstrably nonliving things or exclude certain particular living organisms. Crystals such as snow or pyrite grow but are not biological in nature; offspring of separate but related species such as mules (offspring of a donkey and a horse) are almost invariably unable to reproduce, yet clearly are living.

Some biologists lean toward a definition that incorporates the concept of *Darwinian evolution*, defined broadly to mean reproduction, variation of characteristics from one generation to another, and natural selection whereby some individuals with specific traits gain an advantage over others and hence are more successful in producing offspring. In this context, one working definition of life might be "a self-sustained chemical system capable of undergoing Darwinian evolution," as devised by University of California biologist Gerald Joyce and colleagues.

There are two major drawbacks to this definition: First, it has become clear that, although species do evolve, the classical Darwinian concept of natural selection is only one factor that comes into play in such evolution. Second, the definition may be unnecessarily narrow in that

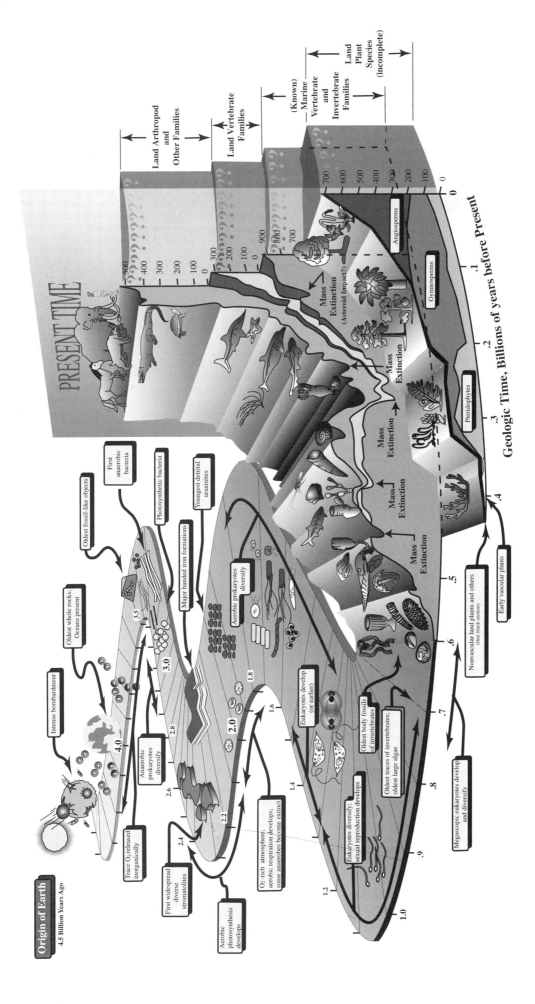

Figure 12.1. Schematic history of life on Earth, showing where key milestones in the history of life likely occurred relative to geologic events on Earth. Beginning at the Vendian-Cambrian diversification of life (chapter 18), the rise and fall with time of the number of families of land and marine creatures is depicted.

"life" on other planets might not undergo Darwinian evolution, but might still involve biochemical reactions resembling those on Earth; non-Darwinian evolution might have occurred in the very earliest, primitive organisms on our planet as well. The definition also excludes "artificial life," experiments in computer information replication described in chapter 13, but could easily allow inclusion of such experiments by replacing the phrase "chemical system" by "material system," as has been suggested by NASA Ames planetary scientist Chris McKay. Finally, a more general definition of life – perhaps too general in that it might apply to some nonliving systems – is "a system that possesses the ability (*homeostasis*) of maintaining form and function through feedback processes in the face of changing environments."

What is required to maintain terrestrial life? Many different things are required for different forms of life, but the essentials are organic (carbon-based) molecules for structure and processes, liquid water as an energy and information transporting medium, and a source of usable energy (most often from the Sun, but Earth's heat can be a source as well).

12.2.2 Basic Structure of Life

All known life-forms live on Earth and are based on the same small set of molecular units and chemical reactions. Four types of essential molecules are *organic* (contain carbon and hydrogen) and account for most biological processes and structures: *proteins*, *nucleic acids*, *carbohydrates*, and *lipids*. Carbohydrates are molecules in which the hydrogen and oxygen atoms form a whole number (that is, 1, 2, 3 ...) of water molecules. Some classes of carbohydrates (*sugars*) are produced by plants using sunlight as an energy source, and water and carbon dioxide as the raw materials. This process, *photosynthesis*, led to fundamental changes in Earth's atmospheric composition early in its history, as we see in chapter 17.

The molecules that provide the primary structural material for life, as well as contribute crucially to its functioning, are called *proteins* (from the Greek word *proteios*, or primary, hence "primary substance"). Proteins are long chains (or *polymers*) of relatively small molecular units (*monomers*), called amino acids. An example structure of an amino acid is shown in figure 12.2. The "R" group distinguishes the particular amino acid – it could be hydrogen or methyl (CH_3) or more complicated combinations of hydrogen, carbon, and oxygen. Of the vast variety of possible amino acids, only about 20 are found to be the building blocks of the major proteins of life.

Long-chain proteins fold into tight bundles, which give rise to the physical and chemical behaviors associated with

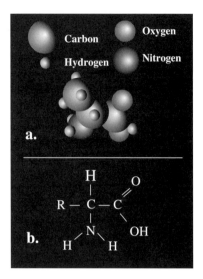

Figure 12.2. (a) Atomic structure of the amino acid alanine, used in proteins. (b) Schematic structure of many amino acids, including most biological ones, where "R" represents a functional group of atoms that defines the particular amino acid.

particular proteins. A typical protein chain may contain from about 50 to 1,000 amino acid molecules strung together. The total number of possible proteins is vastly more than the relatively few (of order 100,000) that actually occur in terrestrial life. Of those that do occur in cells, some play a role in defining the cellular structure, some act to transport or store molecular compounds, and others act as catalysts to control the rates of biochemical reactions; the latter are called *enzymes*.

Proteins cannot make copies of themselves; in the absence of some directive agent or template, the faithful production of proteins from the simpler amino acids would not occur in cells. Nucleic acids are molecules that form the building blocks of the templates, which we consider next.

12.2.3 Information Exchange and Replication

The information-carrying and replicating (or *genetic*) components of terrestrial life are types of nucleic acids called DNA (deoxyribonucleic acid) and RNA (ribonucleic acid). DNA molecules are long double chains normally twisted into a helical structure. The side rails of the double chains consist of a string of alternating sugar and phosphate molecules. Sugar is a simple carbohydrate. Many common sugars, such as glucose, have the chemical formula $C_6H_{12}O_6$; others have slightly different ratios of carbon to hydrogen and oxygen. Phosphate molecules, or phosphate groups, form high-energy bonds in living systems; a phosphate group involves phosphorous and, for example, would have the formula PO_3H_2.

The cross-ties of the DNA chain consist of pairs of four different types of *bases*: adenine (A), thymine (T), guanine

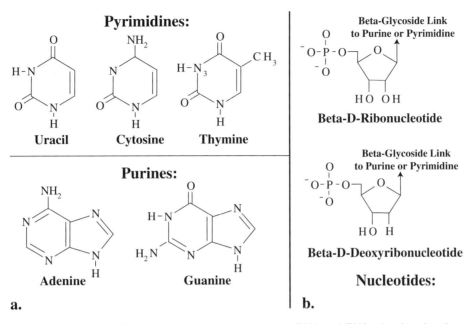

Figure 12.3. (a) The five types of nucleic acid bases in DNA and RNA, showing the characteristic ring structure. (b) Two types of nucleotides are produced from the bases: (top) a ribonucleotide that is the foundation for RNA and (bottom) a deoxyribonucleotide that is the foundation for DNA. Empty vertices correspond to carbon paired with zero or one hydrogen atoms; double lines indicate two shared pairs of electrons. Redrawn from Mason (1991).

(G), and cytosine (C). These bases consist of carbon, nitrogen, oxygen, and hydrogen in complicated ring structures (figure 12.3). The combination of a base with the sugar and phosphate backbone is called a *nucleotide*.

The pairing of the nucleotide bases is restricted: A with T and G with C. Thus, the two sides of the chain (*conjugates*) are redundant to each other because, from the letter on one side, you know what the letter on the other side must be. In replication, the net result is that the two sides of the chain are split, with each side reconstituting (through the mediation of enzymes) its conjugate, resulting in two copies of the original DNA.

12.2.4 Formation of Proteins

Protein synthesis is governed by DNA, through the intermediation of RNA. The synthesis begins when DNA, instead of replicating to make new DNA, transcribes RNA. RNA differs from DNA in two aspects: The sugar is of a different form, and the nucleic acid base uracil (U) is present in place of thymine (T). These are relatively minor structural changes in the molecule (figure 12.3), a fact that we return to in chapter 13 as we consider the origin of the genetic code.

Thus, a chain of RNA contains a long sequence of molecular monomers chosen from among the four nucleic acid bases A,U,G,C. This chain of monomers can be "read" as a sequence of three-letter "words" constructed from a four-letter "alphabet." Each three-letter word is called a *codon*. Some examples of words are **GUA**, **AAG**, **UGA**. The number of possible words is $4 \times 4 \times 4 = 64$.

Each codon codes for a specific amino acid; thus, the sequence of codons in an RNA molecule (which, remember, is ultimately derived from the sequence in the original DNA) specifies a sequence of amino acids. This amino acid sequence constitutes the synthesized protein. A particular amino acid generally is coded for by more than one codon, because 64 codons are available for the 20 amino acids commonly used in terrestrial biology.

The actual protein synthesis is a bit more complicated, with *messenger RNA* carrying the protein-structure information from the DNA, *transfer RNA* attaching to specific amino acids and aligning them based on the messenger RNA sequence, and *ribosomal RNA* (located in a cellular structure called the ribosome) receiving the ordered amino acid sequence (ferried by the transfer RNA) and acting as a catalyst for final assembly of the amino acid chains. Other RNA molecules assist in DNA replication and in the construction of the messenger RNA. This diverse range of roles for a single kind of molecule makes tempting the proposal that, at some time in the distant past, RNA was central to the genesis of life as we know it. By contrast, DNA, which is not terribly dissimilar to RNA, has a very specialized function as a record of the genetic information of the individual organism and (in separate DNA strands) of certain structures in the cell. This essential but much more limited role compared to that of RNA suggests that DNA is a subsequent, derived molecule.

A length of DNA that carries the genetic information that is ultimately expressed as a single protein is called a *gene*. The genetic code is the complete sequence of nucleotides in DNA which determines the form and function of an organism's proteins. All living organisms on Earth that have been examined use DNA and RNA to record and express the proteins of which they are made.

12.2.5 Mutation and Genetic Variation

The replication process sketched above operates with high fidelity. Errors are rare but occur. These errors are called mutations. Such errors, changes in the structure of the DNA, may have a variety of causes such as chemical impurities in the environment or radiation (ultraviolet photons or particle radiation). Other errors or changes may be a result of accidental mixing or crossover of DNA chains in normal cells. Mutations give rise to changes in organisms. This genetic variation is usually harmful but sometimes not.

Such variation forms the biochemical basis for the evolution of one species from another, via natural selection within a given environment or through environmental changes in the ecosystem itself. The large-scale pressures for the evolution of species are discussed later, but, without the imperfection and vulnerability in the genetic code that allows changes (both good and bad), such evolution would not be possible, or too slow to be relevant to the history of life on Earth.

12.3 THE BASIC UNIT OF LIVING ORGANISMS: THE CELL

With the exception of viruses and virons, which are essentially strands of DNA or RNA sheathed in proteins and which cannot survive independently of other organisms, all Earth life is organized into cells. These structures provide a boundary or membrane for separating the outside environment from the internal one where biochemical reactions occur, and house the DNA and RNA genetic machinery for replicating the particular organism.

Two basic types of cells exist today on Earth (figure 12.4). *Prokaryotic* (from the Greek "pro" for before and "karyon" for nut, hence seed or nucleus) cells are exemplified by the bacteria, and are fairly simple bags of *cytoplasm* (a salt water medium containing proteins)

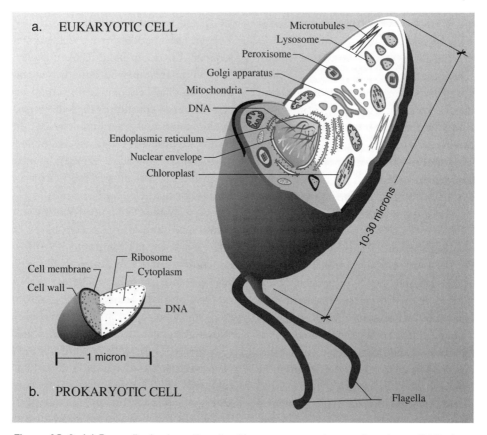

Figure 12.4. (a) Generalized eukaryotic cell, with structures and organelles shown. (b) Prokaryotic cell (a bacterium).

in which are contained a loop of DNA and *ribosomes*, which are structures hosting the RNA for protein production. The cell walls are composed of sugars and *peptides*, chains of amino acids that are shorter than the full-fledged proteins. These simple cells are as small as 10^{-6} meters, are able to store energy by either aerobic or anaerobic processes, and divide by a simple splitting process. Most can move by means of attached protein strands called *flagellum*.

Eukaryotes (from the Greek "eu" for good, hence true, and "karyon" for nucleus) are much bigger, more complex cells. The DNA is housed in a *nucleus*, and other structures called *organelles* also occupy the cytoplasmic space. These include *plastids* (the commonly known green ones being *chloroplasts*) in plants, which are the sites of photosynthesis (see section 12.4), and *mitochondria*, within which respiratory processes take place. Most eukaryotes are obliged to utilize oxygen in their sustaining processes; there are a lesser number of anaerobic eukaryotes. Furthermore, the cells are 10 to 100 times larger in diameter than prokaryotic cells, and reproduce by somewhat more complex processes which ensure the presence of nucleus and organelles in each new cell.

Clues to the origin of eukaryotic cells lie in the ability of most to take advantage of oxygen and the resemblance of organelles in size and structure to bacteria. They seem to be later arrivals in the history of life, though how late is controversial, and in chapter 18 we consider their origin.

12.4 ENERGETIC PROCESSES THAT SUSTAIN LIFE

Chapter 13 considers life as a phenomenon of non-equilibrium thermodynamics, driven and sustained by substantial flows of energy. Life on Earth primarily utilizes the Sun for energy, with heat from Earth's interior as a secondary source. To appreciate the coupling of life to such energy sources, we must understand how they are utilized by living organisms on Earth.

12.4.1 Common Metabolic Mechanisms

Energy for living processes requires a usable raw material and suitable chemical reactions to store energy in chemical bonds which then can be utilized by the organism. *Fermentation* and *respiration* are the most common metabolic processes used by organisms today (figure 12.5). In each case, energy is stored by the organism in bonds involving the element phosphorus. Molecules such as adenosine triphosphate serve as the storage medium through their phosphate bonds; a single phosphate bond stores 7.3 kilocalories of energy, a large amount. (Biochemists use the unit calorie, but so do nutritional scientists: Confusingly, the nutritional "calorie" listed on a cereal box is 1,000 of the physicists' calorie, or 1 kilocalorie. Only the physicists' calorie is used in this book.)

In fermentation, which is practiced by bacteria, the sugar, glucose, is split into two molecules of pyruvate. Two phosphate-bonds worth of energy are used to break the bond, but the resulting reaction produces a total of four phosphate-bonds worth of energy. The pyruvate then is converted into ethyl alchohol and carbon dioxide, or into lactic acid (depending on the type of bacteria) as waste products. Net energy gain is two phosphate-bonds or 14.6 kilocalories of energy per glucose molecule.

Respiration takes advantage of the presence of free oxygen (O_2) in Earth's atmosphere to extract much more energy from the glucose molecule than fermentation can. Pyruvate again is produced as in fermentation, but instead of immediate conversion of pyruvate into waste products, a complex series of chemical reactions with six oxygen molecules leads to the production of carbon dioxide, water, and 34 additional phosphate-bonds worth of energy. The net result is 36 phosphate-bonds, or 263 kilocalories, worth of energy. A number of biological catalysts, that is, enzymes, are required to mediate and control the *citric acid* cycle that produces the additional 32 phosphate bonds. Although some bacteria do undertake respiration, eukaryotes take the greatest advantage of this process. As we discuss in chapter 18, the onset of oxygen in Earth's atmosphere was likely the enabling factor for the dominance of multicellular eukaryotic life, with its specialization of cells and high degree of mobility. Confined to only one-eighteenth the amount of energy per glucose molecule, as is the case in fermentation, living processes would be much too sluggish to sustain macroscopic animals.

12.4.2 Photosynthesis

The source of glucose and other sugars used in metabolic processes must lie in an energy-collecting process. Without some means to create such sugar, limitations of food supply for metabolic processes would be far more severe than they actually are. *Photosynthesis* is the production of sugars from water and carbon dioxide, using sunlight as the energy source. Chemically the reaction (in plants) is $6CO_2 + 6H_2O \rightarrow C_6H_{12}O_6 + 6O_2$, where the sugar (glucose) appears as the first compound on the right side of the equation. Energetically, sunlight charges a natural battery in the plant: A molecule called *chlorophyll* is able to donate an electron upon absorption of photons. The source

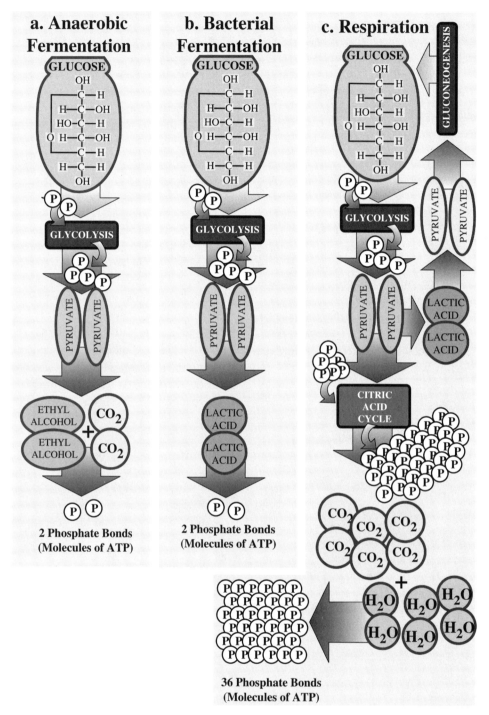

Figure 12.5. Comparison of the common metabolic processes of fermentation and respiration. Various molecules involved in the process are shown in light grey, but only glucose's structure is shown. Other boxes with white lettering refer to processes; gluconeogenesis is the production of glucose. Phosphate bonds are symbolized by a P. Note that fermentation sequesters for the organism, in the end, two phosphate-bonds worth of energy; respiration sequesters 36 such bonds. Modified from Gould and Gould (1989).

of the electron in plants and most photosynthesizing bacteria is a water molecule. The electrons so liberated then are used to drive the formation of high-energy phosphate bonds which, in turn, the plant uses to produce sugars.

There are several varieties of chlorophyll and chlorophyll-type molecules utilized by different photosynthesizing organisms. Modern plants employ water as the electron source, and produce molecular oxygen as a waste product. Some bacteria use a less efficient cycle in which the chlorophyll-type molecule is the electron source itself, and the electron then is returned to the donor molecule. This more primitive *cyclic* photosynthesis does not produce molecular oxygen, and captures less energy from a given amount of sunlight than does plant photosynthesis. One type of cyclic photosynthesis, as an example, begins with hydrogen sulfide and ends with sulfur: $2H_2S + CO_2 \rightarrow CH_2O + H_2O + 2S$.

Some bacteria that conduct cyclic photosynthesis are in fact intolerant of oxygen. Others switch between oxygen-free photosynthesis and respiration. As discussed in chapter 19, the rather late occurrence of large amounts of molecular oxygen in Earth's atmosphere suggests that the less efficient cyclic photosynthesis dominated early on, and that the oxygen-producing form of photosynthesis was a later innovation.

12.5 OTHER MEANS OF UTILIZING ENERGY

Respiration and photosynthesis are currently the predominant means of producing storable energy from sunlight and then utilizing that energy for biological processes. However, other mechanisms are employed by organisms that are not in environments where they can photosynthesize or gain access to photosynthesized sugars. *Chemisynthesis* is employed by organisms that live in the environment around deep-sea volcanic vents, where hot, hydrogen sulfide-rich waters pour out of newly formed ocean crust (figure 12.6). Such waters, compared to the colder, sulfide-poor adjacent regions, have an abundant supply of *free energy*. This term refers to a source of energy that can be utilized readily to do some form of work, such as sustain biological processes, or can be stored in high-energy phosphate bonds.

One readily available means to extract energy from the vents is to combine hydrogen sulfide with oxygen to form sulfur dioxide with production of energy. Such a process

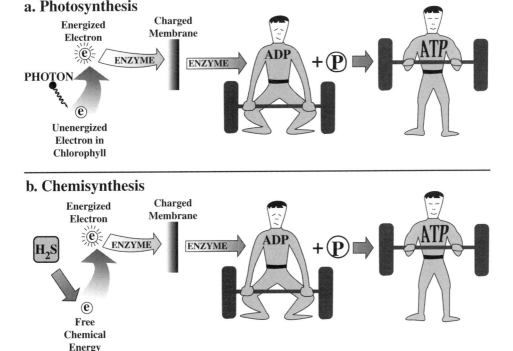

Figure 12.6. Comparison of photosynthesis and chemisynthesis: Both allow organisms to sequester useful energy for cellular functions in the form of the phosphate bonds of the molecule adenosine triphosphate (ATP). However, photosynthesis derives energy from sunlight whereas chemisynthesis garners energy from reduced molecules (such as H_2S) not in chemical equilibrium with their surroundings. ATP is symbolized as the energized weightlifter; the preceding molecular form that does not have the energetic phosphate bond is adenosine diphosphate (ADP). Redrawn and modified from Gould and Gould (1989).

is possible in an ocean that has free oxygen available, but would not work on the primitive, pre-oxygen-rich Earth. Other biochemical cycles which use sulfur but not oxygen are conducted by some prokaryotic organisms, but these capture much less energy than the oxygen-driven cycles. As with fermentation, chemisynthesis without free oxygen was the hallmark of a rather sluggish primitive biota.

12.6 ELEMENTAL NECESSITIES OF LIFE: A BRIEF EXAMINATION

12.6.1 Why Carbon?

Why is the element carbon the basis for biochemistry? Of all elements, carbon possesses the greatest tendency to form covalent bonds and, in particular, has a remarkable tendency to bond with itself. These characteristics reside to some extent with all elements near the center of the periodic table (figure 2.6), for which there is not a strong tendency to favor donation over acquistion of electrons. However, carbon is most distinguished in this regard as a small element (few electron shells) and being positioned in the central column IVA of the table. It readily forms a variety of long-chain, sheet, and ring structures, many of which play important roles in the basic biological molecules, as seen in figures 12.2 and 12.3. In addition, carbon is the fourth most abundant element in the cosmos (after hydrogen, helium, and oxygen), being made readily from helium fusion in stars.

The element silicon has chemical-bonding properties very similar to those of carbon, being one row below the latter in the periodic table. Silicon also forms chain, sheet, and ring structures with nearly (but not quite) the same ease and variety as does carbon. It is not nearly as abundant as carbon in the cosmos because it is the product of a later stage of fusion reactions achieved only in massive stars, but, after oxygen, it is the primary constituent of the crust of Earth. The similarities in silicon's properties to those of carbon and its high abundance in geologic materials are responsible for the lithification of biological remains in the form of fossils, as discussed in chapter 8. It is natural to ask whether silicon might be the elemental basis for biology on another planet in another planetary system, perhaps one on which conditions are not quite suitable for carbon-based life because of higher temperatures, or where carbon-bearing molecules were not supplied to the planet's surface, for whatever reason.

The biologist A. Cairns-Smith has proposed that, on the early Earth, layers of crystalline silicates might have served as the basis for a very primitive kind of life, which served in turn as a sort of template for, and gradually evolved into, carbon-based life. Certainly, some clay minerals have a surface upon which organic molecules can adhere, in favorable positions, which might have encouraged the formation of replicative molecules similar to RNA (or perhaps RNA itself). Cairns-Smith's proposal is more radical, however, and represents a kind of inverse fossilization in which the end result is a vigorous and robust carbon biochemistry.

Because the evidence for any very primitive silicon-based life surely has vanished, such speculations must remain just that. However, consider that, in natural chemical systems, very small intrinsic advantages can be magnified into very large effects. In addition to forming stronger bonds with itself (C–C) than can silicon, carbon is more versatile in its bonding properties, and hence can form a wider variety of structures. The *information* content, therefore, inherent in carbon-based biochemistry is larger than that available using any other single element as a basis. The American chemists J. Feinberg and R. Shapiro suggest the analogy of an alphabet. By way of example, the English alphabet has 26 characters, from which are made words and sentences. Other human languages may have hundreds or thousands of symbols. A computer, on the other hand, uses two basic states or characters (usually symbolized as 0 and 1) to carry information. A computer can record the same information in its two-letter alphabet as an English writer would in his or her 26-letter code, but the computer must use longer words and longer sentences to record the same concept or amount of information.

Biochemistry in which the number of possible compounds is fewer than in a carbon-based system, by analogy, would have to utilize those fewer compounds to build sufficient functional structures and information-carrying molecules to survive and perpetuate its particular kind of chemistry. Organisms based on such a biochemistry might build essential structures more slowly and less efficiently, reproduce less frequently, and perhaps be limited to more narrowly characterized environments than is carbon-based life. In 1961 Carl Sagan noted that silicon cannot form the changeable and versatile side chains that are crucial to the proteins and nucleotides that characterize carbon-based life. Thus there might well be some environments that favor silicon over carbon but, in a general sense, carbon-based biochemistry would more likely be fruitful and multiply, gain access to and utilize a variety of sources of free energy in its environment, and respond robustly to drastic environmental changes, than would its silicon-based counterpart.

There is one environment, however, in which silicon-based life is in a sense extant on Earth today. The term *artificial life* has been co-opted by information scientists and physicists to refer to computer programs that move bits of information around within the silicon-based processors and memories of modern computers and, in doing so, simulate the biological processes of complexification, growth, and reproduction. The significance of such programs is highly controversial, and most laboratory biologists do not regard the resulting transfers of energy and information within the computers as equivalent to living processes. We return to artificial life and its lessons for terrestrial carbon-based biology in chapter 13.

12.6.2 Why Water?

The third most abundant element in the cosmos is oxygen, and it occurs primarily in two molecular forms: carbon monoxide (CO) and water (H_2O). In our own solar system, the former is rare, presumably because its high volatility discouraged the trapping of large amounts in solid material. Water ice, on the other hand, was a primary "mineral" in the primordial solar system, and Earth may have had access to large amounts of it through comets. Water is thus abundant on Earth and elsewhere in the solar system, but its importance for life is not due solely to abundance: The properties of water are crucial.

Water exists as a liquid over a temperature range particularly suited for organic reactions – not so cold that reaction rates are too slow to sustain biological processes, and not so hot that organic bonds are too readily broken. Water's existence as a liquid from 273 K up through the critical point at 647 K (beyond which only a single, gaseous/fluid state is stable) is unusually broad and is not significantly overlapped by many other abundant molecular species. (Liquid water in an open container is not stable in our atmosphere above 373 K – the boiling point – because the vapor pressure of water exceeds the ambient atmospheric pressure at sea level. For liquid water to exist above that temperature on Earth, it must be confined to a pressure vessel, such as a pressure cooker.)

This important property of water is due mostly to the somewhat unusual bonding mechanism between water molecules, in which the hydrogen atoms from one water molecule form bonds of modest strength with those of another water molecule. This mechanism of *hydrogen bonding* produces much stronger bonds than those seen in similar-sized molecules such as methane and carbon dioxide, for which the liquid phase occurs at much lower temperature, and much weaker bonds than in silicates, which melt at much higher temperatures. The hydrogen bond itself, caused by small residual positive charges at the hydrogen end and negative charge at the oxygen end of each molecule, is also responsible for making liquid water a good conductor of electricity and a good solvent for materials of biological importance.

Water functions as a medium within which biochemical processes take place, and as a liquid through which nutrients can be transported in a controlled fashion. Cellular membranes are important boundaries within which the salinity and acidity of the water are carefully regulated by biochemical processes, so that nutrients can be properly absorbed through the membrane and waste products extruded. In complex organisms with circulatory systems, blood is essentially a liquid-water medium packed with cells of various kinds and with a multitude of functions; the liquid nature of the blood enables transport over large distances to be rapid and efficient.

Most biologists would argue that, of all the requirements of life, the need for liquid water or a similar liquid medium is paramount. In its absence, the ability to selectively pass nutrients into loci of biochemical activities (cells, in the terrestrial case) and to remove undesired products of biochemical reactions is extraordinarily limited. It has been suggested that other liquids, such as ammonia-water solutions or hydrocarbon liquids, could be substitutes in low-temperature environments. However, the properties of ammonia-water solutions are such that certain kinds of organic reactions on which terrestrial life depends are prohibited; Shapiro and Feinberg suggest that weaker chemical bonds, such as those involving nitrogen, would tend to be more important. Such biota might be less robust than terrestrial life, and if the ammonia-water solution were near its freezing point (as low as 176 K), organic reactions would be much more sluggish.

12.6.3 Is Free Oxygen Essential?

One common misconception is that life on Earth requires molecular oxygen (O_2) for its existence and propagation. In fact, as discussed above, molecular oxygen and associated respiration processes are only one kind of energy-producing mechanism that sustains life. Many simple organisms not only do not use oxygen in metabolic processes, but are poisoned by it. Because molecular oxygen reacts readily with many organic compounds and breaks bonds in such compounds, organisms that tolerate the large amounts of free oxygen in Earth's atmosphere had to evolve protective mechanisms to avoid undesirable oxidation of organic molecules. Additionally, the presence of free oxygen made utilization of nitrogen as a nutrient more difficult. However, the very large amount of energy-storing phosphate bonds that can be created using molecular oxygen outweighed the disadvantages, and complex cells and multicellular life became possible as O_2 abundance in the atmosphere increased over time (see chapters 17 and 18).

Strategies for searching for inhabited planets around other stars focus on very sensitive spectroscopy to detect the presence of molecular oxygen in the atmosphere. Remember, however, that a negative result does not indicate the absence of life, but rather indicates the absence of massive photosynthetic processes that produce large amounts of molecular oxygen. Life utilizing anaerobic processes, such as fermenting bacteria or the sulfur metabolizers at deep-sea vents, might still be present. This does not mean such strategies are flawed; the detection of molecular oxygen is feasible with large interferometers planned for the next century, and hence a worthy goal particularly if one is interested in finding advanced life. However, there might well be a multitude of planets throughout the cosmos on which life exists but, for whatever reason, remains in a stage in which less-efficient anaerobic metabolisms predominate.

12.7 SOLAR SYSTEM SITES FOR LIFE

Although liquid media other than water are possible sites for life, a well-constrained search for solar system environments in which life might have arisen should focus on liquid water – because we know that life arose at least once in such a medium! In our own solar system, there are four environments within which liquid water is known to be stable today or was likely stable for long periods in the past: Earth, Mars, the interior of Europa, and the water clouds of the giant planets. Saturn's moon, Titan, although too cold for liquid water to be stable, is rich in organic molecules and may be an important site for certain kinds of organic reactions that precede life. Also, large impacts on Titan may have provided energy to melt its ice crust and provide liquid water for relatively short periods of time.

12.7.1 Atmospheres of the Giant Planets

The atmospheres of the giant planets represent the most speculative site for life, one proposed by Carl Sagan and others some decades ago. As discussed in chapter 11, no solid surface exists except at enormous depths in the interiors of these gaseous bodies. The water clouds lie below a layer of ammonia clouds, which in turn lie below methane clouds in the cases of cold Uranus and Neptune. Living organisms there might be composed of structures that allow them to cycle in depth from the relatively warm water clouds up to the ammonia clouds, absorbing sunlight at the higher levels and various trace organic constituents at a variety of altitudes. Although one cannot rule out such a biota, the initial evolution of such organisms to a point of sophistication such that they could safely cycle in altitude without sinking into excessively hot depths is an open issue.

12.7.2 Interior of Europa

The interior of Europa, a satellite of Jupiter just 10% smaller than our own Moon, is a less speculative venue for life. Spectroscopic studies of its surface from Earth reveal that water ice is an important or predominant component. Radar signals bounced off of Europa from Earth are reflected back with very high intensity, like a mirror, again indicating the predominance of water ice. Photographs taken by *Voyager 2* in 1979 reveal a bright surface covered in cracks, which themselves are only slightly darker than the surrounding surface. So bright is Europa's surface that most of the sunlight hitting it is reflected, and the average surface temperature is below 110 K, colder than the darker surfaces of Ganymede and Callisto.

Europa's surface layers may be cracked water ice at very low temperatures, but the density of this moon tells an intriguing story. At 3.0 g/cm^3, Europa cannot be composed entirely of water ice, which has a density close to 1 g/cm^3 (varying somewhat with pressure). To match the density with water ice and rock requires a moon composed of 80% rock and only 20% ice by mass. The rocky component has embedded within it the radioactive isotopes of potassium, uranium, and thorium and, as described in chapter 10 for Earth, the decay of such isotopes produces heat.

Add to this source of heat one other: tidal heating. Both Io and Europa have orbits that are slightly noncircular and are maintained that way by the mutual gravitational pulls of Io, Europa, and Ganymede against each other. These pulling motions are effective because the orbital periods of the three satellites are simple multiples of each other – the period of Europa is twice that of Io, and that of Ganymede twice Europa's. So, like a child on a swing pumping his or her legs in synchroneity with the period of the swing, these satellites tug gravitationally on each other and keep their orbits noncircular. This, in turn, means that even though each moon keeps one face approximately toward Jupiter all the time – as does our Moon to Earth – there is a small amount of twist as each moon varies in its orbital speed. The twisting is enough to cause frictional rubbing of rocks against each other in Io, leading to extraordinary heating that has melted its rocky interior and produced spectacular volcanic eruptions viewed by Voyager. Europa, 50% farther from Jupiter than Io is, suffers some tidal heating, but much less than Io experiences.

Models of the interior heating of Europa by radioactive isotopes and tides indicate that, if Europa developed a liquid water mantle early in its history, *liquid layer could be sustained today*. The heating rate is not enough to create a liquid mantle today out of solid ice, and so, the existence of a water mantle rests on melting in the past. The presence of cracks filled with a slightly darker material than the icy plains is consistent with liquid water and carbon-bearing materials welling up from deeper down, and the paucity of craters suggests that at least softening of the ice has occurred relatively recently in the geologic past. The cracks themselves might remain thin enough to allow some biologically useful sunlight to penetrate to the liquid ocean beneath.

Beginning in 1996, the sensitive electronic camera aboard the *Galileo* Jupiter orbiter imaged the surface of Europa in much greater detail than could *Voyager*. Preliminary study of the images yields additional circumstantial evidence that a liquid water layer exists beneath the ice. These include an enormous variety of cracks of different degrees of freshness, areas where cracks have been cut or buried by flows of liquid or warm ice, pieces of crust that have tilted upward as if floating on a layer of liquid water beneath, and craters in the ice displaying softened edges consistent with a thin ice crust (figure 12.7). Europa, then, might have a liquid water ocean lying beneath a frigid surface, and a source of heating that maintains the liquid state. Europa is almost certainly not just made of water ice and silicates, but likely acquired some carbon-bearing, nitrogen-bearing, and other compounds during its formation early in the history of the solar system. It is not too much of a stretch, then, to speculate that in the mantle ocean of Europa, perhaps near areas with particularly high heating (due to a concentration of radioactive isotopes, for example) lies a modest biota.

12.7.3 Titan

Saturn's largest moon, Titan, is bigger than the planet Mercury. It has an atmosphere that has a surface pressure of 1.5 atmospheres and is mostly nitrogen. Methane is the next most abundant gas in the atmosphere. Titan is so far from the Sun that the surface temperature is 95 K. This is so cold that methane and similar molecules may exist on Titan's surface as rivers and seas. Most of what we know about Titan comes from the *Voyager* missions, which could remotely sense the atmosphere but could not photograph Titan's surface through its permanent atmospheric haze. We know little about the surface, but it may be an intriguing mimic of Earth, with methane substituted for water. The action of solar ultraviolet photons creates a complex organic chemistry in Titan's atmosphere, forming the haze and many different molecules from methane and nitrogen.

Titan is a complex, planet-sized chemical factory that almost assuredly has no extant life because of the low temperatures. Exploration of this world is of high interest, however, because much of the chemistry going on in the atmosphere and on the surface may give us insight into organic chemistry on the earliest Earth. Although Titan's atmosphere is probably chemically much more reducing (that is, rich in hydrogen) than the early Earth's, it is likely to be a better analog to the Hadean Earth than is our present, oxygen-rich, biologically dominated planet. Occasional impacts or volcanic eruptions might melt the crust and produce pools of liquid water and ammonia for some thousands of years or more; within these transient pools interesting prebiotic chemistry might occur.

There is no way to reproduce fully in the laboratory the large and complex organic chemistry on Titan. The joint United States-European robotic mission to explore Saturn and Titan, called Cassini/Huygens, will drop a probe into Titan's atmosphere (figure 12.8). What will be learned about Titan in this mission, in the year 2004, may tell us some important things about the complicated chemistry postulated to have preceded life on Earth.

12.7.4 The Mars of Today and Yesterday

The American *Viking 1* and *Viking 2* robotic landers operated on the surface of Mars beginning in 1976 and extending into the early 1980s. Among the experiments were four designed to determine whether life existed in the upper few centimeters of soil collected by each lander. Each of the experiments was designed to test for a particular kind of energy-generating mechanism, such as photosynthesis, fermentation, and chemisynthesis. Soil was activated through the addition of potential nutrients, and then the experiments monitored release of gases that might indicate biological activity.

Some of the experiments showed positive results, but in combination, the experiments weighed against biological activity in the Martian soils. The positive activity was best explained as reaction of a nonbiological, highly oxidized component in the soil reacting with the gases and liquid nutrients supplied by the experiments. The death-knell for a biological explanation came from the *gas chromatograph/mass spectrometer* experiment, a device that can detect very small amounts of organic molecules. The amount of organic molecules at the two *Viking* landing sites, in the soil, was less than one part per million, and probably at the part-per-billion level for more complex

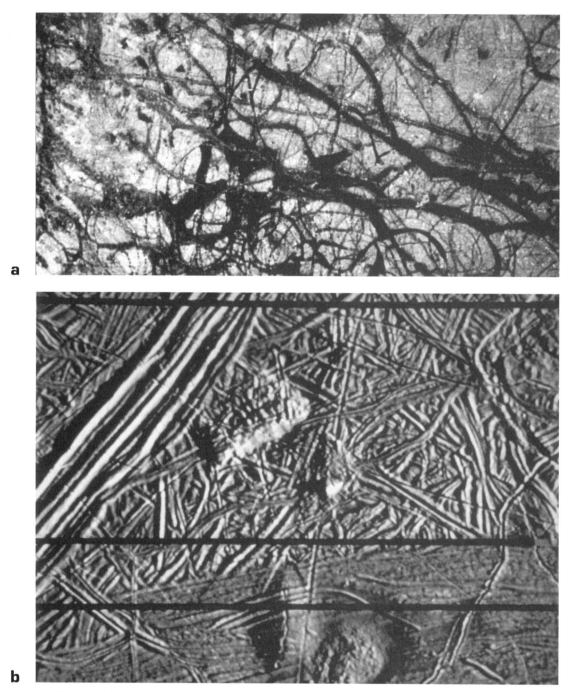

Figure 12.7. *Galileo* views of Europa: (a) Large-scale view. (b) Close-up image showing details around one of the crack systems. Courtesy of NASA Jet Propulsion Laboratory.

organics. With so few organics, carbon-based life evidently was not present, and speculation about, for example, silicon life was not profitable because no evidence for such life could be obtained from the experiments.

The two *Viking* lander sites, on high plains, were particularly poor candidates for sustaining living organisms, and areas near canyon bottoms, where the thin atmosphere has higher pressure, might be better. Nonetheless, current Martian conditions – an atmosphere of carbon dioxide with a total pressure less than a hundredth that at the surface of Earth, and surface temperatures generally well below the water freezing point – are not promising for life. The search for life in protected enclaves will be extremely challenging by robotic means, and piloted expeditions are decades away. Hence, the focus of research into life on Mars has shifted to finding evidence of past life.

12.7 SOLAR SYSTEM SITES FOR LIFE   147

Figure 12.8. Stages in the climax of the Cassini/Huygens mission: (a) Spacecraft fires engines to break into orbit around Saturn. (b) Orbiter releases Titan probe.

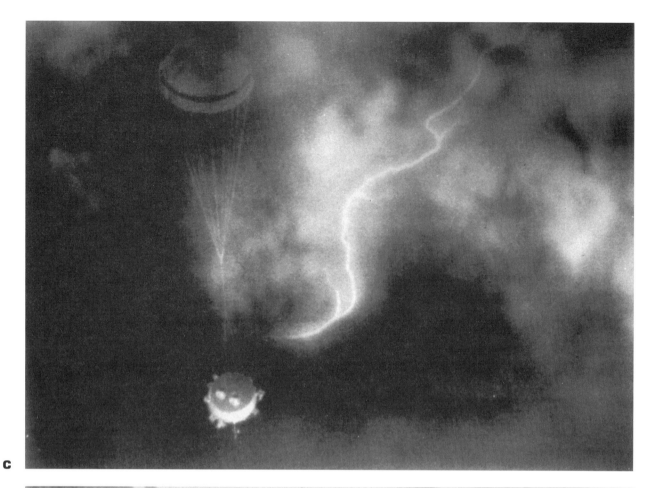

c

d

Figure 12.8. (Continued) Stages in the climax of the Cassini/Huygens mission: (c) Probe descends through the atmosphere of Titan, collecting scientific data. (d) Probe nears impact on the surface. Courtesy of European Space Agency and NASA Jet Propulsion Laboratory.

Evidence for ancient clement conditions on Mars is described in chapter 15, and consist of now-dry valley networks and outflow channels that appear to have been carved by liquid water (or debris carried by water) and more-controversial evidence in the form of possible glacial features and dried lake beds. Clearly something different happened on Mars in the past than today, and ancient conditions appear more promising than those at present. What further buoys the hopes of searchers for evidence of past Martian life is the presence of living organisms on Earth in extraordinary places. Submarine hot springs at mid-ocean ridges are, in the absence of sunlight, a rich abode of life. The Dry Valleys of Antarctica sport lakes whose surfaces are frozen over year-round, but in which a variety of microbial organisms thrive. Bacteria have been found living in rocks kilometers under the surface of Earth. All these sites could plausibly have had their analog on early Mars, and in chapter 15, we place possible life in the context of the history of Mars and the times when such environments might have existed.

To conduct the search for fossil evidence of life on Mars will be a daunting challenge, because such life very likely was microbial and not large, complex eukaryotic organisms (see chapters 15 and 18). However, even microbial life leaves behind signatures. The isotopic ratio of carbon-12 to carbon-13 in organic molecules may be increased from the baseline value by cycling through biological systems, and hence looking for evidence of unusual isotopic patterns in trapped gases in Martian rocks is one avenue. Researchers at the University of California have found that relatively primitive life-forms can have an effect on the kinds of mineral structures that precipitate from seawater on Earth, and the same effect might be preserved in rocks on Mars. Finally, the macroscopic evidence of earliest abundant Earth life is the lithified remains of bacterial colonies, the stromatolites of chapter 10. Because evidence for such organisms stretches back to the oldest fossil record, examination of sedimentary Martian rocks for such patterns also should be undertaken.

In August 1996, NASA geologist David McKay and colleagues brought the search for Martian life to the attention of millions of people around the world with an astonishing assertion: that they found evidence for relict biological activity in a meteorite thought to have come from Mars. Meteorite ALH84001, found in Antarctica in 1984, is one of 12 SNC meteorites thought to have been from material blown off of Mars by an impact, and onto a collision course with Earth. These samples contain trapped gases which, in their chemical and isotopic signatures, are identical to the present atmosphere of Mars as sampled by the *Viking* mission. Work over the past decade by a number of researchers has shown that it is plausible for a large impact to gouge out a portion of the Martian crust and send some of it toward Earth.

ALH84001 is an old igneous rock, with a radioisotopic age of 4.5 billion years. It is therefore from some of the earliest Martian crust. The meteorite was heated again about 4.0 billion years ago by a strong shock, possibly a nearby impact. Globules of carbon-bearing minerals called *carbonates*, described in more detail in chapter 14, were found in the rock. These formed later than the rock itself and are tentative evidence for the presence of liquid water flowing through the rock, tentative because carbonates can under some circumstances form in the absence of liquid water.

The age of the carbonate formation in the rock is highly uncertain, with different groups estimating ages from 3.6 billion years (from potassium-argon dating) to 1.4 billion years (from rubidium-strontium isotopes). In either case, these ages are much older than the time when the rock was blown off of Mars. This time is estimated by examining tracks made by cosmic rays on the surface of the rock exposed to space. The abundance of unusual isotopes of noble gases made by cosmic-ray collisions give a residence time in space of between 10 million and 20 million years. Once on Earth, isotopes such as carbon-14 and others of boron and chlorine, also made by cosmic-ray hits, begin to decay; their abundances indicate that ALH84001 has been on Earth only 13,000 years. So, the impact and delivery of this rock to Earth were much more recent than the formation of the carbonates, indicating that the carbonates were formed when the rock was in the Martian crust. Supporting this is the fact that the organic (carbon-bearing) content of the meteorite increases toward the center of the rock, suggesting that at least some of the carbon-bearing material is from Mars.

McKay and colleagues went a step further to propose three lines of evidence that are consistent with biological activity. First, ring-shaped carbon-bearing molecules called *polycyclic aromatic hydrocarbons*, or PAHs were found near the carbonate globules. Although PAHs can be formed in nonbiological environments, including interstellar space, from which their spectra are detected by sensitive Earth-based telescopes, the structure of the Martian PAH molecules differs from those seen to date from nonbiological sources. However, McKay and colleagues have not shown that biological activity necessarily forms such PAH structures either, and so, by themselves, the PAHs do not argue strongly for biological activity.

The second line of evidence comes from the association of two mineral types, magnetite and greigite, in the carbonate globules. The presence of these two mineral types together requires very alkaline water, an unusual geological situation, or the mediation of biological activity. Further, the texture of the mineral grains resembles those produced by bacterial processes on Earth. Although these seem to point strongly toward biology, McKay and colleagues caution, as do others, that some nonbiological processes can produce the textures seen; further, the identification of the mineral greigite is tentative.

Most controversial are images, constructed by bombarding electrons into the sample, of very small structures near the globules – about 10 to 100 times smaller than terrestrial bacterial. They look like microbial forms (figure 12.9), and McKay and colleagues argue that they could be evidence of life. Biologists and geologists today argue about the possible existence of simple cells some 10 times smaller than bacteria that are indirectly inferred to be present in Earth rocks. These *nannobacteria* are a speculative and as yet unproven form of terrestrial life, and their invocation as support for the Martian microbe interpretation has raised more controversy. To show that the forms in ALH84001 are cells requires finding cell walls, which as yet cannot be seen in the images.

The reader should be skeptical of the interpretation of the ALH84001 data in terms of biology. McKay and colleagues argue that the timescales are right – if the carbonates are 3.6 billion years old, then the evidence in the *Viking* images of a more clement Mars at the time are consistent with life beginning then. But the carbonates could be younger.

Further evidence against Martian life in ALH84001 came in 1998 when scientists from the University of California and University of Arizona analyzed the "organic" carbon (that is, the carbon not in the carbonates) from ALH84001 in more detail. From several lines of investigation, including measuring the carbon-14 to carbon-12 ratio, they concluded that most or all of this carbon is terrestrial, not Martian – though they confirm that most of the carbonate phase is likely from Mars. The ambiguities of interpretation associated with the claim of evidence for life in ALH84001 illustrate the great challenge of finding evidence for microbial life from subtle chemical clues and images at very small scales. It will be even more difficult to perform such searches on the surface of Mars; perhaps the best strategy is to locate promising sites on Mars where life may have existed, and then return samples back to Earth.

The search for extinct life on Mars by robotic means must be part of a larger effort to thoroughly characterize the histories of water, carbon compounds, and other biogenically important elements on Mars. A properly-conducted series of expeditions will enable an understanding of when and how life could have begun on Mars or, if

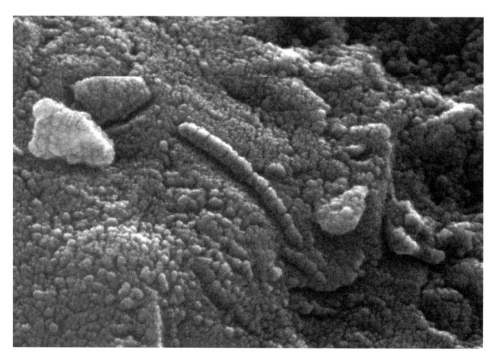

Figure 12.9. Image taken through an electron microscope of a portion of the ALH84001 meteorite, by David McKay and colleagues. The shapes that resemble bacteria are actually 10 to 100 times smaller than known Earth bacteria, and do not show evidence of cell walls.

no evidence of life appears, why it never occurred. The missions of the U.S. *Mars Surveyor* orbiter and *Pathfinder* lander beginning in late 1996 marked the start of such a series which, over the next decades, will begin to answer these haunting questions.

12.7.5 Earth

Earth is home to an abundant variety of biological organisms, all of which are carbon-based and utilize the nucleic-acid-based molecules DNA and RNA to record and transfer the blueprint for structures and metabolic processes. With the exception of viruses and the even-simpler virons, both of which rely on DNA-based cells for their continued existence, all Earth life is organized through the fundamental unit of the cell. Two types of cells exist: prokaryotic (simple) and eukaryotic (complex) cells. Earth life uses a variety of processes to capture energy from the Sun and Earth's interior and to store it in sugars and then high-energy phosphate bonds to power production of proteins and replication of genetic materials. Most types of eukaryotic cells must employ molecular oxygen in metabolic processes because of their high demand for energy; free oxygen, in turn, is produced by photosynthesis in plants and certain bacteria.

At the most fundamental level, Earth life can be divided into three groups: *bacteria* or *eubacteria* (true bacteria), *archaea* (literally, old ones), and *eukarya* (organisms composed of eukaryotic cells, including fungi, plants, and animals). By comparing the resemblances between the RNA in various organisms, one can construct a tree on which the relative distances between organisms indicate their degree of commonality, as shown in figure 12.10. The organisms of least familiarity to us are the ones that live in the most unusual environments, the archaea. The three groups of prokaryotes here are the *methanogens* which produce

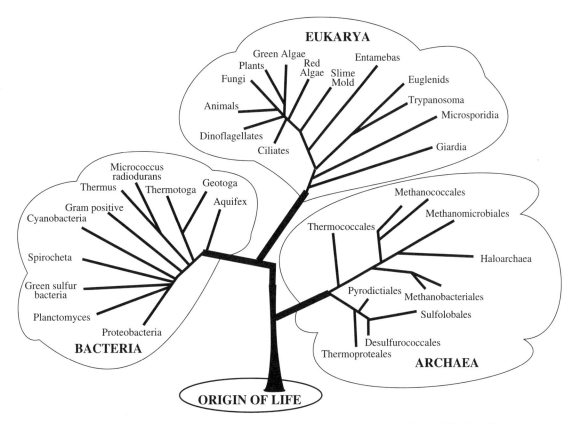

Figure 12.10. The *phylogenetic tree*, or the organization of the major kingdoms of life. The kingdoms are shown grouped within the three domains of life: bacteria, archaea, and eukarya. The tree is constructed by comparing the sequences of nucleotides on RNA molecules in organisms belonging to each of the kingdoms shown. The longer the branch along which a kingdom is located, the greater the difference in genetic sequences from other kingdoms. The differences presumably have resulted from mutations and consequent evolution which has led to the divergence of life-forms seen today. Hence, the animal kingdom is most closely related to the fungi, and much less so to amoeba with ciliae. A few minor kingdoms are omitted for clarity. Figure constructed with data from Forterre (1995) and Pace (1997).

methane in an anaerobic energy cycle, the *halobacteria* which survive in extremely salty environments, and the *sulfothermophiles* which live in high-temperature environments such as submarine vent areas and rely on sulfides for their energy source.

There is a deep evolutionary gap between these cells and the true bacteria, and it has been popular to argue (hence the name) that the archaea are the most primitive organisms on Earth. However, the French biologist P. Forterre and others have asserted that the archaea show evidence of being later adaptations to unique environments rather than progenitors of other organisms. In particular, the transcription of RNA in archaea is done in a fashion more similar to that in eukaryotes than in bacteria, and eukaryotes are almost certainly cells evolved at late times, because of their reliance on oxygen (with a few exceptions). To trace back the origin of life through the phylogentic tree of figure 12.10 may not be possible because the "universal ancestor" of us all may not be with us today, even among the simplest life-forms. Thus is the origin of earthly life a daunting question, one that we tackle in the next chapter.

12.8 QUESTIONS

a. Can you think of physical evidence that living processes require a creative spark beyond the chemistry of the nonliving world?
b. Why does DNA use only four distinct nucleic acid bases? Could there be an advantage in a system using six or even eight, such bases?

12.9 READINGS

12.9.1 General Reading

Sagan, C., and Druyan, A. 1992. *Shadows of Forgotten Ancestors: A Search for Who We Are*. Random House, New York.

12.9.2 References

Bada, J.L., Glavin, D.P., McDonald, G.D., and Becker, L. 1998. A search for endogenous amino acids in Martian meteorite ALH84001. *Science* **279**, 362–365.

Chyba, C., and McDonald, G.D. 1995. The origin of life in the solar system: Current issues. *Annual Review of Earth and Planetary Sciences* **24**, 215–249.

de Duve, C. 1995. The beginnings of life on Earth. *American Scientist* **83**, 428–437.

Forterre, P. 1995. Looking for the most "primitive" organism(s) on Earth today: The state of the art. *Planetary Space Science* **43**, 167–177.

Gould, J.L., and Gould, C.G. 1989. *Life at the Edge: Readings from Scientific American*. W.H. Freeman and Company, New York.

Jull, A.J.T., Courtney, C., Jeffrey, D.A., and Beck, J.W. 1998. Isotopic evidence for a terrestrial source of organic compounds found in Martian meteorites Allan Hills 84001 and Elephant Moraine 79001. *Science* **279**, 366–369.

Mason, S.F. 1991. *Chemical Evolution*. Clarendon Press, Oxford.

Morrison, R.T., and Boyd, R.N. 1973. *Organic Chemistry*. Allyn and Bacon, Boston.

Mojzsis, S.J., Arrheniun, G., McKeegan, K.D., Harrison, T.M., Nutman, A.P., and Friend, C.R.L. 1996. Evidence for life on Earth before 3,800 million years ago. *Nature* **384**, 55–59.

Pace, N.R. 1997. A molecular view of microbial diversity and the biosphere. *Science* **276**, 734–740.

Reynolds, R.T., Squyres, S.W., Colburn, D.S., and McKay, C.P. 1983. On the habitability of Europa. *Icarus* **56**, 246–254.

Sagan, C. 1961. On the origin and planetary distribution of life. *Radiation Research* **15**, 174–192.

Shapiro, R., and Feinberg, G. 1995. Possible forms of life in environments very different from the Earth. In *Extraterrestrials: Where Are They?* (B. Zuckerman and M.H. Hart, eds.). Cambridge University Press, 2nd ed.

Taylor, M.D. 1960. *First Principles of Chemistry*. D. Van Nostrand, Princeton, NJ.

Treiman, A.H. 1996. Fossil life in ALH 84001? *Lunar and Planetary Information Bulletin* **80** (Summer), 2–6.

13 THE ARCHEAN EON AND THE ORIGIN OF LIFE: II. MECHANISMS

13.1 INTRODUCTION

Having covered in the previous chapter some of the basics of present-day living organisms and considered the limitations of life in terms of terrestrial and extraterrestrial environments, the present chapter addresses some of the issues surrounding the origin of life. We begin with general considerations about living processes and their relationship to the natural laws that govern the workings of the universe. In particular, self-organization seems to be a property of complicated physical systems, and computer simulations of such systems suggest the kind of bootstrapping necessary to build well-controlled biochemical processes from simpler suites of chemical reactions. We then move to more specific ideas about how life might have begun and examine the issue from two somewhat different points of view: that the origin of life lay in the primitive mimicking of cellular processes (the vesicle model) or that the essential point of origin lay in an RNA or slightly more primitive genetic-coding molecule (the RNA world).

13.2 THERMODYNAMICS AND LIFE

Thermodynamics, introduced in chapter 3, is the study of energy transfer in macroscopic systems. A fundamental principle that governs the transfer of matter and energy in both natural and artificial systems is called the second law of thermodynamics. It is most precisely expressed mathematically but, in words, it says that the capacity for a system to do useful work (move something) decreases with time, unless usable energy is pumped into it. Looked at another way, systems tend to become more disordered with time. An engine takes a hot fluid and converts this to motion of some kind; the waste heat is dumped into a cold reservoir. Over time, the hot reservoir cools, the cool one warms, and the efficiency of the engine to keep the moving parts moving decreases. Only by heating up the hot reservoir (adding some energy from outside) can the engine be maintained at a given efficiency; but then the device heating up the reservoir must itself be stoked or it will run down, too.

A measure of the ability of a system to do work on its environment is called *entropy*. The more entropy, the more disordered the system and the lower the system's capacity to do work. The entropy of a system increases with time unless an external device is applied to pump more usable energy into it. But then, that external device gains entropy. In fact, a fundamental principle of thermodynamics is that the total entropy of the universe is increasing with time. One situation in which entropy is not increasing, but is constant instead, is that of a system in equilibrium with its surroundings; this is a state in which there is no tendency to exchange matter or energy with the surroundings. Can work be done in such a state? No – hence systems in which entropy is not increasing can do no work. A big tank of warm water can do no work if it is in contact with an environment at the same temperature and pressure, that is, it is in equilibrium with its suroundings. A corollary, then, of the second law is that, given enough time, the entire universe will come to equilibrium, entropy will increase no more, and there will be no further change in anything.

The second-law concept, along with other laws and corollaries, is fundamental to the design of machines in our civilization and is observed repeatedly in the natural world around us. A chemical reaction converts fuel in a camping stove to carbon soot and various gases, releasing heat and driving the system toward a more disordered state; while it does so, food may be cooked. After the fuel is converted to gas and soot, it is hard to use these products to heat after-dinner coffee! Snow taken from the frigid heights of a high mountain down to the surrounding lowlands melts quickly; once melted, it requires a refrigerator (powered by electricity) to make snow or ice again. An egg may be scrambled and cooked to taste using various sources of energy; to reconstitute the original egg from the scramble is a bit more challenging.

What is the physical origin of the second law? It arises from the fact that the macroscopic (large-scale) world that we see actually is composed of a very large number of microscopic particles, and *a given condition of the large-scale world represents an enormous number of possible states in which the microscopic particles may find themselves*. The most probable state of the macroscopic system is the one that expresses the largest number of configurations of the microscopic particles under the given physical conditions – this is the state of equilibrium. Macroscopic systems evolve toward equilibrium as the microscopic particles underlying them wander into more and more accessible states – and hence become less characterizable, or organized, in the process. Equivalently the entropy increases as a macroscopic system evolves toward equilibrium.

It has been said that life is a counterexample to the second law of thermodynamics. Life from seed organizes itself into a more complex state: It grows, becomes stronger, and propagates its own seed. In the fossil record over time, scientists see a remarkable flowering of complex forms from an original rather limited and primitive set of species. Admittedly, all things die, but in the growing and complexification, is life violating the second law? *No!* Life is very definitely providing us with a "living" example of the workings of the second law, and is churning out entropy like crazy! Here, it is necessary to elaborate on a corollary of the second law. If a system is brought only a little ways away from equilibrium, it will move back to equilibrium in such a way as to produce entropy as slowly as possible. But brought far from equilibrium, the system could operate in very many complicated ways as it moves back toward equilibrium, and the production of entropy speeds up.

This complex behavior as a system is moved far from equilibrium is easily seen in simple everyday life. Open a bottle of soda (or beer) and tilt the bottle just enough to start letting liquid out. This system is not in equilibrium because the liquid can fall out of the bottle (and do work). It will come out smoothly and predictably. Now open a bottle and turn it upside down; this configuration is much farther from equilibrium. The complex behavior of the liquid as it "glug, glug, glugs" out of the bottle is more interesting and less predictable. And, more entropy is produced.

Systems brought far from equilibrium are at the root of not only complicated behavior, but organized behavior in natural systems. This is a crucial concept! The heating of plains during summertime produces a disorganized stirring of the air around it. The rising air, if it is moist enough, condenses water as it reaches higher, colder altitudes. This condensing process adds more heat to the system and allows the air to rise more rapidly: The system is very far from equilibrium. The eventual result is a beautiful, organized column of cloud, which looks like a giant piece of cotton candy. But even further, the end result of this process – the thunderstorm – produces discharges of electricity that carry enormous power. Surprisingly, a good fraction of the energy of a thunderstorm, originally the energy of the sunlight reaching the ground, is expended in lightning flashes: rather organized electrical energy that is perfectly capable of doing work. Therefore, from the simple, seemingly random process of heating of the ground and transfer of heat to the air, a highly organized system has been created. But entropy is being produced rapidly in this process.

Life is a set of complex physical systems in which chemical and energetic sources are held far from equilibrium. Life produces large amounts of entropy, but because it is so far from equilibrium, organized and complex behavior is to be expected. Living processes do not violate thermodynamics; they simply are maintained far from equilibrium states. This statement can be quantified by comparing living systems with chemical systems that are much closer to equilibrium. The chemistry and energetics of individual living processes are generally well understood today; what is difficult for anyone to comprehend is how these processes interact to produce forms of extraordinary complexity and, in at least one case, sentience.

The understanding of a living organism as a set of interlinked systems held far from equilibrium is essential to approaching the question of how life formed. If life is a natural product of the early evolution of Earth, then the essential characteristics of life must be an outgrowth of early Earth systems that were (i) far from equilibrium and, as a result, (ii) self-organizing and (iii) self-complexifying.

It is easier to hold a small system far from equilibrium than a big one, and so, the isolation of a portion of the environment from the whole likely was involved in the origin of life. This isolated piece of the environment had to have a source of energy, and it had to operate in such a way as to build increasing complexity up to that of the genetic coding molecules RNA and DNA. The pathway to such an entity is not known at present. Little pieces of the puzzle appear to be understood, but they are little indeed.

13.3 THE RAW MATERIALS OF LIFE: SYNTHESIS AND THE IMPORTANCE OF HANDEDNESS

That the early Earth was rich in the building blocks of life is now generally accepted. After the chaos of the first few hundred million years, a liquid ocean was stable on the cooling crust of the Hadean Earth. Comets striking Earth would have delivered methane, ammonia, carbon dioxide, carbon monoxide, nitrogen, and other more complex molecules. This atmospheric soup, exposed to ultraviolet light or the electrical discharges of lightning bolts, would have been converted into even more complex organic molecules and nitrogen-bearing species called *nitriles*. In a classic set of experiments in the 1950s, Stanley Miller and Harold Urey showed that, if the atmosphere was rich in methane (CH_4) and ammonia (NH_3) and contained not too much carbon dioxide, spark discharges (simulating lightning) could produce amino acids, the building blocks of proteins, from the gas and water.

The laboratory synthesis of amino acids seemed to promise a quick experimental resolution to the origin-of-life question, but two problems arose in the years subsequent to those experiments. First, the atmosphere in the Miller-Urey flask contained primarily reducing gases, an atmosphere with large amounts of methane as is found today on Saturn's moon Titan. However, models and geochemical data suggest, rather definitively, that the predominant molecule in Earth's early atmosphere was carbon dioxide, as is the case today for Mars and Venus. Although organic molecules such as methane likely were present, as was ammonia, they were far less abundant than assumed in the Miller-Urey experiment. As the amount of hydrogen-bearing organic molecules relative to carbon dioxide is decreased in the Miller-Urey experiment, the amount of synthesized amino acids plummets. It is possible that enough hydrogen-bearing molecules existed to make some amino acids in the early Earth's environment, but probably not in the extreme quantities manufactured in the original Miller-Urey synthesis.

Subsequent analyses of meteorites revealed that they contain small amounts of amino acids as well, apparently synthesized on parent asteroids or in space. Although the different types of amino acids was much broader than we find in living systems today, the essential lesson is that, on the early Earth, chemical molecules up to the complexity of amino acids were made or supplied readily. We might therefore imagine the early Earth's ocean as a soup of organic molecules, including amino acids, drifting from hot environments to colder environments (for example, from submarine vents to the surface), occasionally being disrupted by impact events that still were frequent, and then reforming in the atmosphere or being supplanted by more material delivered by smaller impactors.

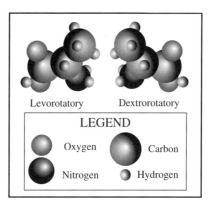

Figure 13.1. Left- and right-handed enantiomers of the amino acid alanine. The left-handed type is referred to as *levorotary* or L-alanine; the right-handed is *dextrorotary*, or D-alanine.

The other important problem has to do with the *chirality* or sense of handedness of the amino acid molecules. Figure 13.1 shows two versions, or *enantiomers*, of the amino acid alanine. Each contains exactly the same number of elements with the same types of chemical bonds, and yet they are the mirror image of each other. A molecule that is not superimposable on its mirror image is *chiral*. When a molecule with a definite sense of handedness reacts chemically with one that is symmetric (or otherwise does not have a particular handedness), the left- and right-handed amino acids have similar properties. Likewise, the chemical properties of an interaction between two left-handed molecules or two right-handed molecules are the same. However, neither of these interactions is the same as when a left- and right-handed molecule are interacting with each other. Hence, *the handedness of biological molecules such as amino acids or nucleotides plays a role in their functionality*.

Earthly life has the remarkable property that virtually all proteins are constructed only from left-handed amino acids, whereas the nucleic acids RNA and DNA utilize only right-handed sugars in their structures. Terrestrial organisms cannot utilize right-handed proteins (with a few exceptions) or left-handed sugars in their biochemical processes; they would starve to death if such wrong-handed materials were the sole food source. Yet, nonbiologically-produced amino acids such as those in meteorites and the Miller-Urey experiments are a roughly equal mixture of left-handed (L) and right-handed (D) molecules (there is one meteorite in which there is a modest excess of left-handed amino acids). Furthermore, chemical production of polymers such as proteins or nucleotides does not prefer a particular handedness when the starting molecules are a mixture of L- and D-enantiomers.

In the remainder of this chapter, we use the term chirality to indicate a strong sense of handedness (left or right) in a particular molecular species. *Heterochiralic* means that comparable amounts of left- and right-handed, non-superimposable, molecules with a given chemical formula are present, and *nonchiralic* means that the molecule can be superimposed on its mirror image. Chirality is a property; enantiomers are the left- and right-handed versions of a molecule that exhibits chirality.

An important consequence of chirality in biological molecules is that a nonbiological mix of L- and D-type amino acids occurring in meteorites, or in a flask after irradiation (Miller-Urey synthesis), represents more of a problem than a solution to life's origin. How could a particular handedness be selected by prebiological, or primitive-biological, chemistry? We return to this issue later in the chapter because it stands as one of the major challenges to theories of life's origin in which RNA plays an early, primary role.

13.4 TWO APPROACHES TO LIFE'S ORIGIN

From here, the road to take is far from certain. The synthesis of amino acids is a far cry from the construction of complex, self-replicating molecules that carry enough information to construct proteins from amino acids. Two approaches to the origin of life from the soup of organics are explored here. In the first, one argues that certain structures could form spontaneously, capable of isolating parts of the chemical soup from the environment. These *vesicles*, if the right chemicals were present, could have become little factories of increasing chemical complexity, eventually growing and splitting in two but still lacking a reliable genetic code for reproducing the chemical activity within them. The other approach focuses on the genetic code, in particular RNA, which might have been synthesized in the environment of the chemical soup, and once synthesized, multiplied and co-opted vesicles to form cells. Neither approach yet tells a convincing story, but both have led to some tantalizing suggestions as to how life could have arisen from complex, energetic chemical systems.

13.5 THE VESICLE APPROACH AND AUTOCATALYSIS

We consider first the vesicle approach. One of the crucial properties of certain biochemical substances is their ability to enable or speed up reactions, without themselves being expended. This chemical effect is called *catalysis*; some biological catalysts are called enzymes. Catalysis is a common feature in many nonbiological chemical systems, and is essential in biology. A special kind of catalysis is *autocatalysis*, in which a product of a reaction acts as a catalyst in its own production.

Autocatalysis is a process that can lead to complex behavior. Beginning with two chemical substances that tend to react with each other, and supplying enough such *reactants* to maintain vigorous chemical reactions, progressively more complex molecules can be built up in the soup, including molecules that catalyze certain reactions. The key is a continuous supply of reactants, and a source of energy, that is, the system must be maintained far from equilibrium.

Even more significant is that complicated autocatalytic systems, as simulated in computers, have the capacity to increase their level of organization over time. If several sets of autocatalytic cycles are in operation in the same environment, they have the possibility of producing complex chemical species that can couple the sets together and create further organization and complexity. These self-organizing chemical systems increase their network of reaction steps and become more organized so long as energy is available to hold them far from equilibrium (figure 13.2).

Scientists have argued that perhaps such autocatalytic sets brought amino acids and other molecules to increasing levels of complexity until primitive proteins and other structures were produced. To do this, the chemical system must have been held far from equilibrium, which means isolating the system from the surrounding environment and enabling energy and reactants to be pumped in and products to be removed. This is where vesicles play a role. In a watery (*aqueous*) medium, certain simple molecules spontaneously form *bilayer membranes* that partition off an "inside" from an "outside." Biological materials such as egg white will do this. Simpler organic molecules called *lipids*, which need not be produced biologically, will do likewise. So, in the early Earth's oceans, it may have been that small environments, microns or less in size, were partitioned off spontaneously by simple organic molecules.

Within the interior of a vesicle, the environment was partially isolated from the outside. It then would be possible to create a small system, out of equilibrium, within which complex chemical reactions, perhaps autocatalytic sets, could have occurred. To do so, we must specify two essential ingredients, namely, energy and a pathway for introducing reactants and removing products.

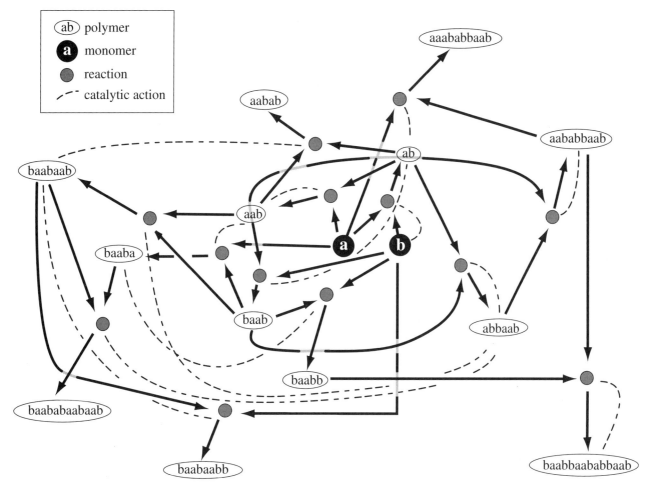

Figure 13.2. Autocatalytic network of 15 chemical species, simulated by computer. The starting point is with two *monomers* (single molecules or atoms as opposed to polymeric chains), labeled a and b. Each dot is a reaction that causes simpler reactants to combine to make more complicated products, to which the arrows point. A polymer composed of two b's in the middle capped on each end by an a would be labeled **abba** in the figure. Some reactions are catalyzed by certain polymers indicated by the dashed lines. Eventually in this system, every polymer will have its formation catalyzed by some other polymer in the system. Such a fully autocatalytic set exhibits a stability similar to that achieved in biological metabolisms. Redrawn from Bagley and Farmer (1992) by permission of Addison–Wesley.

The spontaneously formed vesicles would not themselves be a source of energy, except for the heat given off by chemical reactions inside them. This heat, though, represents increase of entropy and is not available to do work inside the vesicle. Likewise, heating from the outside is possible but would merely tend to equalize the inside and outside temperatures – not a promising start for bringing our environment away from equilibrium. One novel suggestion that has been made recently is that certain simple organic molecules, attachable to the vesicle, may have had the capacity for capturing light energy from the Sun and using it to ionize parts of the vesicle membrane. Although speculative, such molecules would *transduce* energy from the Sun and make it available as chemical energy (via the ions formed from the membrane) inside the vesicle.

What about the transfer of reactants and products? This also appears to be possible through a particular set of complex organic molecules that could have acted as channels, filtering some substances through and excluding others. Some preliminary experiments have suggested the possibility that such an attribute might be developed on the vesicles through molecules available in the organic soup of the early ocean.

Although still hypothetical, we have conjectured a vesicle machine that can be charged up, transfer molecules in and out, and serve as an isolated environment for autocatalytic reactions – all of this using molecules plausibly

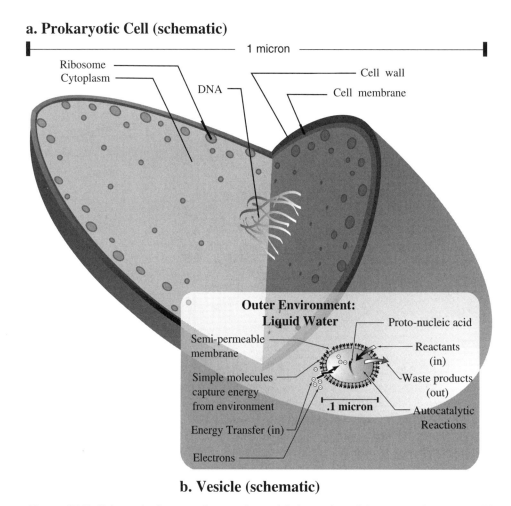

Figure 13.3. Schematic diagram of a putative vesicle in a primordial ocean environment, with molecules attached to transduce energy from sunlight or heat. Superimposed for size comparison is a modern prokaryotic cell.

available in the prebiotic environment (figure 13.3). The purpose of the machine is simply to make molecules of higher and higher complexity. Whether such molecules eventually would move toward proteins and enzymes is completely unclear. The principle, though, is simple: Hold it away from equilibrium and let it get more complex!

13.6 THE RNA WORLD: A SECOND OPTION

Although vesicles appear to be a natural and compelling structure for evolving complicated chemical factories, they lack a detailed, formalized set of instructions for producing molecules of the complexity of proteins, and for reliably reproducing themselves. A vesicle that has split off from another, and floated off to a slightly different environment in the early ocean, might well become host to completely different chemical cycles, which fail to sustain autocatalytic sets. Living forms are able to continue their chemical processes from one generation to the next. They also exhibit the ability to self-regulate their internal processes in the face of environmental changes that would completely alter or shut down nonbiological chemical cycles – a capability called *homeostasis*.

13.6.1 The Promise: RNA as Replicator and Catalyst

A second school of thought holds that the formation of life had as its essential step the formation of RNA from natural chemical processes. The RNA would then function as a primitive form of life unto itself, existing perhaps in the early ocean and quickly co-opting protective structures such as lipid-like vesicles. The proponents of this view prefer RNA over DNA because it plays various central roles in all cells today. Not only does it synthesize proteins, but it also primes DNA for replication. In modern cells, DNA is required to produce and regulate RNA, but this may

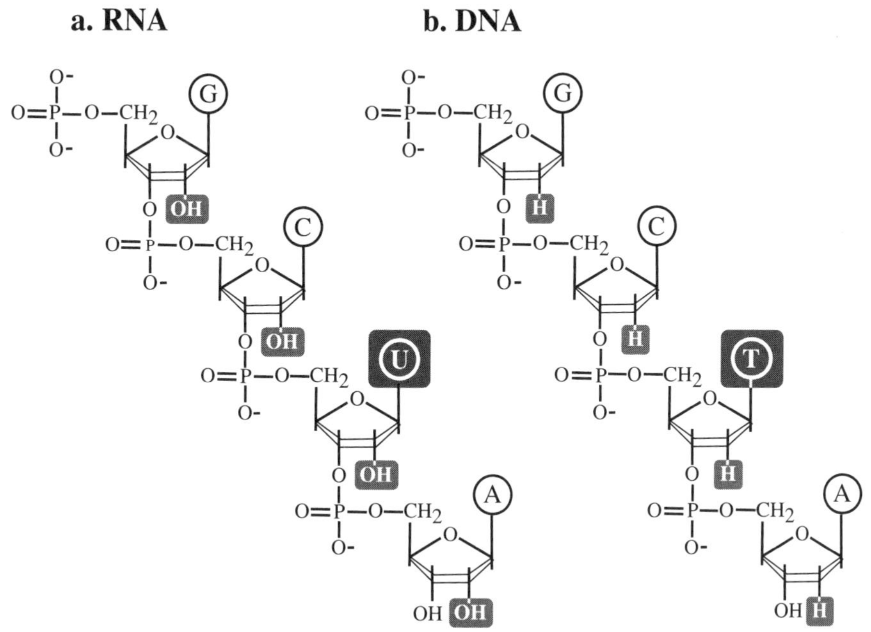

Figure 13.4. (a) Primary structure of the RNA molecule. Nucleotides are labeled within circles by the first letter of their name (for example, G for guanine). Other letters are elements (O for oxygen, P for phosphorus, etc.). Carbon atoms occupy the unlabeled vertices, in accordance with common chemical convention. The straight lines show *single* and *double* bonds, which reflect the number of electrons shared. (b) Structure of a DNA molecule. The differences between RNA and DNA are highlighted on the two structures.

be a later refinement in the evolution of life. In fact, the modification of an RNA molecule to make a single strand of DNA is a relatively minor chemical step (figure 13.4). There is little argument among biologists that RNA came before DNA.

What made the notion of an RNA world so attractive was the discovery by T. Cech, S. Altman, and colleagues that RNA molecules can act as catalysts, participating in and speeding up the production of nucleic acid sequences and other biological molecules. Although biologically occurring RNA molecules are rather weak catalysts, these abilities can be enhanced and expanded in the laboratory, by splicing and reproduction of selected sequences on the RNA chain of nucleotides. The resulting modified RNA structures are sufficiently impressive catalysts that it is possible to imagine early RNA as biological catalyst in place of the present-day proteins that function as enzymes. Thus, RNA could have been both the reproductive and the catalytic molecule of the very early stages of life on Earth, acting in autocatalytic cycles to sustain a primitive biology.

13.6.2 The Problem: Invention of RNA

Most serious is how to put the jigsaw-puzzle RNA together in the first place. To understand whether RNA could have been synthesized in the absence of pre-existing biological molecules, it is convenient to consider the three fundamental chemical parts of an RNA molecule: (i) the nucleic acid bases A, G, C, and U; (ii) a *phosphate group* that contains the element phosphorus and serves as the connector of each of the bases; (iii) the sugar ribose that functions as the binder or backbone of the molecule, so that each nucleic acid base is bound to a sugar and these in turn are attached to each other by the phosphate groups. (As noted in chapter 12, each unit composed of a ribose, a phosphate group, and a nucleic acid base is called a nucleotide; the polymer composed of a string of nucelotides is an RNA molecule).

The production of nucleic acid bases by nonbiological means appears to be understood at least in the case of adenine (A), cytosine (C), and uracil (U); there does not seem to be any fundamental hurdle in eventually making guanine (G). Somewhat more difficult is the understanding of how a phosphate group would tend to attach to the right position of a ribose molecule to provide the necessary chemical activity; the same challenge is present in attaching the nucleic acid bases to the ribose. However, one might imagine a random assortment of nucleotide-type molecules, those of which that happened to be configured like an RNA nucleotide possessing a chemical advantage.

The real problem lies in the synthesis of ribose, with the right chirality. Carbohydrates possessing the formula $C_nH_{2n}O_n$, where n represents a number 1, 2, 3, ...,

Figure 13.5. The disaster of heterochirality (mixing L- and D-sugars in nucleotides): (left) a normal DNA molecule built of nucleotides of a single chirality; (right) a DNA molecule built with *just one* nucleotide of the opposite chirality (i.e., all D-ribose except for one nucleotide with L-ribose). The one-defect DNA is forced into a much looser structure. The strain in the chemical bonds created by trying to force L- and D-nucleotides together causes bond breakages elsewhere in the structure. The result is a much more open, loose DNA molecule, which is very fragile and thus cannot carry out its templating and replicating functions before falling apart. A similar problem faces the synthesis of RNA from a heterochiralic soup of nucleotides. Photographs reproduced from Avetisov et al. (1991) by permission of American Institute of Physics.

including the sugar ribose, are readily manufactured by reacting formaldehyde ($n\text{CH}_2\text{O}$) with itself. A catalyst is required to initiate the reaction, but this is not a problem. The problem is that ribose is not particularly preferred over other sugars. Hence, an autocatalytic cycle designed to produce large amounts of carbohydrates from formaldehyde will not preferentially make ribose.

One novel suggestion that has been made is that clay minerals may have been involved to concentrate ribose. Clay minerals have ordered surfaces that could form templates, forcing organic molecules that bind to their surfaces to form certain structures. Although no synthesis of RNA has occurred this way, the suggestion is in the right direction: Force molecules away from randomly defined patterns to a subset of structures that might allow ribose to form preferentially. This suggestion is also geologically consistent because, with the formation of an ocean in the Hadean era, the environment would have become suitable for formation of clays.

However, a difficulty is encountered here. The ribose produced must have the correct handedness or chirality; on Earth, D-sugars are exclusively involved in living processes. Production of a mixture of D- and L-sugars produces nucelotides that do not fit together properly, producing a very open, weak structure that cannot survive to replicate, catalyze, or synthesize other biological molecules. In fact, the synthesis of the RNA molecule itself is interrupted by mixing nucleotides of different chirality; only in a controlled laboratory experiment or theoretical model can such an assemblage be realized (figure 13.5).

To create a properly functioning RNA molecule out of a batch of heterochiral L- and D-sugars is a daunting challenge. Two approaches have been pursued. The first is to consider precursor molecules with function similar to RNA but which are much easier to synthesize. The second approach is to understand how prebiological or very primitive biological processes could have selected a particular chirality and allowed its dominance, and hence permitted RNA. In a sense, these two approaches are linked; some precursor chemistry must have operated out of a heterochiral soup prior to the concentration of the D-sugars.

Considering the first approach, it is possible to imagine substituting another sugar for ribose in making RNA, and in particular a sugar that is symmetric and hence nonchiralic. Possible sugars suggested by University of California biologist G. Joyce include glycerol; others have suggested additional candidates such as glucose. Candidates proposed are generally ones that could have been fairly easily synthesized on the early Earth by nonbiological processes, and as sugars, they are capable of binding a nucleic acid base and a phosphate group. However, the properties of the resulting pseudo nucleic acids can be very different. Some have much more flexible structures than RNA, leading to a much greater chance of breakup and hence replication or catalysis failure in a fluctuating environment. Others are too stable, and may not catalyze.

Finally, many of these substitutes allow not only complimentary pairing (A with U, G with C, as in chapter 12) but also other pairings (A with C, A with A, G with U, etc.). Under such conditions, the genetic template that sustains a particular kind of chemistry and set of structures is quickly lost after just one generation.

Other possibilities have been conceptualized. For example, amino acids are found readily in meteorites and synthesized under early Earth conditions; could they substitute for sugars as the RNA foundation? Indeed, one can synthesize a backbone composed of glycine (an amino acid) attached to a nitrogen and hydrocarbon unit. This then can attach to a nucleic acid base and a phosphate group to form a nucleotide. The resulting structure is a *peptide nucleic acid*, or PNA. PNAs have been synthesized and shown both to be sturdy and to produce pairing of complementary nucleic acid bases (i.e., A with U, G with C). They could therefore serve as a replicator molecule. However, three open questions remain: Can PNA function as a catalyst? Can one actually induce the polymerization of the amino acid with the nucleic acid bases to make PNA in a plausible prebiotic setting? Is PNA subject to the same chiral restrictions that RNA is? (Recall that many amino acids exhibit chirality.)

The second approach is to understand how the dominance of D-sugars and L-amino acids took place on the early Earth from an initially heterochiralic soup. One would first look to some innate preference for one or the other handedness in the environment or the nature of the molecules. The environment itself yields small effects which tend to select out one or the other sense of handedness, but some means to amplify the selection must be found. Interestingly, at the subatomic physics level, there is a very small preference for right-handed sugars and left-handed amino acids, the current state on Earth. Such a preference is so small, however, that it cannot *by itself* lead to the distillation of L-amino acids and D-sugars from a heterochiralic soup. Recent analysis of the Murchison meteorite indicates a significant overabundance of some L-amino acids relative to D-amino acids, but the origin of this imbalance and its possible connection to prebiotic chemistry have not yet been explored.

If there is an answer to the development of preferred chirality at the dawn of life, it might well lie in the propensity of complex physical systems to self-organize, as discussed at the very beginning of this chapter. Mathematical models of autocatalytic systems in which polymer production takes place in a heterochiralic environment show that, under certain conditions, the system can exhibit a transition in which the symmetric treatment of D- and L-molecules is broken, and the system rapidly evolves toward a single kind of chirality. No simple physical preference is at work here; instead it is the intrinsic *chaotic* nature of a complicated physical system, held far from thermodynamic equilibrium, that leads to such a self-organizing property. Certainly an autocatalytic system, enclosed in a special environment that allows energy flow and reactants (nutrient) in and products (waste) out, is a candidate to exhibit such behavior. It is in the necessity to invoke such a behavior to make chirally-sensitive molecules such as RNA that we might find the combination of the vesicle world and the RNA world to be a requirement for the formation of life.

13.7 THE ESSENTIALS OF A CELL AND THE UNIFICATION OF THE TWO APPROACHES

What are the essentials of a cell? Operationally, they are

(i) a dynamic membrane that exhibits fluid-like, flexible motion;
(ii) a set of embedded, membrane proteins that capture energy-bearing molecules (*metabolites*) from the environment, and transport them into the cell;
(iii) a set of enzymes that break down the metabolites and use the breakdown products to construct more membrane, more enzymes, and more genetic material (RNA/DNA);
(iv) a genetic string, RNA coded by DNA, which encodes for the set of enzymes;
(v) a genetic program, DNA primed by RNA, consisting of the set of triggering relations between the various genes. The program will cause the cell to grow, duplicate the genetic-string DNA, and eventually divide when it has gotten large enough, resulting in two cells that will continue to metabolize, grow, and divide.

The chemical vesicle factories embody in a primitive way properties (i) through (iii); the RNA world covers (iv) and (v). Neither model yields all five properties. It may be that if the origin of life occurred as a natural chemical process on Earth, the first step was the formation of the autocatalytic vesicles, which were short-lived and formed over and over again in different varieties over millions of years – chemical experiments that failed repeatedly. However, at some point a vesicle system exhibited the property of producing polymers of a dominantly single chirality, either sugars or amino acids, and within this system the production of an RNA, a PNA, or other nucleic acid structure (ONA) was enabled. ONA varieties with catalytic capability became coupled into the autocatalytic networks of some vesicles, and a subset of these used the energy and catalytic properties of the sets to

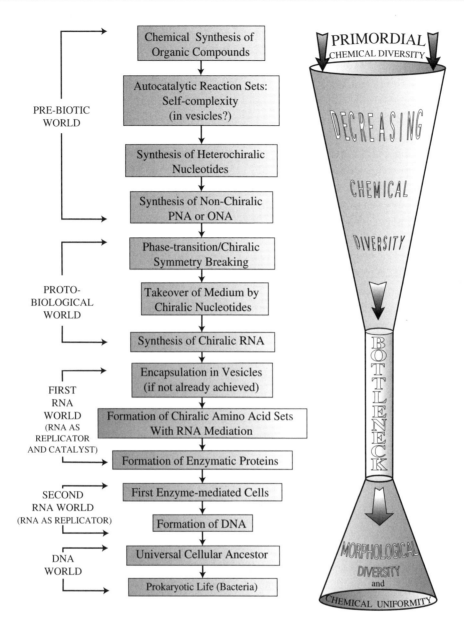

Figure 13.6. One possible schedule of the steps by which life formed. In the left-hand sequence, RNA appears before encapsulation in vesicles, although as the text argues, the reverse might well have been the case. The bottleneck in the origin of a chemically uniform, morphologically diverse biology from a chemically diverse terrestrial environment is illustrated on the right, aligned with the chemical/biological steps on the left. Adapted from Cloud (1988).

reproduce. Such symbiosis, which is a theme in the evolution of life, could have represented the very primitive precursor to a biological cell. All of the ingredients, (i)–(v), of a cellular structure capable of maintaining and reproducing itself are present in such an RNA-primed vesicle.

Although from here, the formation of DNA is not well understood either, the jump in complexity from RNA to DNA is not considered by biochemists to be as much of a hurdle. Along the way, in some RNA-driven vesicle, DNA may have arisen, and the universal cellular ancestor of all earthly life was born.

Figure 13.6 suggests two ways to look at the origin-of-life issue. One, on the left, is to try to list the steps, in order, by which life began; this approach is fraught with dissent because we still do not know whether vesicles, RNA, ONA, chirality, or other precursors came first. [For example, Nobel Laureate C. de Duve notes that the development of energy storage in phosphorus-bearing molecules such as *adenosine triphosphate* (ATP) is yet another problem that requires the identification of simpler precursor molecules.] The other approach is to recognize that biology represents a self-controlled selection of a subset of possible molecules out of an enormous range of possibilities. DNA, for example, has four kinds of letters and about 1,000,000 base pairs (ladder rungs) per molecule. The number of possible varieties of DNA molecules then is $4^{1,000,000}$, or 4 followed by one million zeros. *It is the role of enzymes, biological catalysts, to suppress the random nature of chemical reactions so as to preserve and ensure a particular suite of biological molecules at the rate needed to sustain the production and replication of the whole system.*

In this regard, it is perhaps most instructive to view the right-hand sketch in figure 13.6. The prebiotic Earth was a system of high chemical diversity, but with an environment that tended to select certain elements and naturally occurring molecules as preferred in increasingly complicated autocatalytic chemical systems. Reaction sets that straddle the barrier between biology and chemistry were still chemically diverse, and likely limited in size and capability to interact with the environment: morphologically (appearance-wise) simple. It is the bottleneck of chemically diverse, morphologically simple, protobiotic chemical systems that lies at the crux of understanding how life began. As one kind of reliable protobiochemistry took hold, the chemical diversity of protobiology plummeted, but the success of the system was such as to allow the blossoming of a great diversity of morphologies that functionally allowed different kinds of interactions with a changing environment. Today, biological processes have co-opted most of the available carbon and oxygen in the atmosphere, ocean, and continental surface of Earth, so that the chemistry of these elements is largely limited to the rather uniform and specific biochemical processes that sustain life. Most or all of the other planetary environments in our solar system may never have crossed this bottleneck, but how close they came is an intriguing question (figure 13.7).

13.8 THE ARCHEAN SITUATION

True life in the Archean, that which left stromatolites and other faint records, consisted of the most primitive type of cells called prokaryotes. The more complex eukaryotes were apparently not present in the Archean, and we tackle their origin in chapter 19 in the context of the formation of an oxygen-rich atmosphere during the Proterozoic eon. At least in terms of structural complexity of the container of living processes, it does not seem like such a long step from the vesicles of the theorists to the bacterial prokaryotes of the Archean.

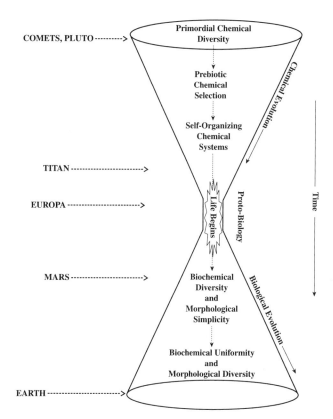

Figure 13.7. Another look at the transition from chemical to biochemical evolution. Conservative guesses as to where various planetary bodies lie on the hourglass are indicated. The cold distant bodies of the outer solar system – Kuiper Belt comets and Pluto – store organic molecules relatively unaltered from interstellar processes. Titan's surface may be host to organic chemistry that on occasion exhibits self-complexifying processes. Europa, with a possible liquid ocean, might support life, but agreeable conditions may be so short-lived that this body always stands just on the threshold of life's origin. Mars may have had conditions early in its history, and in brief episodes thereafter, capable of sustaining a primitive biota. Only Earth, in our solar system, has an atmosphere, ocean, and crust that play host to an extensive biochemistry expressed in the great diversity of life-forms we see today.

When did the prokaryotes form? Certainly before 3.5 billion years ago, based on fossil evidence of photosynthesizing bacteria in Australia, more than perhaps 3.9 billion years ago from isotopic evidence, but after the formation of the liquid water ocean, 4 billion years ago or earlier. The limiting factor may have been the rate of large impacts: If the early Earth environment was rendered unstable by a high frequency of such impacts, there would be no chance for robust vesicles (or RNA) to form. Until the timescale for forming self-sustaining vesicles is understood, we can make no reliable judgment as to when the environment became sufficiently stable. Sometime after the formation of the water ocean, and perhaps sooner rather than later, life appeared on Earth.

13.9 QUESTIONS

a. Imagine a planet with two well-developed biota, one able to synthesize left-handed sugars and use right-handed amino acids, the other synthesizing right-handed sugars and using left-handed amino acids. What kinds of competition might ensue in such a situation? Is it intrinsically unstable, i.e., will one form of life win out?

b. If indeed RNA was the initial genetic encoder and DNA developed later, do you think that some viruses might be remnants of that earlier epoch? Why or why not?

13.10 READINGS

Avetisov, V.A., Goldanskii, V.I., and Kuz'min, V.V. 1991. Handedness, origin of life and evolution. *Physics Today* 44(7) (July), pp. 33–41.

Bagley, R.J., and Farmer, J.D. 1992. Spontaneous emergence of a metabolism. In *Artificial Life II*, (Langdon, C.G., Taylor, C., Farmer, J.D., and Rasmussen, S., eds.). Addison-Wesley, Redwood City, California, pp. 93–140.

Brown, M.W. 1993. Fossils shorten time frame for life to have developed. *New York Times* 142 (April 30), p. 1.

Chyba, C.F., and MacDonald, G.D. 1995. The origin of life in the solar system: Current issues. *Annual Review of Earth and Planetary Sciences* 24, 215–249.

Cloud, P. 1988. *Oasis in Space: Earth History from the Beginning*. W.W. Norton, New York.

de Duve, C. 1995. The beginnings of life on Earth. *American Scientist* 83, 428–437.

Engel, M.H., and Macko, S.A. 1997. Isotopic evidence for extraterrestrial non-racemic amino acids in the Murchison meteorite. *Nature* 389, 265–267.

Forterre, P. 1995. Looking for the most "primitive" organism(s) on Earth today: The state of the art. *Planetary and Space Science* 43, 167–177.

Levy, S. 1992. *Artificial Life: A Report from the Frontier Where Computers Meet Biology*. Vintage Books, New York.

Morrison, R.T., and Boyd, R.N. 1973. *Organic Chemistry*. Allyn and Bacon, Boston.

Pace, N. 1996. New perspectives on the natural microbial world: Molecular microbial ecology. *ASM News* 62, 463–470.

Schwartz, A.W. 1995. The RNA world and its origins. *Planetary and Space Science* 43, 161–165.

14 THE FIRST GREENHOUSE CRISIS: THE FAINT EARLY SUN

14.1 THE CASE FOR AN EQUABLE CLIMATE IN THE ARCHEAN

There is ample evidence that the Archean Earth possessed liquid water. The existence of metamorphosed sedimentary rocks from this period, as discussed in chapter 11, require erosion by liquid water and deposition in a lake or marine environment. The presence of life itself, recorded through isotopic signatures and fossil evidence, also implies liquid water. As discussed in chapter 12, we know of no living thing today that can get by without water. Many don't require oxygen (and are poisoned by it), but all require liquid water.

Figure 14.1 summarizes constraints arguing for Earth's mean temperatures being above the melting point of water during the Archean. In chapter 15, we explore the case for a Martian climate, at the time of Earth's Archeon eon, which was warmer than at present (either continuously or episodically). In total, the evidence on Earth and Mars points to planetary climates at least as warm as those experienced today. Surprisingly, as we now show, such climates impose rather strong constraints on the nature of the Archean atmospheres of the Earth and Mars – provided our understanding of the evolution of the Sun is correct.

14.2 THE FAINT EARLY SUN

Simple reasoning about the physics of hydrogen fusion indicates that the Sun was cooler in the past than it is at present. As the Sun converts hydrogen to helium, the mean atomic weight of the atoms in the core goes up (helium is four times heavier than hydrogen), whereas the number of atomic nuclei goes down (four hydrogen nuclei having combined to make one helium nucleus). As the core evolves toward a state of heavier but fewer atomic nuclei, it compresses, forcing the density up. The compression of the core toward higher density, in turn, increases the average temperature of the material as the energy of compression is converted into the random energy of collisions of nuclei.

Finally, the rate of fusion is very sensitive to the temperature, such that small increases in temperature lead to a large increase in the rate of fusion. This is the case because fusion requires a threshold collisional speed in order to allow protons (hydrogen nuclei) to overcome their intrinsic repulsion and bind together (see chapter 3). Hence, a small increase in the mean speed of collisions (small temperature increase) leads to a very much larger percentage of collisions in which the hydrogen nuclei can fuse to form helium. Therefore, over time, the Sun has gotten more luminous. Computer models predict that, at the time of the early Archean, 3.8 billion years ago, the Sun's luminosity was 75% of the present value, that is, it was 25% dimmer than at present. By the end of the Archean eon, 2.5 billion years ago, the Sun's luminosity was 82% of the present-day value (figure 14.2).

With less sunlight streaming to Earth in the past, the surface would have been colder than at present. The surface temperature of Earth's oceans today, averaged over their surface and over a year, is 288 K. A very rough guide to what the surface temperature would be for lower solar luminosities (all else kept the same) is given by scaling the temperature to the fourth root of the solar luminosity. (Such a scaling derives from the way in which photons are emitted from a surface that is heated at a given rate.) Hence, at 82% of the solar luminosity, the mean surface temperature is $288 \times (0.82)^{1/4} = 274$ K; for 75% of the solar luminosity, the surface temperature becomes 268 K, below the freezing point of water.

The sensitivity, though, is actually greater than this, because as Earth cools, the atmosphere cannot hold as much water vapor, and this dryness leads to an even lower

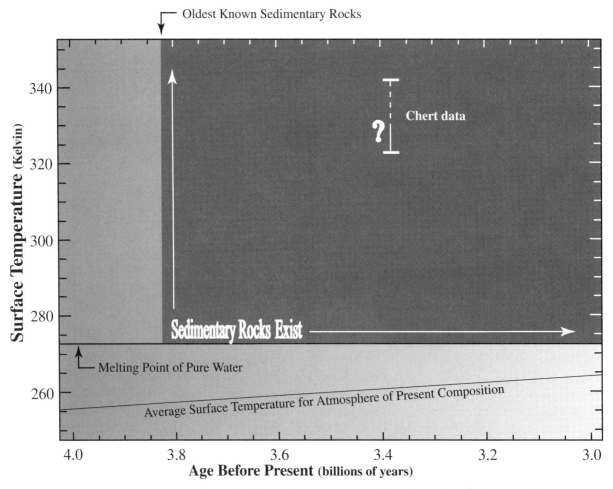

Figure 14.1. Some constraints on Earth's surface temperature during the early to mid-Archean. The line marked "Average Surface Temperature for Atmosphere of Present Composition," derived from figure 14.2, shows what happens if today's atmosphere is combined with the fainter Archean sun: surface temperatures lie well below freezing. The chert data described in chapter 6 suggest ocean surface temperatures at 3.4 billion years ago much higher than today's. A more robust but looser constraint is the appearance of metamorphosed sedimentary rocks in the geologic record after 3.85 billion years ago, indicating that widespread liquid water was present and hence that the global mean surface temperature was above the melting point of water.

temperature through the *greenhouse effect*, which we describe in the following sections. Work by Pennsylvania state atmospheric scientist James Kasting indicates an earth surface temperature of 255 K at the start of the Archean. Such an earth could not have had a stable, liquid water ocean. What kept the oceans from being frozen? To answer this question, we need to consider how the atmospheric greenhouse effect works.

14.3 THE GREENHOUSE EFFECT

It is a sunny midsummer day and you have parked your car, windows closed, in the asphalt parking lot of your favorite shopping center for an hour of shopping. Upon leaving the building, you notice that the outside air temperature is warm but not broiling. Once the car door opens, though, that familiar blast of heat greets you from the hellishly torrid interior. What happened?

The glass of your car windows and windshield allows plenty of sunlight to get in – glass is transparent at optical (or visible) wavelengths, where most of the Sun's energy is emitted. When the visible photons from the Sun hit the seats, dashboard, and other parts of the passenger interior, they are partly absorbed and then re-emitted as infrared photons, lower in energy but more numerous. Some of the solar photons are instead reflected by bright surfaces and exit through the windows but, even for whiter color schemes, much of the sunlight is absorbed.

The automobile's glass is not transparent at infrared wavelengths: Infrared photons are absorbed by the glass

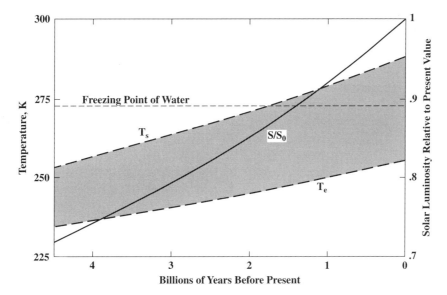

Figure 14.2. The faint early sun problem. Plotted as a function of time before present are Earth's surface temperature (T_s), its effective temperature (T_e), and the luminosity of the Sun relative to its present value (solid curve). The temperature values are to be read on the left-hand axis; the luminosity refers to the right-hand axis. The surface temperature assumes an atmospheric composition through time identical to the present one, just to illustrate the problem. Under this restriction, Earth's mean surface temperature remains below the freezing point of water (dashed horizontal line) for the first 3 billion years of Earth's history. Reproduced from Kasting (1989) by permission of Academic Press, Inc.

on their way out, and partly re-emitted back into the car again. In effect, the heat, in the form of infrared photons, has trouble getting out. (If human eyes were sensitive to light in the infrared rather than the optical, automobile glass would not appear transparent to us.) This situation is not a stable one, because there must be energy balance between the inside and the outside air of the car. As more sunlight streams in and infrared photons are hindered in getting out, the temperature inside the car rises. This increases the flow of infrared photons out of the car, as the inside air temperature rises more and more above the outside value.

Eventually, a balance is reached where the temperature difference between the inside and the outside of the car is enough to balance the free flow of visible photons in, and the arrested flow of infrared photons out. The *greenhouse effect*, then, refers to the increase in temperature of the air caused by the greater difficulty that infrared photons have moving outward to cooler regions, relative to the ease of movement of visible photons. Its efficiency depends on the property of the medium through which the visible and infrared photons move. If they are allowed through the material with equal ease, there is no resulting elevation of air temperature.

For a planetary surface like Earth's, the role of the glass is played by the atmosphere. Sunlight streams down through the atmosphere, some of it reflected by clouds, but a good fraction reaching the surface. The visible photons are absorbed by the ground, people, trees, buildings. The increased vibrations of the molecules in all these things, engendered by the solar photons, cause emission of heat, in the form of infrared photons, upward through the atmosphere. But the atmosphere, despite being nearly transparent to sunlight (except in cloudy regions), impedes the progress of these infrared photons. They are absorbed and re-emitted many times at many altitude levels in their trip back up through the atmosphere. Photons that are absorbed get re-emitted not only directly upward, but in all directions – up, down, sideways. Thus some infrared photons, about half in each absorption/re-emission event, find themselves moving downward again – against the flow of energy from the warmer ground to the cooler upper air. The result is an impediment to the outward flow of heat energy (figure 14.3).

As in the car, there must be a balance of incoming solar energy to outgoing thermal energy. To achieve this, the temperature of each layer in the lower atmosphere must go up, to compensate for the infrared photons turned around and headed downward again. The lower the layer in the atmosphere, the more the temperature increases, because the lower layers are denser and absorb photons more

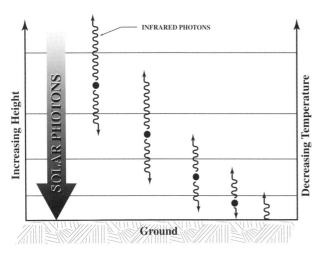

Figure 14.3. Paths of solar photons through the atmosphere (straight arrow) and emitted infrared photons (squiggled lines).

effectively than the thin upper layers. So, the absorption of infrared photons causes the gradient of temperature – its change with altitude – to become steeper. The steeper the temperature gradient, the more efficient the flow of infrared photons outward, until energy balance is achieved.

The result is an increase in the temperature at Earth's surface by about 33 K over what we would get if the atmosphere freely let infrared photons escape. For Venus' massive carbon dioxide atmosphere, the temperature *increase* is 500 K!

The actual situation in the atmosphere of Earth is more complicated, as shown in figure 14.4. Clouds and surface ice very efficiently reflect roughly one-third of the solar photons back outward, so that they never contribute to the warming of Earth's surface and atmosphere. About one-fourth of the solar photons are absorbed directly by clouds or the atmosphere itself. Also, photons usually do not move directly vertically in the atmosphere; there is a spread of photon directions over all angles in the sky, and clouds in particular are very effective at changing the directions of photons (*scattering* of light).

The infrared emission also is somewhat complicated. Because the ground gets several degrees or more warmer than the air immediately above it, turbulent air motions or *convection currents* are set up. The bulk motion, or convection, of the rising warm air moves heat from the ground upward. Convection continues at altitudes well

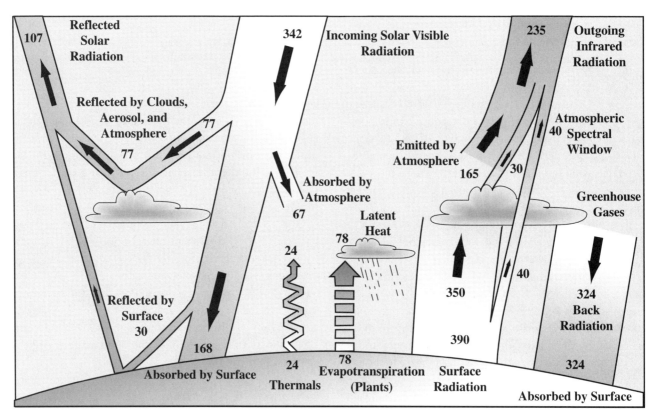

Figure 14.4. Processes affecting the movement of energy in Earth's atmosphere. Solar visible photons are sketched on the left part of the figure, infrared photons on the right, and other kinds of energy transfer in the middle. Numbers show the flux of energy, in watts per square meter, involved in each process; the amount of incoming solar radiation in a square meter per second is an average value for the globe. Latent heat refers to release of heat as water condenses to form clouds; evapotranspiration is the movement of water through plants combined with evaporation back into the atmosphere. Modified from Trenberth et al. (1996).

away from the ground: The rising warm air is replaced by cooler air, and the warm air loses heat through radiation of infrared photons in the more transparent higher layers of the atmosphere. In contrast, a hot automobile interior does not cool by convection because the air inside cannot move a significant distance upward; in this respect it is an imperfect analogy for our atmosphere. Therefore, use of the term greenhouse effect for a planetary atmosphere such as Earth's is something of a misnomer, because greenhouses (like a car's interior) cannot cool by convection.

Turbulent motions in our atmosphere are also responsible for forming clouds: As air is raised in the atmosphere, it cools and, as it cools, it becomes less capable of retaining water as a gas. Eventually, the water vapor condenses (much as it does on a cold drink glass) to form clouds. The process of condensation releases some energy in the form of heat, and this must be included in the atmospheric energy balance as well; it is called *latent* heat. Clouds are an important part of Earth's atmospheric greenhouse process because they can cool the air (by reflecting sunlight) and heat it (by absorbing infrared photons). The amount of water vapor that the air can hold is a sensitive function of temperature, so that small amounts of warming of Earth can lead to large changes in cloud cover. Also, the rainfall produced by clouds is the principal source of water for land-based life.

Other complications include the daily and seasonal variations in sunlight, the effects of continental landmasses and ocean currents in changing the temperature and moisture content of the atmosphere. These, together with the spinning of the Earth on its axis, generate the very dynamic atmospheric phenomena we call weather. For Earth as a whole, however, the net heating process that warms the surface above freezing is absorption of sunlight at or near the ground, and emission of infrared photons the upward escape of which is impeded by absorption in the atmosphere.

14.4 PRIMARY GREENHOUSE GASES

Gases that cause the greenhouse effect must absorb photons in the mid-infrared part of the spectrum, corresponding to wavelengths of light of roughly 10 to 50 microns. The major constituents of the air – oxygen and nitrogen – are poor absorbers. More important are two trace species: water (H_2O) and carbon dioxide (CO_2). Water vapor makes up on average no more than several percent of the molecules in the atmosphere close to Earth's surface, but is highly variable over time and location, depending on the surface temperature and meteorological conditions.

Carbon dioxide gas cannot condense to form clouds in the atmosphere, and hence is uniformly distributed around Earth at 0.03%. Methane, nitrous oxide, man-made *chlorofluorocarbons*, and other trace gases contribute as well.

Figure 14.5 shows the wavelengths at which these gases absorb photons. Also given are curves that represent the amount of energy at each wavelength contributed by the Sun, and the distribution of wavelengths at which the energy from the Sun's heating of the ground and air comes out. Note that there is almost no overlap between the wavelengths of solar photons reaching Earth and photons emitted by Earth's warmed surface and atmosphere. (See chapter 3 for a discussion of the relationship between the temperature of a body and the wavelength distribution of light it emits.) This nice separation of photons into "solar-optical" and "terrestrial-infrared" makes possible the straightforward conceptualization of the greenhouse effect described above.

14.5 IMPLICATIONS FOR EARTH DURING THE FAINT-EARLY-SUN ERA

Returning now to the early Archean, over 3.5 billion years ago, if the amount of sunlight at that time was less than today, with no change in atmospheric composition, the surface of Earth would have been below the freezing point of water. But rocks dated at 3.8 billion years ago are metamorphosed from sediments, which strongly suggest that running liquid water was present; other information such as chert data corroborates this. So, the simplest way to obtain an elevated temperature for Earth in the face of a cooler sun is to increase the amount of greenhouse-absorbing gases. Water vapor is controlled by the atmospheric temperature (the higher the temperature, the more evaporation from the ocean), and so, we are not free to simply invoke more water. Carbon dioxide, however, is not so linked and, given that carbon is abundant in continental and oceanic sediments, it is plausible to explore models in which carbon dioxide was more abundant in the atmosphere than at present.

Figure 14.6 is from a calculation by Kasting and Ackerman, using the physics of the greenhouse effect described above, to show the effect of increased carbon dioxide abundance on the surface temperature of Earth. The calculation assumes that there is always 1.0 atmosphere of nitrogen pressure at Earth's surface. The present CO_2 mixing ratio in the atmosphere is 0.0003 in the figure. As more carbon dioxide is added to the atmosphere, surface temperature goes up. The effect is amplified because, as the surface temperature of Earth goes up, more water

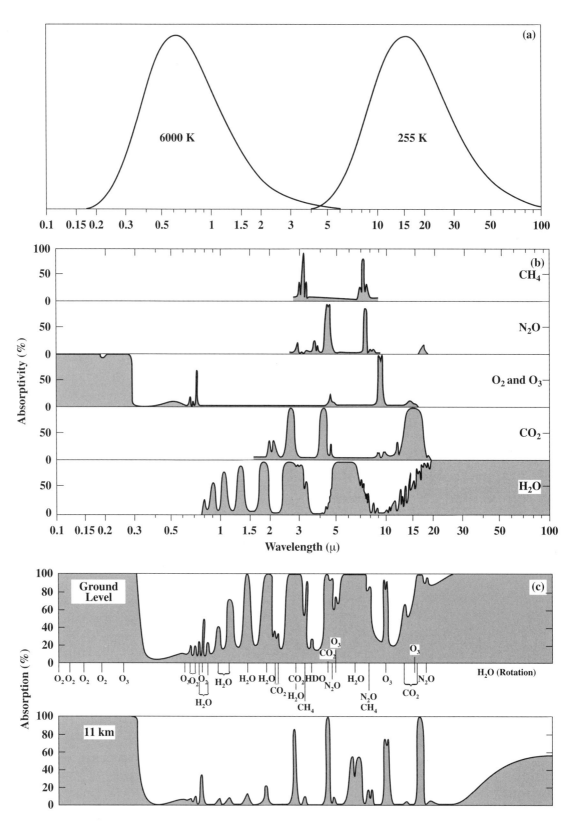

Figure 14.5. (a) Approximate wavelength distribution of photons emitted by the Sun and Earth. (b) Amount of absorption of photons as a function of wavelength, by various gases, through the whole Earth's atmosphere (from the surface upward): 100% absorption means that the atmosphere is completely opaque at the particular wavelength; 0% means the atmosphere is completely transparent. (c) Combined absorption for all gases at sea level, and 11 km above Earth. Notice that, at the surface, the atmosphere is mostly opaque to photons except at optical (0.3–0.7 micron) wavelengths. At a much higher altitude, with less of the gases overhead, the atmosphere is more transparent at infrared wavelengths than it is at Earth's surface. Modified from Peixoto and Oort (1992) by permission of Springer-Verlag.

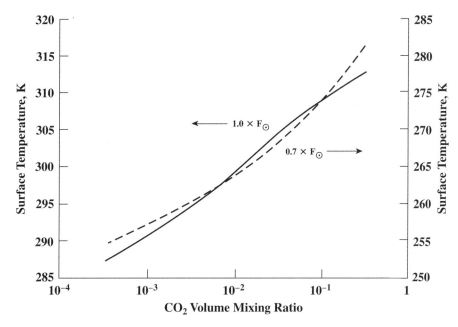

Figure 14.6. Temperature as a function of carbon dioxide abundance in the atmosphere. Solid line is for present-day solar luminosity (scale on left); dashed line assumes a solar luminosity 70% of the present value (scale on right). The air at sea level is assumed to contain one atmosphere pressure of nitrogen, hence, a carbon dioxide mixing ratio of 0.5 means that gas has a pressure of 1.0 atmosphere. Reprinted from Kasting and Ackerman (1986) by permission of the American Association for the Advancement of Science.

is evaporated from oceans and goes into the atmosphere, where it further enhances the greenhouse absorption and adds to the temperature increase.

For comparison, two curves are shown, corresponding to the run of temperature with carbon dioxide abundance for the present-day solar luminosity, and for a value 70% of the present, appropriate for the opening of the Archean eon. To obtain an average Earth temperature above the freezing point of water in the Archean requires a carbon dioxide abundance 1,000 times the present value, corresponding to carbon dioxide being one-fourth of the total atmosphere. To obtain a temperature equal to that today, 288 K, requires that carbon dioxide be the dominant gas in the atmosphere, with a pressure well above 1 bar (the current total pressure in Earth's present, mostly nitrogen, atmosphere).

What is the amount of carbon available today that could have been carbon dioxide in the past? An estimate of the carbon buried in sediments on the ocean floor and continents as well as cycling through the upper crust of Earth could yield at least a 60-bar-pressure CO_2 atmosphere, not much less than that of Venus today (see chapter 15). Hence, there appears to be enough carbon locked in the crust today to maintain above-freezing temperatures in the Archean, if indeed it was in the form of atmospheric carbon dioxide.

Somewhat troubling, however, are the chert data described in chapter 6. If those data are correctly interpreted, the average ocean temperature in the Archean was higher than today, pushing the required carbon dioxide abundance to even larger, perhaps implausible, values. Because the interpretation of the chert data in terms of surface temperature is highly controversial, what is needed is an independent determination of the carbon dioxide abundance. Although this is not available for the early Archean, it is available for the late Archean. This determination comes from paleosols.

14.6 PALEOSOLS AND THE CARBON DIOXIDE ABUNDANCE

As rocks are attacked by water during erosion and weathering processes, elements embedded within the crystal structures of the minerals become mobile and may move to other locations. They may form new chemical compounds, and the nature of these compounds can be a sensitive function of the ambient conditions – including the atmospheric composition. *Paleosols*, weathered rocks (that is, soils) that are preserved through burial and hardening (without extensive metamorphism), may preserve an indirect record of the composition of the atmosphere at the time the weathering occurred.

At present, molecular oxygen is the second most abundant gas (21% of the total) in our atmosphere. Its presence and reactivity have a profound effect on rock weathering; in particular, iron readily combines with oxygen to form various iron oxides. Rocks occurring prior to about 2 billion years ago show evidence of ferrous iron (FeO) being mobilized by weathering processes. This relatively oxygen-poor compound of iron requires that the atmospheric oxygen abundance at the time be very low, a fact of profound importance that we consider in chapter 17. What is important here is that the FeO is dissolved readily in water, mobilized, and becomes available to combine with other materials, the choice of which is sensitive to the atmospheric composition.

In particular, some iron-rich basalts, weathered to soils in oxygen-poor conditions (Archean to early Proterozoic time), lost much of their iron through reaction with water. Some of the iron probably ended up in the local groundwater table, but some reacted with silicates to form iron silicates such as *greenalite* [$Fe_3Si_2O_5(OH)_4$]. The formation of such silicates would not be possible if large amounts of carbon dioxide existed in the atmosphere. The carbon dioxide, diffusing into the soils, would force the formation of an iron carbonate such as *siderite* ($FeCO_3$). Thus, the absence of iron carbonate in paleosols formed in oxygen-poor conditions sets a limit on the abundance of carbon dioxide.

Analysis by Harvard geologist R. Rye and colleagues of paleosols from Canada, formed 2.5 billion years ago, shows no evidence for iron carbonates; instead iron silicates are present. Thus, an upper limit on the carbon dioxide in the atmosphere at the time of formation should be obtainable. The formation of iron carbonates is sensitive to temperature, and the carbon dioxide concentration in the soils was not exactly the same as that in the atmosphere. Despite these uncertainties, it is possible to use laboratory experiments and theoretical calculations to estimate a carbon dioxide upper limit. Rye and colleagues obtain a maximum of 3% carbon dioxide in the atmosphere at the time of the paleosol formation, or 100 times the present value. This upper limit is obtained by assuming a soil temperature of 300 K; lower temperatures yield lower upper limits because of the dependence of the chemistry of the iron carbonate formation on temperature.

Note carefully that the number derived is an upper limit to the carbon dioxide concentration – it could have been much lower. Nonetheless, what is important about the result is that even 3% carbon dioxide would have been insufficient to produce a surface temperature of 300 K, given the amount of solar luminosity (80% of the present value) some 2.5 billion years ago. Assuming a lower surface temperature does not help, because that in turn lowers the carbon dioxide upper limit from the paleosol data.

Could it be that enhanced carbon dioxide was not the sole contributor to a thicker greenhouse atmosphere? Other gases, particularly methane, might have been present to contribute additional infrared absorption; the lack of molecular oxygen tends to stabilize methane. Another possibility is that the models for computing greenhouse warming in the Archean are not accurate. Pennsylvania State climatologist G. Jenkins examined the climatic consequences of low molecular oxygen during the Archean. Although O_2 is not an important greenhouse gas, its low abundance implies that ozone, O_3, was rare in Earth's upper atmosphere, where it presently serves the important function of blocking ultraviolet radiation from the Sun. Lack of ozone would have cooled the upper atmosphere, forcing water to condense out as high-altitude clouds. These clouds might have contributed to a warming of the surface, perhaps by a few degrees relative to models that ignore clouds.

The uncertainty associated with high-altitude water clouds in the Archean greenhouse model reflects the general difficulty of understanding the role of clouds in climate. It still might be necessary to invoke additional gases to contribute to the greenhouse effect and offset the weak Archean Sun. Nonetheless, the abundance of carbon on Earth makes enhanced carbon dioxide the favorite of most climatologists as the primary mechanism for maintaining clement surface temperatures. Analysis of older paleosols, perhaps early Archean when carbon dioxide must have been well above the threshold to trigger iron carbonate formation, would be an important test. Such analysis must await the discovery of very ancient paleosols. However, regardless of whether there is experimental corroboration of enhanced carbon dioxide, a crucial question that must be answered is: If elevated levels of carbon dioxide existed, where did it go? What was the mechanism for evolving the atmosphere from CO_2-rich to CO_2-poor? It turns out that the answer lies in a cycle that is tied to the fundamental geologic process of plate tectonics, a cycle that continues today and that is partly mediated by life.

14.7 CARBON DIOXIDE CYCLING AND EARLY CRUSTAL TECTONICS

14.7.1 Basic Carbon-Silicate Weathering Cycle

How did a thick carbon dioxide atmosphere dwindle away over time? Carbon dioxide is taken out of the atmosphere by weathering of rock, and by plants and bacteria during photosynthesis. Much of the biologically-trapped carbon continues to cycle through living organisms at the surface. However, some of the carbon dioxide ends up in

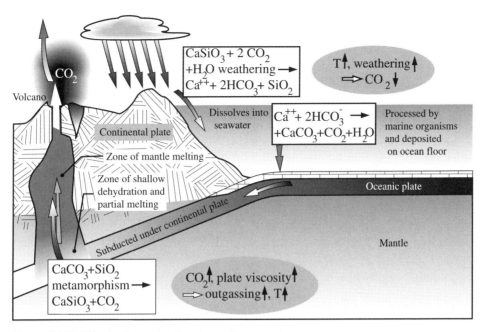

Figure 14.7. Weathering cycle of carbon, silicate, and water.

shell-forming organisms, which die and drop to the ocean floor, effectively removing the carbon from circulation in the biosphere. The chemistry of the weathering process, which depends critically on rainwater, goes as follows:

The breakdown of silicate rocks by the weathering action of rainwater is very efficient, because CO_2 gas dissolves in the rainwater to make a weak acid that can attack the rock chemically. This yields, among other products, silicon dioxide (SiO_2, the basic molecule of which quartz is made), bicarbonate ions (HCO_3^-, where the negative sign indicates that the molecule is negatively charged), and doubly charged ions of calcium (Ca^{2+}). These ions are quite reactive, and are used by shell-forming organisms to make calcium carbonate ($CaCO_3$) shells. The shells, along with silica (opal) shells made by other organisms, are preserved as thick layers of sediment on the floors of lakes, seas, and the ocean. The coral reefs are perhaps the most spectacular example of structures formed by deposition of calcium carbonate.

How long would it take this process to remove all carbon dioxide from the present atmosphere? Calculations show that, at current weathering rates and with the present mass of biota in the oceans, the removal time is less than one million years – only 0.02% of the whole history of Earth. With most of the carbon dioxide gone, the oceans would freeze over very quickly. Something else must happen, both today and in the past, to release the carbon dioxide from the calcium carbonate and return it to the atmosphere.

That "something" is plate tectonics. The ocean floor is continually being recycled back into the mantle at subduction zones, and the carbonate-laden sediments are carried with it. Much of the ocean-floor material subducted at trenches is melted at temperatures well over 1,000 K. Carbon-bearing materials, such as the calcium carbonates, react with silicates at these high temperatures to make calcium silicates ($CaSiO_3$) and CO_2 gas. The gas makes its way back to the surface not at the midocean ridges, but right at the subduction zones where volcanism is occurring. Mount St. Helens, Mount Pinatubo, and other active volcanoes belch CO_2 gas from deep within the subducted plates, resupplying the atmosphere.

Carbon is cycled through the atmosphere into the ocean and onto the seafloor, only to be subducted and returned to the atmosphere. This cycle, shown in figure 14.7, is limited by the time it takes ocean floor (once formed at midocean ridges) to be subducted. At current plate-tectonics spreading rates, the cycle takes about 60 million years; in other words, any given carbon atom in atmospheric carbon dioxide typically will form carbonate, be subducted, and then released again as carbon dioxide gas in a time of order 60 million years.

14.7.2 Negative Feedbacks in the Carbon-Silicate Cycle

Michigan geophysicist J.C.G. Walker proposed that this carbon-silicate-weathering cycle might well act as a stabilizing influence on Earth's climate. Because the cycle requires liquid water to dissolve carbon dioxide gas and to effectively weather rock, a fully frozen Earth would have lost the erosive portion of the process. Carbon dioxide gas would not be lost to ocean floor, while more carbon – previously cycled into the crust, or derived from deeper mantle rocks – would continue to accumulate in the atmosphere. This would raise the temperature through the greenhouse effect, until the oceans could melt and liquid precipitation became possible again. (These are two separate conditions; rainfall requires somewhat higher temperatures than does melting the oceans, the freezing point of which is lowered by salts.)

Conversely, higher atmospheric temperatures increase the evaporation from oceans, the amount of cloud, and hence rainfall rates. Higher temperatures also favor rainfall over snow at high latitudes and elevations. These have the net effect of increasing the rate of weathering by rainfall, and hence removal of carbon dioxide from the atmosphere. Sedimentation rates are higher, but the rates of subduction and carbon dioxide outgassing are not affected. As a result, carbon dioxide abundance in the atmosphere decreases, weakening the greenhouse effect and lowering temperature.

The carbon-silicate cycle possesses what is called a *negative feedback* loop in which changing the conditions in one direction tends to cause the system to move back in the other direction. This is characteristic of physical systems that are stable and provides at least a partial explanation as to why Earth's climate has remained in the temperature range allowing liquid oceans over geologic time. Absent such a feedback, changes in the Sun's luminosity, in the rate of spreading of plates, and in the amount of volcanism, as well as disasters such as giant impacts (chapter 18) could have moved Earth's climate out of the range in which life is sustainable.

Life itself has played a role in altering carbon loss rates: The development of soil-forming microorganisms some 3 billion years ago accelerated the trapping of carbon dioxide in soils and hence may have led to a net decrease in carbon dioxide levels in the late Archean atmosphere. The development of calcareous plankton shifted most of the deposition of carbonate to deep ocean rather than to shallow, continental-shelf environments, hastening the transport of carbonates to subduction zones and increasing the rate of reintroduction of carbon dioxide to the atmosphere. These and other evolutionary changes in Earth's biosphere have thus caused shifts in climate to which life has had to adjust through the formation of new species (chapter 18). The Britist scientist James Lovelock, and others, have even proposed that life acts to control the environmental feedback processes to maximize the habitability of Earth. This controversial "Gaia" hypothesis is thought-provoking but not required necessarily to explain the stability of Earth's climate. Perhaps instead, life is a somewhat meddlesome passenger largely along for the ride.

14.7.3 The Carbon-Silicate Cycle During the Archean

Although the carbon-silicate cycle seems to be a good candidate for explaining how Earth's climate has been stabilized over time, it is necessary to think carefully about how it operated during the Archean, when conditions were different from today. Smaller amounts of continental mass exposed above sea level could have reduced the efficacy of silicate weathering, which is the step that determines the rate of carbon dioxide sequestration. More rapid recycling of crust at the time would have led to a faster return of carbon dioxide to the atmosphere, with less stored in sediments on the seafloor. Both of these somewhat speculative differences between the Archean and more modern times mitigate in favor of leaving the available CO_2 in the atmosphere, rather than locked in sediments, during the Archean.

The high CO_2 abundances required to sustain Archean temperatures above that of the water freezing point led to a potential instability in the early Archean: The Sun's luminosity was low enough that the feedbacks in the carbon-silicate cycle might not have worked to bring Earth out of an ice-covered state, if it fell into one during that time. The reason for this lies in a phenomenon that we discuss for Mars in chapter 15 – cold temperatures and high carbon dioxide abundances would have caused carbon dioxide clouds to form, with the cooling effect of these clouds short-circuiting the gas's ability to warm the surface. This is a problem unique to the Archean because, for higher solar luminosities, less carbon dioxide is required to drive Earth out of a global ice age, so that carbon dioxide cloud formation is not an issue. However, some models suggest that carbon dioxide clouds might, under certain conditions, actually warm the surface (chapter 15). Whether the Archean Earth ever dipped into a deep ice age remains conjectural.

14.8 A BALANCE UNIQUE TO EARTH, AND A LINGERING CONUNDRUM

The history of Earth's atmosphere has been one of declining carbon dioxide abundance from the Archeaneon onward. Geologic evidence for glaciation in the early Proterozoic and the lack of iron carbonates in paleosols are important constraints, as is glaciation in the late Precambrian (figure 14.8). The decline has been slow because of the buffering effects of the carbon-silicate cycle, operating on a planet with liquid water and plate tectonics.

If Earth did not possess liquid water, erosion of rock would be extraordinarily slow. On Venus, the surface pressure is 90 atmospheres of carbon dioxide. The total

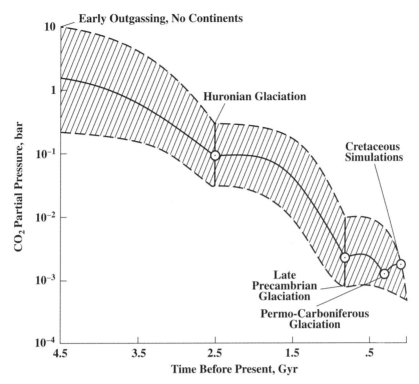

Figure 14.8. History of the CO_2 partial pressure in Earth's atmosphere based on various lines of evidence regarding Earth's surface temperature. Adapted from Kasting and Toon (1989) by permission of University of Arizona Press.

amount of carbon dioxide in Venus' atmosphere is not very different from the total equivalent of carbon dioxide locked up in various forms in Earth's crust. As discussed in chapter 15, Venus lost whatever water it had early in its history, forcing all of the carbon dioxide to remain in the atmosphere. The resulting massive greenhouse has pushed the surface temperature of Venus to 730 K, much too high for liquid water to exist. Venus is trapped outside the carbonate cycle with no way to rid itself of the carbon dioxide that keeps the surface so warm.

Mars has a thin atmosphere, and its surface is too cold to support an ocean. Evidence (chapter 15) shows that early Mars had episodes in which liquid water existed on its surface, but the crust of Mars shows no evidence of plate recycling – it is a small planet with little internal heat and hence lethargic tectonics. No recycling means that the carbonates formed from carbon dioxide essentially will never yield the greenhouse gas again. The liquid water on early Mars would have encouraged weathering, carbonate formation, and decreasing carbon dioxide – until the temperature of the surface got so low that the water froze and the weathering process ground to a halt.

In this "Goldilocks" view, Earth has two unique things that keep the cycle going: (i) liquid water to make carbonates from carbon dioxide and (ii) a vigorous recycling of the ocean crust, which releases much of the carbon dioxide

back into the gas phase. Earth is far enough from the Sun to enable liquid water to exist without catastrophically vaporizing and escaping, the fate of Venus discussed in chapter 15. Earth also is much larger than Mars, which is too small to have the tectonic recycling needed to keep carbon dioxide in the atmosphere. Nonetheless, the atmospheric supply of carbon dioxide on Earth has continued to decrease slowly over time, from as much as 10 atmospheres in the Hadean or Archean, through to the present value of 0.0003 atmospheres. And, in spite of the increasing output of the Sun, Earth's climate generally has gotten cooler over time: The first glacial episode for which evidence exists was just after the close of the Archean. In these cold glacial episodes, less rainfall has meant that less carbon dioxide is lost from the atmosphere, allowing buildup of the gas to warm the climate again and encouraging faster subsequent loss of carbon dioxide from the atmosphere. Therefore, the overall march toward less CO_2 and cooler climes appears to be a theme of Earth history up to the present.

Although the story presented above seems tidy, not all issues are resolved. It remains a serious problem that carbon dioxide alone may not have been enough to explain warm temperatures on Earth in the weak glow of the faint early Sun. Explaining warm episodes on Mars, as we show in chapter 15, exacerbates the problem. Louisiana astronomer D. Whitmire and colleagues have proposed a quite different solution: that the early Sun was not low in luminosity after all. To get around the seemingly simple solar physics that led to the faint-sun problem, they suggest that the Sun was more massive 4 billion years ago and has lost that mass through expulsion of hydrogen gas in the form of a wind.

The present solar wind, a tenuous medium of protons and other ions, is much too weak to remove the mass required to create a brighter early sun, and so, the hypothesis must rely on the notion that the mass loss was much higher earlier in the history of the solar system. The viability of the idea rests on observing such mass loss from stars similar in age to the Archean Sun elsewhere in the galaxy, an observation that currently is very difficult. Nonetheless, the proposal itself reminds us that the keys to understanding the history of Earth lie buried not only in our own planet, but in our planetary neighbors, in the Sun, and in the neighboring galactic regions illuminated by the burning of billions of other suns.

14.9 QUESTIONS

a. The presence of carbon recycling on Earth, as a buffer against the faint early Sun, and excessive temperatures later, might strike some as a kind of "just right" story, such that few planets other than twins of Earth could sustain life. What other kinds of processes could keep a climate habitable for life?
b. Is there any limit on planets much more massive than Earth sustaining life? Could a rocky body 10 times Earth's mass sustain life? What might be the problems?

14.10 READINGS

14.10.1 General Reading

Williams, G.R. 1996. *The Molecular Biology of Gaia*. Columbia University Press, New York.

14.10.2 References

Houghton, J.T. 1977. *The Physics of Atmospheres*, 1st ed. Cambridge University Press, Cambridge, UK.
Jenkins, G.S. 1995. Early Earth's climate: Cloud feedback from reduced land fraction and ozone concentrations. *Geophysical Research Letters* 22, 1513–1516.
Kasting, J.F. 1989. Long-term stability of the Earth's climate. *Palaeogeography Palaeoclimatology Palaeoecology* 75, 83–95.
Kasting, J.F. 1997. Planetary atmosphere evolution: Do other habitable planets exist and can we detect them? In *The Search for Extra-Solar Terrestrial Planets: Techniques and Technology* (M. Shull, H. Thomson, and A. Stern, eds.). Kluwer, Dordrecht, pp. 3–24.
Kasting, J.F., and Ackeman, T.P. 1986. Climatic consequences of very high CO_2 levels in the Earth's early atmosphere. *Science* 234, 1383–1385.
Kasting, J.F., and Toon, O.B. 1989. Climate evolution on the terrestrial planets. In *Origin and Evolution of Planetary and Satellite Atmospheres* (S.K. Atreya, J.B. Pollack, and M.S. Matthews, eds.). University of Arizona Press, Tucson, pp. 423–450.
Kasting, J.F., Whitmire, D.P., and Reynolds, R.T. 1993. Habitable zones around main sequence stars. *Icarus* 101, 108–128.
Knauth, L.P. 1992. Origin and diagenesis of cherts: An isotopic perspective. In *Isotopic Signatures and Sedimentary Records* (N. Clauer and S. Chandhuri, eds.). Springer-Verlag, Berlin, pp. 123–152.
Moroz, V.I., and Mukhin, L.M. 1977. Early evolutionary stages in the atmosphere and climate of the terrestrial planets. *Kosmicheskie Issledovaniya* 15, 901–922.
Nutman, A.P., Mojzsis, S.J., and Friend, C.R.L. 1997. Recognition of ≥3850 Ma water-lain sediments in Greenland and their significance for the early Archean Earth. *Geochimica Cosmochimica Acta* 61, 2475–2484.
Peixoto, J.P., and Oort, A.H. 1992. *Physics of Climate*. AIP Press, New York.
Press, F., and Siever, R. 1978. *Earth*. W.H. Freeman and Company, San Francisco.
Rampino, M.R., and Caldeira, K. 1994. The Goldilocks problem: Climatic evolution and long-term habitability of terrestrial planets. *Annual Review of Astronomy and Astrophysics* 32, 83–114.
Rye, R., Kuo, P.H., and Holland, H.D. 1995. Atmospheric carbon dioxide concentrations before 2.2 billion years ago. *Nature* 378, 603–605.
Trenberth, K.E., Houghton, J.T., and Meira Filho, L.G. 1996. The climate system: An overview. In *Climate Change 1995: The Science of Climate Change* (J.T. Houghton, L.G. Meira Filho, B.A. Callander, N. Harris, A. Kattenberg, and K. Maskell, eds.). Cambridge University Press, Cambridge, UK, pp. 51–65.
Whitmire, D.P., Doyle, L.R., Reynolds, R.T., and Matese, J.J. 1995. A slightly more massive young sun as an explanation for warm temperatures on early Mars. *Journal of Geophysical Research* 100, 5457–5464.

15 CLIMATE HISTORIES OF MARS AND VENUS, AND THE HABITABILITY OF PLANETS

15.1 INTRODUCTION

Earth at the close of the Archean, 2.5 billion years ago, was a world in which life had arisen and plate tectonics dominated, the evolution of the crust and the recycling of volatiles. Yet oxygen (O_2) still was not prevalent in the atmosphere, which was richer in CO_2 than at present. In this last respect, Earth's atmosphere was somewhat like that of its neighbors, Mars and Venus, which today retain this more primitive kind of atmosphere.

Speculations on the nature of Mars and Venus were, prior to the space program, heavily influenced by Earth-centered biases and the poor quality of telescopic observations (figure 15.1). Thirty years of U.S. and Soviet robotic missions to these two bodies changed that thinking drastically. The overall evolutions of Mars and Venus have been quite different from that of Earth, and very different from each other. The ability of the environment of a planet to veer in a completely different direction from that of its neighbors was not readily appreciated until the eternally hot greenhouse of Venus' surface and the cold desolation of the Martian climate were revealed by spacecraft instruments.

However, robotic missions also revealed evidence that Mars once had liquid water flowing on its surface. It is tempting, then, to assume that the early Martian climate was much warmer than it is at present, warm enough perhaps to initiate life on the surface of Mars. However, the difficulty of sustaining a warm Martian atmosphere in the face of the faint-early-sun problem of chapter 14 remains a daunting puzzle, one that is highly relevant to the broader question of habitable planets beyond our solar system. What is the range of distances from any given star for which liquid water is stable on a planetary surface and life can gain a foothold?

In the temporal sequence that Part III of the book has been following, we stand near the end of the Archean eon. By this point in time, the evolution of Venus and its atmosphere almost certainly had diverged from that of Earth, and Mars was on its way to being a cold, dry world, if it had not already become one. This is the appropriate moment in geologic time, then, to consider how Earth's neighboring planets diverged so greatly in climate, and to ponder the implications for habitable planets throughout the cosmos. In the following chapter, we consider why Earth became dominated by plate tectonics, but Venus and Mars did not. Understanding this is part of the key to understanding Earth's clement climate as discussed in chapter 14.

15.2 VENUS

15.2.1 Origin of Venus' Thick Atmosphere

The atmosphere of Venus contains somewhat more nitrogen than does that of the Earth: 3 atmospheres of pressure instead of 0.8 atmosphere. More striking, however, is the enormous surface pressure of 90 atmospheres of carbon dioxide. The consequence of Venus' massive atmosphere is an enormous greenhouse effect: Even though the clouds of Venus' upper atmosphere, largely sulfur compounds, reflect much more sunlight away than do the clouds of Earth, Venus has a surface temperature of 730 K. In other words, even though the surface of Venus receives less sunlight than does the Earth's surface, the temperature at Venus' surface is above the melting point of lead. Liquid water is not stable on the surface or anywhere in the atmosphere; gaseous water vapor is only 10 parts per million of the atmosphere. Oxygen is not abundant either, with a pressure of 0.002 atmosphere, one-hundredth that in our atmosphere.

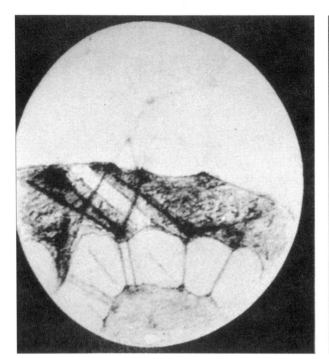 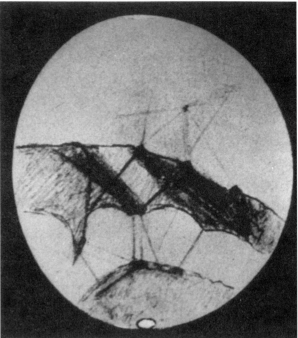

Figure 15.1. Prior to the use of photographic and electronic detectors, maps of Mars sketched by hand typically showed unnaturally straight lines, a result of atmospheric turbulence that blurred telescopic images and caused the merging of irregular dark features. Such lines were considered by some as a sign of intelligence. At the turn of the twentieth century, the American astronomer Percival Lowell interpreted these illusory features as vast canals bringing water from the Martian polar caps to the parched equatorial deserts, a grander version of what was actually undertaken at the time in the Arizona and California deserts south and west of his high plateau observatory.

How Venus came to this state is still a subject of heated debate. Venus is almost the same size as Earth, of similar density (and hence internal composition), and somewhat nearer to the Sun. One clue is the close correspondence of the amount of carbon dioxide in Venus' atmosphere with the amount of carbon dioxide that could be produced from the carbonates and other carbon compounds trapped today in Earth's crust. If Earth's oceans were to boil away, and the hydrological cycle of rainfall end, recycling of carbonates into the atmosphere might eventually build up a massive carbon dioxide atmosphere on our planet as well. The divergent evolutionary paths that Earth and Venus have taken apparently have to do with the lack, or early loss, of large quantities of water from Venus. Direct measurement of Venus' atmosphere from *Pioneer* Venus entry probes in 1978 revealed a large abundance of deuterium (defined in chapter 2) relative to light hydrogen in the atmosphere of Venus, the ratio of the two being about 150 times that in the oceans of Earth. One interpretation of such an overabundance is that large amounts of water escaped from Venus early in its history; as the water was lost in gaseous form from the atmosphere, the heavier deuterium atoms in HDO and D_2O (versus H_2O) were more likely to be retained. Although alternative models have been proposed (for example, that the high deuterium abundance is a contaminant from impacting comets), the water-loss model appears at present to be the best explanation for the deuterium data.

If Venus did have liquid water early in the solar system's history, the challenge is to understand how it was lost and when. The traditional explanation for the loss lies in the so-called *runaway greenhouse*, featured in many textbooks. Here, the solar heating at Venus' distance from the Sun, coupled with a sufficient amount of initial greenhouse heating from water and carbon dioxide, leads to an unstable situation: Heating causes more evaporation of water from the ocean (because the evaporation rate and the total water vapor content in the atmosphere are very sensitive to the temperature). This higher water content, in turn, increases the atmospheric temperature through the greenhouse effect, which in turn causes more water to evaporate, warming the atmosphere further. The system enters a "runaway," leading quickly to the complete boiling away of the oceans.

Very careful modeling of the early history of Venus shows that at the time, a runaway greenhouse was marginal for that planet. The reason lies again in the faint-early-sun problem. Although today, Venus receives 1.9 times the amount of sunlight that Earth does at the top of the atmosphere (remember much of this is reflected by Venus' clouds), in the earliest period of solar system history the sunlight that Venus received was only 1.4 times that received by Earth at present. Below a certain threshold surface temperature, the greenhouse effect does not evaporate enough water to initiate a runaway.

So how did Venus arrive at its present state? The solution to this puzzle lies in considering the effect of water vapor on the entire atmosphere, as shown in figure 15.2. On Earth today, because the temperature drops rapidly with altitude as the atmosphere thins and becomes more

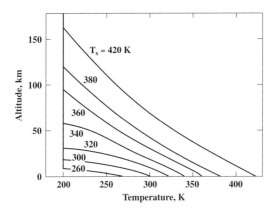

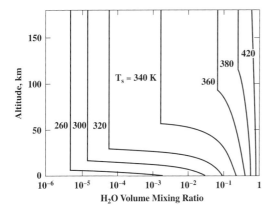

Figure 15.2. A moist greenhouse atmosphere in action. The temperature (top) and amount of water vapor (bottom) are plotted versus altitude for different values of the surface temperature. Each profile is marked with its particular surface temperature, T_s. The water-volume mixing ratio is simply the number of water molecules divided by the number of all molecules (of all chemical species) in the atmosphere at a given altitude. Hence a water mixing ratio of 10^{-3} means that one out of every thousand molecules is water. The stratosphere is simplified in the calculation by assuming that it has a constant temperature of 200 K; in reality, its temperature is not constant. See text for a description of the moist greenhouse loss of water. Reproduced from Kasting (1988) by permission of Academic Press.

transparent to infrared radiation, the amount of water vapor drops very steeply. At about 10 km above the surface lies a boundary between the lower atmosphere, the *troposphere*, and the *stratosphere* above it. This boundary, the *tropopause*, is defined by the altitude at which the temperature stops falling and begins rising at higher altitude as the air becomes transparent to most infrared radiation, and some molecules selectively absorb sunlight in the ultraviolet wavelengths. Above the tropopause, water vapor no longer decreases with increasing altitude; its minimum value is determined by the temperature at the tropopause.

In Earth's atmosphere today, the dropoff in temperature with height leads to a very sharp decline in water vapor with altitude. The water vapor condenses as clouds and these eventually are lost as rain. The Earth's stratosphere is extremely dry today, about as dry as the present bulk atmosphere of Venus. What water vapor does exist in the stratosphere is subject to being broken apart by ultraviolet photons from the Sun to form oxygen (O_2) and hydrogen; because hydrogen is a light molecule, it moves upward in the atmosphere and eventually is lost to space. The ultraviolet radiation is restricted to high altitudes precisely because it is absorbed there by molecules such as water and ozone; the vast majority of Earth's water is protected from such destruction by being resident in the oceans and lower atmosphere.

Consider now what would happen if Earth's surface temperature were increased, simulating what might have happened on Venus if it once had had liquid water oceans. More water vapor is put into the troposphere, allowing formation of more massive cloud decks. Clouds can warm or cool the climate, depending on their altitude, but their formation by condensation always releases heat, which causes the temperature profile to fall more gently with altitude. Because of this effect, the temperature profile for higher surface temperatures declines more gradually than for lower surface temperatures, and the tropopause boundary between the troposphere and the stratosphere moves upward as the surface warms (figure 15.2). More water is admitted into the stratosphere, and eventually large amounts of water are present at altitudes accessible to solar ultraviolet photons. For a surface temperature just 80 K above Earth's current global mean value, the water vapor at high altitudes increases by a factor of 10,000.

In effect, then, a global surface temperature above 340 K "pops the cork" on the water budget of the atmosphere, allowing large amounts of water vapor to flow to altitudes where solar ultraviolet radiation breaks it apart, and the hydrogen escapes. This *moist* greenhouse crisis operates at lower solar fluxes than is required for the runaway greenhouse; for an Earth-like atmosphere with nitrogen and a

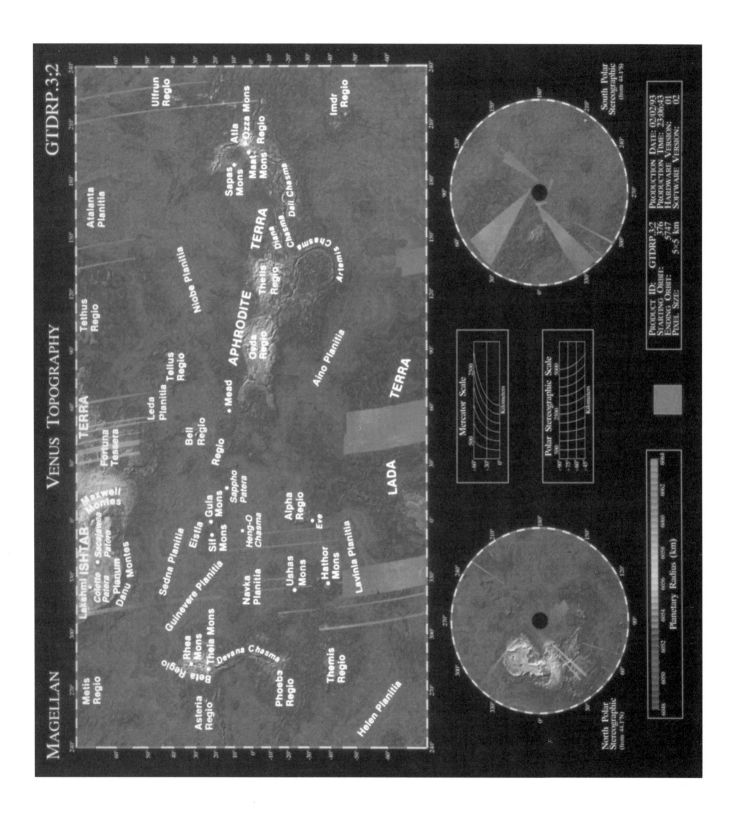

small amount of CO_2, the threshold for the moist greenhouse is just 1.1 times the present solar flux received by Earth. This flux is well below that which was received by Venus during the faint-early-sun epoch, but above that for Earth throughout its history.

We can imagine what happened to Venus early in its history. Possessed of an atmosphere with at least as much CO_2 in gaseous form as Earth possesses today, Venus' surface was above the temperature threshold for the moist greenhouse even when the solar flux was only 70% of its present value. If liquid water did exist on the surface, the atmospheric temperatures were high enough to allow evaporated water to flow freely to the tenuous upper atmosphere. Ultraviolet photons broke up the water molecules, causing most of the hydrogen to be lost and eventually depleting the planet of water. The signature of this lost ocean is with us today in the form of a high Venusian ratio of deuterium-to-hydrogen, because the heavier deuterium tended to be left behind in the atmosphere as hydrogen escaped.

Once bereft of surface water, the die was cast for Venus. Carbon dioxide in the atmosphere had no means of being locked up in surface rocks because liquid water was not available to make bicarbonates. The carbon dioxide that we see today in Venus' atmosphere cannot escape from the top of the atmosphere, cannot be trapped in rocks at the surface, and remains as a massive gaseous memento of the early loss of water.

How quickly could the water have been lost? Observations of young stars suggest that the early Sun put out more ultraviolet radiation than it does today though, as discussed in chapter 14, its overall energy output was lower. Based on the amount of ultraviolet radiation available at Venus from the early Sun, less than 100 million years were needed to remove the equivalent of an Earth's ocean-worth of water. Hence there was little or no time to lock up carbon dioxide as carbonates before the water was lost, and because accretional heat likely was still contributing to a very hot early crust for Venus, most or all of Venus' carbon dioxide complement was likely *never* locked up in the crust. The 90 atmospheres of CO_2 present today probably is close to the original atmospheric abundance, although some of the carbon dioxide could have been added later from the Venusian mantle by volcanoes. The moist greenhouse model has important consequences for the habitability of planets in general, a point we return to in section 15.6.

Figure 15.3. Global topography of Venus from the Magellan radar probe, with various features labeled. Courtesy of NASA/Jet Propulsion Laboratory.

15.2.2 Overview of the Surface of Venus

Although early Soviet and U.S. probes measured the atmospheric composition and temperature of Venus, mapping the geology of the surface was hindered by a global mass of sulfurous clouds at high altitude. First radar from Earth, then radar from two Soviet orbiters *Veneras* 15 and 16 and the U.S. *Magellan* spacecraft have enabled mapping of the surface. A radar mapper functions like a camera that provides its own flash or source of illumination. Photons at radio wavelengths (chapter 3) can penetrate the clouds, and the radar transmits such photons to the ground surface. These are reflected and scattered, and some are received back at the radar antenna. By coding or shaping the transmitted pulse of photons, and taking advantage of the orbital motion of the spacecraft, the received photons can be arranged or mapped by computer into an image of the surface at radio wavelengths. For detailed geologic work, the very-high-resolution Magellan images, collected at Venus from 1990 through 1993, are of greatest use.

The geology of Venus, on a broad scale, looks at first glance like the Earth's with highlands rising out of a lowland plain, akin to continents rising above the ocean floor. However, the proportion of land on Venus that rises above the mean surface elevation is far smaller than on Earth; likewise, there are few long, deep cuts in the crust like Earth's submarine trenches (figure 15.3). Thus the signatures of mature plate tectonics – massive continents and subduction zone trenches – are largely missing. It is as if we were to look at Earth in the Archean eon of time, when plate tectonics was just getting going and continental masses were small. Soviet probes have sampled several regions on the surface; all of the analyses are consistent with basaltic compositions (close to that of Earth-ocean crust). No strong evidence for granites has been found anywhere on Venus.

The surface of Venus contains impact craters. Although the number of these is far larger than on Earth, it is less than that of the Moon and Mars. The number of craters is consistent with a surface that has renewed itself through volcanic flows over geologic time, with the last overall renewal of the surface being perhaps 300 million to 600 million years ago. (Whether the surface is continually or episodically active geologically is addressed in chapter 16.) This is long after the loss of any putative Venusian water ocean, which occurred in the first few hundred million years of Venus' history. Hence, we expect to see no evidence of ancient oceans, and indeed there is no evidence for the past existence of liquid water anywhere on Venus' surface.

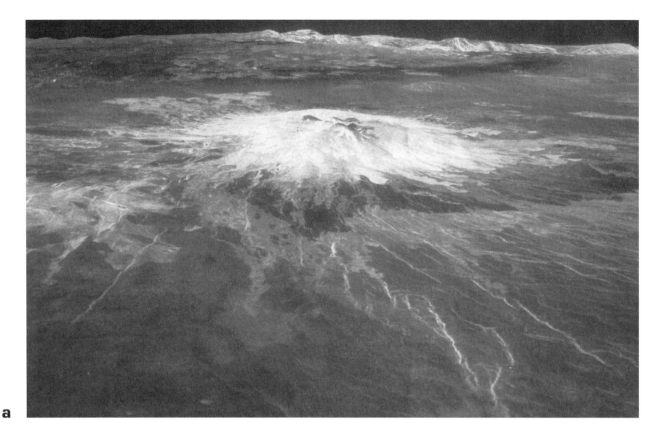

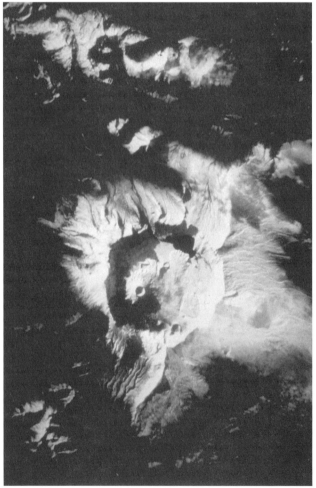

Figure 15.4. (a) Sapas Mons, a 600-km-diameter, 1.5-km-high volcano on Venus, shows no evidence of water erosion; the bright linear features have the form and appearance of lava channels. This *Magellan* radar view exaggerates the vertical extent by a factor of 10. Image courtesy of NASA/Jet Propulsion Laboratory. (b) Aniachak volcano in Alaska, imaged by a Space Shuttle from Earth orbit, is 10 km across and 1.4 km high. Close examination of the caldera and flanks reveals networks of water-carved channels. (Note that this volcano has a new, young cone perched in its caldera.) Courtesy of NASA.

The thick atmosphere prevents small bodies from reaching the surface as well; however, the largest impactors go unimpeded. Because there is no surface water on Venus, craters and other landforms that are not buried in lava erode very slowly. The mean slope of features is therefore larger on Venus than on Earth, and the images of mountain ranges are eerie in their evident absence of water erosion (figure 15.4).

The lack of plate tectonics and its accompanying geologic signatures on Venus is perhaps the most profound difference between Venus and Earth. Remarkably, the presence of water is apparently an important condition for sustaining plate motions, and certainly for the formation of continental masses that are present on Earth in abundance (and may only exist on Venus in one location, if at all). We defer a more detailed development of this idea to chapter 16, where the origin of Earth's plate tectonic geology is explored. Significant and striking geologic differences are apparent on these two planets that should be ridding themselves of the same amount of internal heat; understanding the origin of these differences is perhaps the most important question in Venusian geology.

15.3 MARS

15.3.1 Mars Today

The Martian atmosphere is, in composition, very similar to that of Venus, with carbon dioxide most abundant, nitrogen the secondary constituent, and water and oxygen in minor abundance. Mars' atmosphere is diminutive compared to those of Venus and Earth, however. The surface pressure is only 0.006 of an atmosphere. The thin atmosphere means that Mars has hardly any greenhouse warming. This, combined with its greater distance from the Sun, results in a temperature range from as much as 270 K at the equator to only 150 K at the polar caps. Mars is a true opposite of Venus: a cold dry planet, with air so thin that ultraviolet rays from the Sun penetrate to the surface, effectively sterilizing its uppermost soil.

Mars is so cold that the carbon dioxide atmosphere freezes out seasonally at the poles. The pressure in the atmosphere therefore varies significantly over the Martian year, which is about twice an Earth year. The tilt (*obliquity*) of Mars currently is the same as Earth's; the summer sun shines on one pole, evaporating carbon dioxide and driving it to the winter pole. Mars' axis, however, may undergo large shifts in its obliquity caused by gravitational tugging of the other planets, principally Jupiter; Earth would suffer the same fate were it not for the stabilizing effects of our large Moon. There is some weak evidence in geological features across the Martian surface that past tilt may have exceeded 50 degrees (the current value is 24 degrees).

About one year out of two, heating during the southern hemisphere spring drives large quantities of dust into the atmosphere, allowing more sunlight to be absorbed in the atmosphere and moving dust across the planet. These global dust storms may last for weeks or months.

Water is present today on Mars as ice trapped at one or both polar caps, but probably is more abundant as ground ice trapped in a zone of permanent freezing (*permafrost*) throughout high- and mid-latitude regions of Mars' crust. Water ice also condenses out in the thin atmosphere; storm systems occasionally have been seen in orbiting spacecraft images. The search for life on Mars culminated in the landing of two sophisticated robot laboratories, *Vikings* 1 and 2, in 1976. These laboratories sampled Martian soil and tested for chemical reactions that might indicate living processes. No evidence of life was found in the dry regions to which the landers had been targeted, sites that were chosen to maximize the chances of safe landings. Furthermore, the abundance of organic molecules on the surface was so low as to be undetectable. The thin atmosphere of Mars, with no ozone shield, allows solar ultraviolet radiation to penetrate to the surface and break apart chemical bonds; organic molecules are readily destroyed in such an environment, and much of the hydrogen is lost to space. Additionally, the iron in the soil is combined with oxygen in such a way as to make an extremely reactive mixture that would quickly oxidize organic molecules. The present surface of Mars, at least in the high plains, is an inhospitable location for life.

15.3.2 Geological Hints of a Warmer Early Mars

Unlike Venus, the Mars surface is visible at all times except during dust storms. Cameras sent to Mars on robotic missions have mapped the surface in great detail from orbit and at three landing sites. The geology of Mars has been eloquently summarized by U.S. Geological Survey scientist Michael Carr (1984):

Mars, like the Earth, has had a long history of volcanic and tectonic activity, and its surface has been modified by wind, water, and ice. However, the two planets differ greatly in geologic style. Materials that form at the surface of Earth are continually being cycled deep within the planet as a result of plate tectonics, and are continually being redistributed across the surface as a result of the vigorous action of liquid water. On Mars, no deep recycling occurs; volcanic material brought to the surface simply remains in huge volcanic piles. Moreover, although some water erosion has occurred, the cumulative effect over the life of the planet has been trivial. Once relief is created, it largely remains. The result is a planet on which a record of sustained geologic activity is beautifully preserved in huge volcanoes, enormous canyons, extensive fracture systems, and giant flood channels.

On the large scale, Mars shows no evidence for continents and lowland (ocean-type) basins. The southern hemisphere stands several kilometers above the northern hemisphere; the south is very heavily cratered in contrast to the north. This asymmetry may be the result of a giant impact early in Mars' history; it is not at all what one would get with Earth-style plate tectonics. Two extensive uplands on Mars are sites of past volcanism. The largest one, Tharsis, contains huge shield volcanoes, giant versions of the Hawaiian volcanoes. Again, they are clues to the static nature of the crust: With no lateral movement, magma welling up from the interior keeps spewing out material on the same part of the nonmoving crust, building up huge volcanoes in isolated locations. The *Viking* robotic landers sampled the soils at two widely separated landing sites in the northern hemisphere and found the rocks to be basaltic in composition. The *Pathfinder* lander, arriving in Ares Valles in July 1997, identified one rock with an elemental composition consistent with andesite, which would be suggestive of plate tectonics. However, because only the elemental abundances could be determined on this mission, and not how the atoms are structured in a mineral, the finding was ambiguous: The rock could be an amalgam of basaltic material and more silica-rich debris from an impact.

The lack of plate tectonics likely is caused by the small size of Mars, and hence less vigorous convection than on Earth. The lower amount of heat coming out of Mars results in a thick crust, which resists breaking into plates. The style of geology is more like that of Venus than Earth, but far less active than either.

Evidence for water on Mars abounds. Impact craters appear to have melted ground ice; their peripheries show signs of extensive mudflow. Volcanoes heat the ground and release water; a number of runoff channels reveal that water was melted by the eruptive heat. Most intriguing is evidence for a sustained warmer period on Mars contained in channels and canyons. The evidence is threefold:

(i) Networks of dry channels and valleys are present on Mars. Three basic forms can be identified (figure 15.5):

Figure 15.5. Three types of dried-up channels on Mars, in *Viking* Orbiter images: (a) Outflow channels.

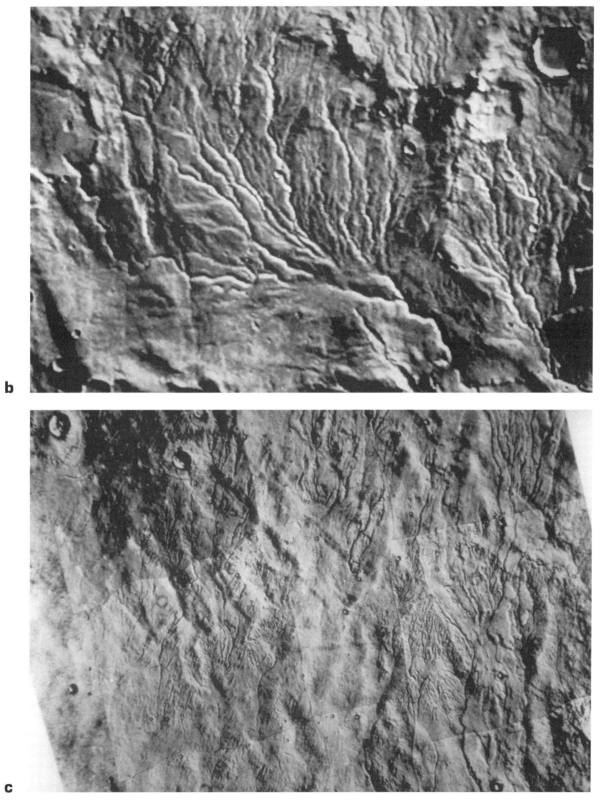

Figure 15.5. (Continued) Three types of dried-up channels on Mars, in *Viking* Orbiter images: (b) Valley networks in the ancient southern highlands. (c) Runoff channels on the volcano Alba Patera. Courtesy of NASA/Jet Propulsion Laboratory.

outflow channels, *valley networks*, and *runoff channels*. The outflow channels appear to have been formed by the very rapid release of large quantities of water, or might have been carved by flows of debris (rocks, mud) mobilized by water. The flows in such channels were sufficiently energetic that they could have been sustained under virtually any atmospheric conditions, including the cold, dry climate existing now on Mars (under which slowly flowing water would quickly freeze and then sublime to water vapor). The wide variation in abundance of craters on surfaces in and around the channels suggests that the channels formed episodically over the history of Mars. The valley networks, on the other hand, have a form that indicates they were carved by slowly flowing liquid water or, alternatively, by collapse of the surface (*sapping*) caused by groundwater flow. The possible sources of the water include melting of buried ice and expulsion to the surface, melting of surface ices, or even precipitation of snow or rain. The valley networks occur primarily, but not entirely, on surfaces that are very heavily cratered, and some of the impacts clearly occurred after the networks were formed. Most are therefore very ancient – dating to the end of the heavy bombardment some 3.8 billion years ago. Because their formation requires conditions very different from those present today (much more restrictive than required for the outflow channels), they could be a record of a time when the atmosphere was thicker and the climate warmer. A few younger valley networks, as well as runoff channels seen on the slopes of some volcanoes, suggest that warm conditions (possibly localized) may have occurred multiple times in Martian history.

(ii) Massive canyon systems formed by geologic processes show evidence of modification by liquid water. The canyons merge into numerous channels that show features caused by the flow of liquid water. Sedimentary deposits within the canyons have been seen on orbiting spacecraft images which suggest the former presence of standing lakes.

(iii) Some geologic features in various areas of Mars appear to have been carved by *glacial* action, that is, the movement of massive amounts of surface ices under their own weight. The features include certain kinds of ridges and troughs that resemble terrestrial landforms carved by glaciers and called *moraines* and *eskers* (figure 15.6). Such interpretation is difficult to make without ambiguity; other nonglacial causes for the features might have been at work. If the glacial interpretation is correct, however, it implies surface conditions in which water ice was stable against rapid sublimation, and hence requires thicker atmospheric conditions.

An early period of warm conditions on Mars, with liquid water, requires a thicker atmosphere of carbon dioxide, perhaps several atmospheres or more of pressure. Because it formed farther from the Sun than did Earth, in a cooler part of the solar nebula, Mars probably started out with at least as much water and carbon dioxide as did Earth. An early thick atmosphere is therefore possible. During such a period, life could have developed. Unlike on Earth, the climate apparently changed because carbon dioxide disappeared and temperatures fell below the freezing point of water, perhaps terminating Martian life. Whether warm conditions occurred in multiple episodes, and how recent the last such episode was, remain controversial. The interpretation of some Martian features as glacial in nature is an important part of the debate, because such features appear to be much younger than the bulk of the valley networks.

The cause of the climate cool-downs might be tied to Mars' small size. On early Mars, carbon dioxide could have been progressively locked up as carbonates in much the same way as on Earth (probably without the mediating step involving life). Mars, however, is much smaller than Earth and therefore has cooled more rapidly than our planet. The result seems likely to be a very thick crust that cannot slide horizontally in the form of recycling plates. The presence of a few massive volcanoes where heat is removed and the lack of typical plate tectonic features attest to this interpretation. Thus, on Mars, there was no plate tectonic activity and hence no significant recycling of the crust: Carbon dioxide locked up as carbonates would have remained that way. Loss of atmosphere by impacts was also important, since the small size of Mars and hence weak gravity (one-third the Earth's) encouraged escape of gases heated by impactors. Whether carbonate formation or impact escape was the more important loss process is a matter of current debate.

Even as surface temperatures fell below the freezing point of water, further (slower) formation of carbonates was possible. The result was the nearly complete loss of atmosphere via escape to space and permanent storage as carbonate sediments. Perhaps changes in Mars' axial tilt or episodes of volcanism temporarily released carbon dioxide from the crust in subsequent warming episodes. Such a model for the evolution of the Martian atmosphere can be tested by searching for carbonates exposed on the surface in selected locations on Mars. This is best done by a combination of orbital surveys, using infrared spectrometers, and robotic visits to promising locations on the surface.

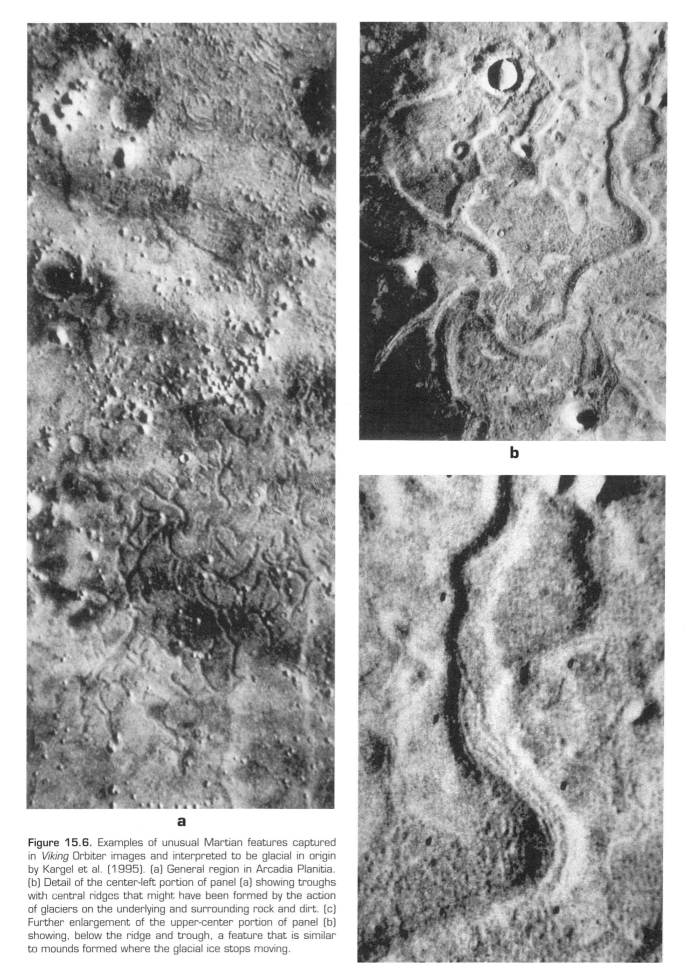

Figure 15.6. Examples of unusual Martian features captured in *Viking* Orbiter images and interpreted to be glacial in origin by Kargel et al. (1995). (a) General region in Arcadia Planitia. (b) Detail of the center-left portion of panel (a) showing troughs with central ridges that might have been formed by the action of glaciers on the underlying and surrounding rock and dirt. (c) Further enlargement of the upper-center portion of panel (b) showing, below the ridge and trough, a feature that is similar to mounds formed where the glacial ice stops moving.

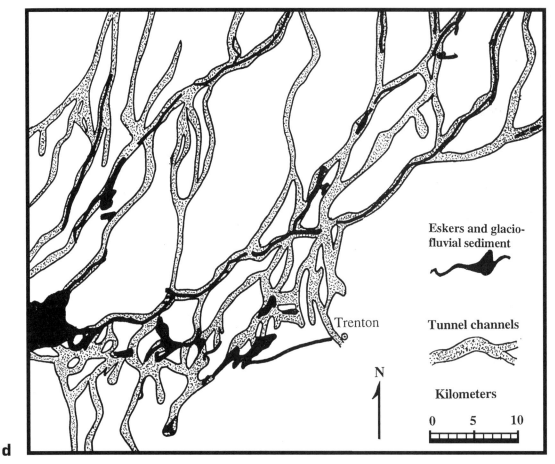

Figure 15.6. (Continued) Examples of unusual Martian features captured in *Viking* Orbiter images and interpreted to be glacial in origin by Kargel et al. (1995). (d) Map of a region north of Lake Ontario on North America, showing glacial trough and ridge systems on a scale similar to the image in panel (c). Figure provided by J. Kargel, U.S. Geological Survey, Flagstaff.

15.4 WAS MARS REALLY WARM IN THE PAST?

15.4.1 Limits to a Carbon Dioxide Greenhouse

The picture of a warm early Mars is drawn by analogy with Earth – a thick carbon dioxide atmosphere sustaining a greenhouse effect in the face of a faint early Sun. Because of Mars' greater distance from the Sun compared to the Earth's – yielding only half the sunshine that Earth receives – a higher carbon dioxide pressure is required to sustain a certain temperature at any given epoch in the Sun's history. At least several atmospheres worth of carbon dioxide, or more, were required for Martian surface temperatures to be above the freezing point of water early in its history.

As shown in figure 15.7, a potentially serious flaw arises for a Martian greenhouse. For progressively smaller amounts of sunlight, one requires a higher carbon dioxide pressure to sustain a given atmospheric temperature. Carbon dioxide, like water, can form clouds, though much lower temperatures are required for a given amount of carbon dioxide to condense than for the same amount of water. It is possible to plot the carbon dioxide pressure for various temperatures at which carbon dioxide cloud formation will occur – the lower the temperature, the lower the pressure at which such clouds will form. An equivalent curve for water determines the altitude at which water clouds form in Earth's atmosphere for particular conditions on any given day. Figure 15.7 shows that CO_2 cloud formation occurs on Mars, for present solar luminosities, when the surface pressure of carbon dioxide exceeds 0.35 bar. This is many times less than the pressure of carbon dioxide needed to warm the surface to the water melting point, and is a direct consequence of Mars' greater distance from the Sun compared with Earth's. The carbon dioxide pressure needed to keep liquid water stable on *early* Mars is higher still

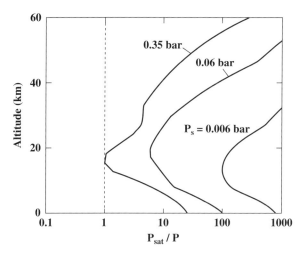

Figure 15.7. Greenhouse problem for Mars. The amount of carbon dioxide in the Martian atmosphere is shown for various values of the carbon dioxide surface pressure. Each profile is the ratio of the saturation pressure of carbon dioxide to the actual pressure as a function of altitude; where this ratio equals one (vertical dashed line), cloud formation occurs. The present-day carbon dioxide pressure near the surface (0.006 bar) leads to an atmosphere that does not produce carbon dioxide clouds (except near the poles). To get a Martian surface warm enough for liquid water, over 2 bars of pressure is needed at the surface (5 bars during the faint-sun epoch). However, for any surface pressure above 0.35 bar, according to the figure, cloud formation will occur. Reproduced from Kasting (1991) by permission of Academic Press.

(since the Sun was fainter), ensuring that CO_2 cloud formation must be considered in models of a warm early atmosphere.

The effect of carbon dioxide clouds on the early Martian greenhouse is not fully understood, because both scattering of sunlight and absorption of infrared energy (heat) may occur within such clouds. Most simple models suggest that the net effect of the clouds is to cool the atmosphere. This, in turn, requires higher carbon dioxide pressures to achieve a given surface temperature, which cannot be obtained because the gas simply condenses into thicker and thicker clouds, and eventually to carbon dioxide snow or rain.

Carbon dioxide cloud formation represents a potentially serious problem for the concept of an early Martian greenhouse. One plausible way to circumvent the problem is to posit other greenhouse gases that enhance the effect of the carbon dioxide. Water vapor is not a candidate, because the low temperatures at which the carbon dioxide cloud formation occurs are such as to keep the atmosphere extremely dry (water condenses out at much lower pressure, for a given temperature, than does carbon dioxide). What is required is a gas that condenses out at much higher pressure, and is a good infrared absorber. Methane (CH_4), ammonia (NH_3), and various compounds of sulfur have been proposed. The first two could plausibly be present only in the very earliest history of Mars, perhaps the first few million years, because they would quickly be broken apart by sunlight or surface reactions to form other species. The sulfur compounds might be more stable, but their effectiveness as greenhouse gases is limited. An alternative way out of the problem comes from the work of Francois Forget of Univ. Curie of Paris and Ray Pierrehumbert of the University of Chicago. They find that clouds made of large particles of CO_2 ice can actually warm the Martian surface. Predicting particle size in clouds is difficult, but it is at least possible that a greenhouse atmosphere on early Mars was supported by CO_2 clouds with this property.

The existence of an early Martian greenhouse to sustain warmer temperatures is a problem that has yet to be satisfactorily resolved, in contrast to the equivalent issue for Earth. Our planet's closer proximity to the Sun ensured that only modest amounts of carbon dioxide were required to sustain equable temperatures, not enough to condense out as carbon dioxide clouds and hence precipitate a potential crisis. The search continues for plausible additional infrared-absorbing gases and better understanding of CO_2 cloud properties, while the skeptics warn that early Mars may not have been warm for extended periods of time, or at all.

15.4.2 Abodes for Life on Early Mars

Where might life have begun and been sustained on early Mars? The answer to this question depends on whether conditions were very different from those at present. If Mars was much warmer, with a thick atmosphere, running or standing water at the surface would have provided similar environments to those on the early Earth for the formation of life. However, if standing or running water was transient or episodic, conditions may have been too unstable for prebiological chemical reactions to develop toward sustained, biochemical systems.

If life did form in lakes at the Martian surface, the advent of freezing conditions would not have meant the immediate end to life. NASA Ames scientist Chris McKay and others have studied lakes in Earth's Antarctic continent that are covered by a layer of ice year-round. They sustain photosynthesizing algae. The liquid region below the ice is maintained in large measure by the warming effects of sunlight, transmitted through the clear ice into the liquid water below. In this respect, the ice-covered lake is analogous to a greenhouse atmosphere. However, the lake liquid also is stabilized by the warming effect of freezing itself, which releases heat. (This is the converse process to the cooling of a drink by the melting of ice cubes.) Calculations and field observations in the Antarctic suggest that such lakes are stable for temperatures as low as 240 K, fully 33 K below the freezing point of water.

Another possible birthplace and abode of life is hydrothermal systems in the early Martian crust, regions where liquid water circulates in the rock and is warmed both by heat flowing from the interior and by the insulating effects of being underground. At Earth's mid-ocean ridges, hydrothermal systems are rich in the chemical and thermal energy needed to support an array of living organisms in the complete absence of sunlight. Such could have been the case on early Mars.

It is difficult to estimate how long either of these two types of ecosystems might have lasted on Mars. The groundwater hydrothermal systems are particularly problematic in this regard; we simply do not understand the details of Martian geologic history sufficiently well to predict where such systems could have been most long-lived.

With regard to the ice-covered lakes, McKay and colleagues argue that they could have been maintained for some 700 million years after mean annual surface temperatures fell below the freezing point on Mars. If, for argument's sake, Mars had a warm climate up through 3.5 billion years ago, then ice-covered lakes might have contained liquid water as late as 2.8 billion years ago. At this time on Earth, life was still solely in the form of single-celled prokaryotes, and the dramatic changes in our atmospheric composition wrought by such organisms (chapter 17) had yet to take place.

We thus expect that any life on Mars would have remained at the single-celled stage at the time of its extinction. Life might have been sustained to somewhat more recent times if the ice-covered lakes provided a bridge between episodes of warmer conditions, triggered by release of groundwater, but such conditions were likely short-lived and geographically isolated. The recent discoveries of bacteria living several kilometers beneath the surface of the Earth, in rocks that allow access to water and nutrients, hint at possible environments for life to exist on Mars today. Perhaps beneath the surface of Mars, warmed by heat sources related to recent volcanism, simple Martian biota carry on.

15.4.3 Searching for Evidence of Life, and the Early Climate

It is a daunting task to explore other worlds, and even more so to search for evidence of microscopic life-forms that might be rare or hidden in inaccessible places. The search for evidence of past Martian life would involve looking for physical evidence (for example, stromatolites in ancient sediments), chemical evidence (ratios of carbon or other isotopes in rocks that are unusual except in biological processes), and mineralogical evidence (types of minerals that are not normally found together, except when formed through the mediation of biological activity). The search for extant life requires techniques to stimulate and detect metabolic activity or (more difficult), to isolate, reproduce, and then study the genetic materials of living organisms. The more exotic a Martian organism is to Earth life, the more challenging is its detection. Such searches must be conducted in regions of Mars where life was most likely to have formed and been sustained, and these are invariably the most difficult and dangerous sites to reach. For cost reasons, the search will be conducted using robotic vehicles.

Although the discovery of evidence for past life on Mars would be a profoundly moving moment in human history, the odds are against such an event in the near future. However, equally important (but less dramatic), is the exploration of Mars to understand the history of its climate and geology. A vigorous campaign to do so will pay off regardless of whether evidence for life is ever found, because the nature and demise of putative warm conditions on early Mars remains one of the great mysteries in solar system exploration. To understand whether and why it happened is to gain the insight necessary to comprehend the remarkable stability of our own planet's climate over four billion years.

15.5 PUTTING A MARTIAN HISTORY TOGETHER

Although speculative, here is a possible interpretation of the spacecraft data from Mars: After formation of Mars and an early epoch of heavy cratering, a period began that was characterized by high carbon dioxide abundance, the possible presence of other greenhouse gases, warm temperatures, and liquid water. This may have lasted for half a billion years, and primitive life might have formed. Over time, the CO_2 in the atmosphere was progressively locked up in surface carbonates or lost by impacts, and atmospheric pressure and temperature decreased. As temperatures dropped below the freezing point, glacial erosion became more important than erosion by liquid water. Perhaps several subsequent warm epochs occurred as chance large impacts or volcanic activity broke the crust, releasing water and carbon dioxide back into the atmosphere and allowing liquid water on the surface. Eventually, however, these reprieves ended. The atmospheric carbon dioxide continued to be progressively locked up in the sediments, and water ice became trapped in polar caps and as permafrost, in a process that continues up to the present cold, dry state. Life, initially retaining a toehold in lakes capped by water ice, or in subsurface hydrothermal vents, eventually ran out of sustainable climates and became extinct or relegated to a few sites deep below the surface.

One possible geologic and climatologic history, compared to that of Earth, is shown in figure 15.8. The specificity of the figure, however, should not be construed to mean that we understand Martian history in detail. In fact, there remains an important controversy about whether early Martian climate was warm and equable over long periods of time, or just in brief episodes. Furthermore, the questions of how late in geologic time the episodes of water or debris flow in the outflow channels extended and the origin of such outbursts are unresolved; such events may be less relevant to the issue of life because they do not imply a global change of Martian climate. In one extreme view, Martian climate was *never* clement, and only impact- or volcanically induced outbursts of water broke the monotony of a dry, cold geologic history. The valley networks seemingly argue against such a view, but some geologists have argued that processes associated with groundwater release and flow could carve such valleys, even under conditions not too different from today's. We will not know for sure until Mars is thoroughly surveyed from orbit and on the surface over the coming decades (figure 15.9 and color plate VI).

15.6 IMPLICATIONS OF VENUSIAN AND MARTIAN HISTORY FOR LIFE ELSEWHERE

The search for evidence of life beyond our solar system is among the most daunting technological challenges imaginable. Direct evidence of life could come in the form of a radio beacon or even a visitation, but only from living forms advanced enough to do so and motivated enough to make contact. Absence of evidence of such contact is not evidence of absence of life-forms, by any means. An alternative approach, to examine neighboring systems for planets of the right size and in the right location for harboring life, will yield interesting results even in the case in which few or no such planets are found (the conclusion then being that Earth is a rare pearl).

The exploration of Mars and Venus, and their scientific results, provide a framework within which to estimate the regions around neighboring stars wherein habitable planets might be found. Habitable is defined by most planetary scientists as being capable of harboring liquid water.

Because evidence exists for liquid water on Mars in the distant past and isotopic data suggest an ocean on primordial Venus, one might argue that the range of habitability around stars like the Sun is from 0.7 to 1.5 AU. However, life must have time to develop and evolve; if we are interested in advanced life-forms we must look for planets with stable climates for liquid water over billions of years. Because the Sun is thought to be typical, most stars should slowly increase their luminosity over time, and hence the zone of "continuous" habitability around each star must be much narrower than the current Venus-Mars range.

We consider for the moment a star like the Sun, that is, of similar mass, composition, and luminosity history. The requirement of abundant liquid water early in a planet's history dictates that the outer edge of the habitable zone be inward of the current Martian orbit because, during the faint youth of the star, any planet at Mars' distance may have the same difficulty in sustaining a greenhouse as Mars had. Computations by Pennsylvania State scientist Kasting suggest that an early carbon dioxide greenhouse is readily sustainable within 1.15 AU of a Sun-like star; this could be extended outward for planets with other greenhouse gases or "warming" CO_2 clouds.

The inner edge of the habitable zone must be much closer to Earth than it is to Venus, because Venus suffered a moist greenhouse loss of water early in its history, and as a solar-type star heats up, planets progressively more distant than 0.7 AU will suffer the same fate. Computations by Kasting and others suggest that planets inward of 0.95 AU will suffer a moist greenhouse crisis for a luminosity of the central star equal to that of the Sun today. Hence, for a planet to sustain a liquid water surface over 4.5 billion years – the current age of the solar system and the length of time for sentient life to develop on Earth – it must be beyond about 0.95 AU from its Sun-like star.

The resulting zone of continuous habitability extends from 0.95 AU to 1.15 AU, that is, a width of 0.2 AU. What is the likelihood of finding a planet in another solar system orbiting at that distance from its parent star? In our own solar system, four rocky planets orbit the Sun between 0.4 and 1.5 AU. The mean spacing of the planets – four planets over 1.5 AU – is about 0.4 AU. If our system is typical, we can say that the likelihood of other planetary systems having a continuously habitable planet is just the width of the required zone divided by the typical (defined by our system) mean spacing, or $0.2/0.4 = 0.5$. This is a probability of 50% – quite high indeed!

Of course, other factors must come into play. Other systems will have planets whose sizes differ from those in

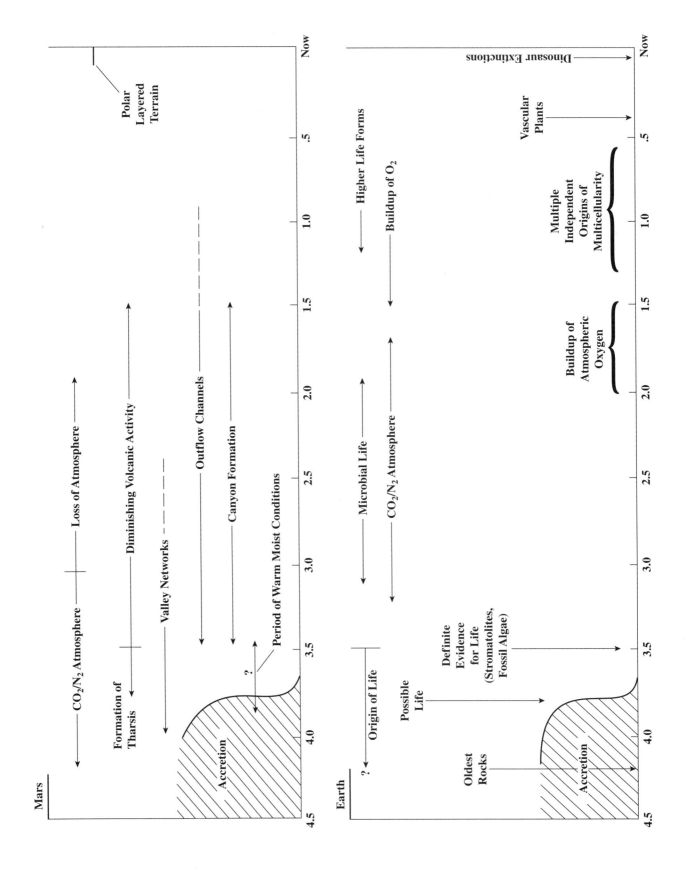

our solar system. Several stars are known to have planets comparable to the mass of Jupiter within the terrestrial planet zone (as defined by our solar system). A system in which a Mars-sized body occupied Earth's orbital position might not have hosted life because such a small planet would not possess the plate tectonic recycling of carbon dioxide needed to sustain an environment stable for liquid water. Conversely, a more massive planet possessing plate tectonics but occupying the orbit of Mars might develop a sustainable greenhouse atmosphere, possibly a bit later than did our Earth as its parent sun increased in luminosity, and then could sustain a clement climate over billions of years.

The mass of the central star itself also determines habitability. More massive stars are more luminous and hence habitable planets must be in proportionately larger orbits than Earth's orbit around the Sun. However, more important, the higher luminosity comes from a more rapid rate of hydrogen fusion, so that massive stars are shorter-lived (chapter 4) and hence provide their planets a much smaller time for biological evolution. Happily, the most abundant stars in the galaxy are the least luminous and longest-lived: the M-dwarfs. They, however, harbor another difficulty: For a planet to gain sufficient warmth from an M-dwarf to be habitable requires that it be as close to its star as Mercury is from the Sun. In that tight orbit, gravitational tugging will force the planet toward a state of *synchronous rotation* where one side faces toward the star – just as our Moon keeps one face toward the Earth. The resulting climate will be vastly different from Earth's, and perhaps not conducive to stable liquid water.

Returning to our own solar system, human technology might expand the habitable zone in the not-too-distant future. A number of futurists have proposed seeding the Martian atmosphere with efficient greenhouse gases such as methane and chlorofluorocarbons, in an effort to warm the surface, release water, and generate conditions that are more Earth-like. Although beyond the reach of current space transportation capability, such a *terraforming* of Mars might be possible for a future generation, which will have to weigh the advantages against the potential ethical dilemmas associated with transforming a vast natural environment.

15.7 THE FINITE LIFE OF OUR BIOSPHERE

The evolution of our Sun has one more consequence for life on Earth. From now to the end of its stable hydrogen-fusing stage, the Sun will continue to increase in luminosity. As it does so, the climate of the Earth will edge closer to the point at which a moist greenhouse is initiated and rapid loss of the Earth's water ensues, as apparently occurred early in Venus' history. A natural delaying tactic is the weathering feedback process described in chapter 14 wherein, as the brighter Sun warms Earth, more rainfall and more erosion will occur, and hence the carbon dioxide budget of our atmosphere will decrease. However, a point will come when rising temperatures cannot be buffered by the decreasing amounts of atmospheric carbon dioxide, and rapid loss of Earth's oceans to space will begin.

Models by Caltech and Pennsylvania State scientists of the Sun's luminosity history and the response of Earth's atmosphere suggest that this crisis will be reached in 1 billion to 2 billion years from now. At that point, if the biosphere has not collapsed already from decreasing amounts of atmospheric carbon dioxide, the lack of liquid water will finally kill off all living organisms. On the other hand Mars will enjoy more clement conditions, if enough water and carbon dioxide are stored in the crust to be partially liberated into a thicker atmosphere by the brighter sun.

Life began on Earth some 3.8 billion to 4 billion years ago, and complex eukaryotic cells appear in the fossil record from 2 billion years ago. Therefore, we are more than halfway through the time period during which life, even complex life, can flourish on Earth. Our time here is not forever. After the brightening Sun drives water, and hence life, from Earth, it will continue to shine by hydrogen fusion for another 2 billion to 4 billion years. For those last several billion years of the Sun's history, Earth's surface might hold a fossil record of its long springtime of clement conditions, during which it teemed with living organisms that eventually looked upward to contemplate the stars.

15.8 QUESTIONS

a. How and where would you search for evidence of past, and present, life on Mars?

Figure 15.8. Comparative geologic and climatologic histories of Mars and Earth, in a view taken by NASA scientists Chris McKay and Carol Stoker. "Formation of Tharsis" is the uplift of a broad Martian highland containing several giant volcanoes. The "Polar Layered Terrain" is a set of alternating layers of dust and ice seen in the polar regions, thought to have formed as the tilt of Mars' pole changed on a frequency of millions of years and altered the pattern of dust and ice (water or carbon dioxide) deposition. From McKay and Stoker (1989).

b. Can you think of any refugia for carbon-based life on present-day Venus?

15.9 READINGS

Amy, P.S., and Waldeman, D.L. (eds.). 1997 *The Microbiology of the Terrestrial Deep Subsurface.* CRC Press, Boca Raton, FL.

Caldeira, K., and Kasting, J.F. 1992. The life span of the biosphere revisited. *Nature* 360, 721–723.

Carr, M.H. 1984. Mars. In *The Geology of the Terrestrial Planets,* M.H. Carr (ed.). NASA SP-469, Washington, DC, pp. 207–263.

Forget, F., and Pierre Humbert, R.T. 1997. Warming early Mars with carbon dioxide clouds that scatter infrared radiation. *Science* 278, 1273–1276.

Kargel, J.S., Baker, V.R., Begé, J.E., Lockwood, J.F., Péwé, T.L., Shaw, J., and Strom, R.G. 1995. Evidence of ancient continental glaciation in the Martian northern plains. *Journal of Geophysical Research* 100, 5351–5368.

Kasting, J.F. 1988. Runaway and moist greenhouse atmospheres and the evolution of Earth and Venus. *Icarus* 74, 472–494.

Kasting, J.F. 1991. CO_2 condensation and the climate of early Mars. *Icarus* 94, 1–13.

Kasting, J.F. 1997. Planetary atmosphere evolution: Do other habitable planets exist and can we detect them? In *The Search for Extra-solar Terrestrial Planets. Techniques and Technology* (M. Shull, H. Thronsoan and A. Stern, eds.). Kluwer, Dordrecht, pp. 3–24.

Kasting, J.F., Whitmire, D.P., and Reynolds, R.T. 1993. Habitable zones around main sequence stars. *Icarus* 101, 108–128.

McKay, C.P., and Stoker, C.P. 1989. The early environment and its evolution on Mars. *Reviews of Geophysics* 27, 189–214.

McKay, C.P., and David, W.L. 1991. Duration of liquid water habitats on early Mars. *Icarus* 90, 214–221.

Moroz, V.I., and Mukhin, L.M. 1977. Early evolutionary stages in the atmosphere and climate of the terrestrial planets. *Kosmicheskie Issledovaniya* 15, 901–922.

Phillips, R.J., and Hansen, V.L. 1994. Tectonic and magmatic evolution of Venus. *Annual Review of Earth and Planetary Sciences* 22, 597–654.

Rampino, M.R., and Caldeira, K. 1994. The Goldilocks problem: Climatic evolution and long-term habitability of terrestrial planets. *Annual Reviews of Earth and Planetary Sciences* 32, 83–114.

Sackmann, I-J., Boothroyd, A.I., and Kramer, K.E. 1993. Our Sun. III Present and future. *Astrophys. J.* 418 457–468.

Sagan, C., and Mullen, G. 1972. Earth and Mars: Evolution of atmospheres and surface temperatures. *Science* 177 52–56.

Figure 15.9. The Mars of today, captured by the Mars *Pathfinder* lander after its successful landing on July 4, 1997, shows hints of earlier times when water rushed across parts of the surface. It also shows the tracks left by the Sojourner rover in its travels across the landing site, the first mobile exploration of Mars. The *Pathfinder* imager was developed for NASA by Peter Smith (University of Arizona) and colleagues. Image courtesy NASA/Jet Propulsion Laboratory and Mark Lemmon (University of Arizona).

16 EARTH IN TRANSITION: FROM THE ARCHEAN TO THE PROTEROZOIC

16.1 INTRODUCTION

The beginning of the Proterozoic eon is set formally by geologists at 2.5 billion years before present. However, the transition between the Archean and the Proterozoic is not a sharp one. From about 3.2 billion to 2.5 billion years ago, rocks with a modern granitic composition made a widespread appearance on Earth. Prior to this time, rocks making up the Archean continents had a composition different from modern granites in several important respects. Beginning around 3.2 billion years ago in what is now Africa, and extending to 2.6 billion years ago on the Canadian shield, large quantities of modern-type granites were produced. We can collect these rocks today and date them by use of radioisotopes. How did the original Archean continents form? Why was there a transition in chemical composition of the rocks roughly halfway through the Archean? What might Earth have been like today if this eruption of new rock types had not occurred? As we see, the transformation wrought on Earth's primitive continents may have been an inevitable consequence of their increasing coverage of Earth's surface.

What might have been inevitable on Earth was apparently difficult or impossible on the other terrestrial planets. No evidence for large granitic masses exists on any other planet. Venus bears two crustal masses that resemble continents, but the details of their geology suggest that they are more similar to primitive Archean continents than to our modern ones and, even then, the connection is a weak one. Venusian geology, taking place as it did on a planet similar in size and composition to Earth, might teach us about the conditions under which Earth's tectonic regime could *not* be achieved.

The formation of continental masses, standing above the level of the seas on an otherwise watery world, had a profound influence on the subsequent history of Earth. The weathering of continental granites provides the essential first step in the transport of atmospheric carbon dioxide into the oceans and sequestration as carbonates. The continents as buoyant regions of the crust probably largely determined the pattern of plate tectonics. They modulate the climate and interior heat flow through cycles of merging into single supercontinents and breaking up into dispersed land areas. Unlike the oceanic crust, which is destroyed and recreated on timescales of a few percent of Earth's history, the continents preserve an ancient record of the geologic and biological history of Earth. Finally, the continents provided a frontier for complex life some half-billion years ago, on which the nature of survival differed drastically from that in the sea.

It is in the Archean and its transition to the Proterozoic that we see Earth, once and for all, diverge in its evolution from that of its sister planets and become the planet with which we are familiar. Understanding how this happened is the challenging subject of the present chapter.

16.2 ABUNDANCES OF THE ELEMENTS IN TERRESTRIAL ROCKS

Table 16.1 summarizes the amounts of the most abundant elements in typical basaltic and granitic rocks, compared to numbers for the mantle of Earth, and for the most primitive class of meteorites (carbonaceous chondrites). Granitic-type composition dominates continental crust, whereas basalt is typical of oceanic crust. Abundances of the major elements vary quite a bit from rock to rock, particularly since continental rock is by no means purely granitic. Hence the numbers in the table are illustrative; what is important are the general trends. The mantle composition is the least certain, because it depends largely on volcanic rocks whose formation in the mantle and subsequent chemical alteration are not precisely known.

Table 16.1 Chemical Composition of Rocks[a]

Element	Chondritic meteorites	Earth's mantle	Basalt	Granite
O	32.3	43.5	44.5	46.9
Fe	28.8	6.5	9.6	2.9
Si	16.3	21.1	23.6	32.2
Mg	12.3	22.5	2.5	0.7
Al	1.4	1.9	7.9	7.7
Ca	1.3	2.2	7.2	1.9
Na	0.6	0.5	1.9	2.9
K	0.1	0.02	0.1	3.2
Other	5.9	1.7	2.7	1.6

[a]Composition in percent by weight, e.g., 33% oxygen means one-third of the weight of the rock is oxygen. Data from Broecker (1985).

Some interesting systematics arise from the table, beyond the dominance of oxygen, the most abundant element in the rocks. Iron is very abundant in the chondrites but much less so in the three types of Earth rocks. This is understood to be the consequence of the formation of a largely iron core very early in Earth's history (chapter 11). Silicon is commensurately more abundant as iron declines. Magnesium, which is very abundant in mantle rock, declines drastically in both basalts and granites. Other elements are more abundant in the crustal rocks than in the mantle or the chondrites. These include aluminum, sodium, and potassium. Finally, between the crustal rocks, some further differences emerge: Basalts contain more iron, magnesium, and calcium than do the granites, whereas sodium and potassium are strongly enhanced in the latter.

If we regard chondrites as roughly representative of the original chemical mix from which Earth formed, it is relatively straightforward to see how the mantle is a residuum of the removal of iron and elements that tend to follow iron. The pattern of the crust, however, seems less certain and, in particular, basalts and granites seem to have followed different paths in their formation. The discussion in chapters 9 and 11 suggest that the crustal rocks ought to be derived in some way from the mantle, and the major elements should provide a guide to that process. To gain some insight into how this might have happened, we must discuss mineral structures and the concept of partial melting of rocks.

In addition to elements that are abundant in crustal rocks, certain trace elements are important indicators of the origin and history of continental rocks. Chief among these are the *rare-earth elements*, or rare earths, which do not dissolve very well in water. They tend, therefore, to stay with the rocks as the rocks are moved around in rivers as sediments. Furthermore, their abundances are an important distinguishing feature between Archean and post-Archean continental rocks. As can be seen by consulting the periodic table (figure 2.6), the rare earths all occupy a single column in the table, but move upward in atomic number by filling an interior (fourth) series of electronic energy levels while the outermost (fifth) is complete (see chapter 2). In consequence, these elements all share certain common chemical properties of diagnostic value in determining the origin and history of continental crust.

16.3 MINERAL STRUCTURE

The structure of minerals represents a scientific discipline in and of itself, and justice cannot be done in the short summary here. Like all chemical substances, the building blocks of minerals are the chemical elements, and the properties of minerals are determined by the types of elements present and how they are bonded together. The abundant elements in minerals are joined by ionic bonds, where electrons are actually transferred from a donator to an acceptor element. For example, in sodium chloride, NaCl, each sodium atom donates an electron to chlorine, and the two types of elements arrange themselves in a regular lattice structure that constitutes the crystalline structure observable on a macroscopic scale. Having donated an electron, the positively charged sodium (Na^+) is packed in between the negatively charged chlorine (Cl^-).

Most significant about ionic bonding is the donation of the electron, which produces ions of very different sizes. As noted in chapter 2, the sizes of the elements across a given row of the periodic table vary slowly, determined by the presence of the electrons. Removal of an electron in ionic bonding greatly decreases the size of the resulting positive ion, whereas the accepting element, becoming a negative ion, increases in size. The *ionic radius* of an element refers to its size in a certain ionic state, that is, having donated or accepted an electron. The *atomic radius* (that is, of the neutral element) of sodium is almost twice that of chlorine, 1.86 Angstroms (1.86×10^{-10} cm) versus 0.99 Angstroms. However, Na^+ has an ionic radius of 0.98 Angstroms, whereas Cl^- is 1.81 Angstroms. Thus, the size situation essentially is reversed in the ionic bonding to form sodium chloride.

The stability of a particular chemical compound, or mineral, lies in part in its structure, which in turn depends on the relative sizes (and hence ionic radii) of the elements. Mineral structures tend to follow a pattern where several negatively charged ions (*anions*) surround a positively

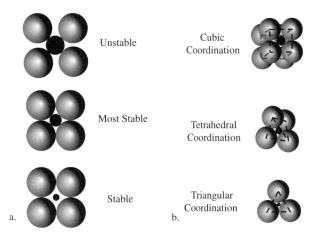

Figure 16.1. (Left) Fit of small, medium, and large cations (black) within mineral structures. (Right) Some examples of typical arrangements or *coordinations* of anions in minerals. Adapted from Press and Siever (1978) and Broecker (1985).

charged ion (*cation*). Oxygen is a very common anion in minerals. In *calcite*, three oxygens surround one carbon (in addition to the presence of the element calcium); in olivine and other minerals, four anionic oxygens (O^-) surround one silicon (Si^{++++}) (in addition to two magnesium or iron cations, which also are present). The ionic radius of the silicon cation is roughly one-fourth that of an oxygen anion; hence, the *tetrahedral coordination* could be predicted essentially from size alone.

The enclosure of an element with a small atomic radius by several with large radii produces a well-packed, closed structure (figure 16.1). However, not all cations and ions fit together. A large cation tends to push the anions outward, allowing a separation between them that is not the energetically most favorable situation. A cation that is too small allows the anions to just touch, but does not interact strongly enough with the anions to stabilize the configuration. Therefore, for any given number and type of anion, cations of particular size *and hence elements of particular ionic radii* optimally stabilize the crystal structure.

Table 11.1 in chapter 11 lists the ionic radii of the elements of table 16.1. The sizes of elements in their cationic form determine their ability to be accommodated in major minerals. For example, aluminum as a cation is similar in size to silicon, and readily substitutes. Magnesium and iron are similar in their ionic radii, and both occur in silicate minerals such as olivine. However, calcium, potassium, and sodium have large ionic radii and cannot substitute for silicon in a four-oxygen structure; they require a different coordination with a larger number of oxygen atoms.

In the structural compatibilities and mismatches of various cations lies much of the reason for the chemical differences between mantle and crustal rocks, and between oceanic and continental crust. The process leading to such differences is the melting of rock beneath the surface of Earth.

16.4 PARTIAL MELTING AND THE FORMATION OF BASALTS

The mantle of Earth is a solid with plastic properties, oozing slowly in a pattern of upwelling and downwelling motions that remove the internal heat of the planet. The solid nature of the mantle has been well established by seismic data and studies of mantle-composition rocks subjected to high pressures in the laboratory (chapter 11). However, volcanism both on continents and at midocean ridges amply demonstrates that some mantle material is being melted virtually all of the time, and this melt rises to form a part of the crust.

Clues to the process by which mantle melting occurs come from laboratory studies of common minerals, combined with some understanding of thermodynamics. On the surface of Earth, we are used to the melting of any given material being characterized by a single parameter, namely, temperature. Water ice melts at 273 K (32°F); common basalts melt at around 1,500 K. Melting, or passage into the liquid state from the solid, is a function also of composition of a mineral and, importantly, pressure. For most materials the liquid form is less dense, or more voluminous, than the solid; that is, the solid sinks in the liquid. Higher pressures compress materials, and therefore favor the solid phase. (The exception is water, for which ice floats on the liquid, and hence increased pressure *lowers* the melting point.)

Imagine now being present in a part of the mantle of Earth that is rising toward the surface, carrying away interior heat. As this part of the mantle rises, temperature drops from high internal values toward the surface value. This brings the mantle material further away from its threshold of melting. However, the pressure also drops, and this tends to push the mantle material toward its melting point. The run of temperature with pressure in the mantle, based on computer models and heat-flow sampling of the surface, is such that the release of pressure is the dominant effect (figure 16.2). That is, the rise of solid mantle material toward lower pressures allows some of the material to melt, in a process called *pressure-release partial melting*.

"Some" is the operative word here. The mantle rock consists of an assemblage of major and minor elements, located in various crystalline structures dominated by oxygen as anions and silicon, magnesium, and iron as cations. Melting does not occur wholesale in such an amalgam; rather, certain combinations of elements tend to

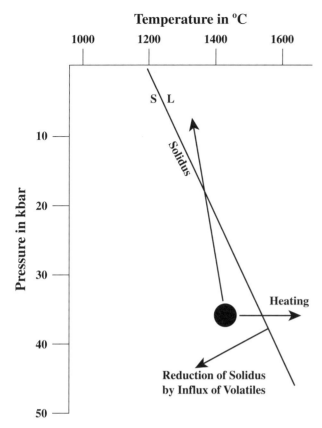

Figure 16.2. Schematic of how melting occurs in the mantle of Earth. The diagonal line is the boundary, or *solidus*, between having all solid (below and to the left) or solid with some liquid (partial melting); it is based on the chemical properties of the mantle rock. Imagine a piece of the mantle at a temperature and pressure given by the circle. Simply heating the sample will slide it across the solidus to the partial melt zone. However, more typical is that mantle material, moving upward, encounters decreasing pressure and temperature shown by the nearly vertical arrow. Although the temperature decreases along the curve, the sample crosses the solidus and melting takes place. The arrow pointing down and to the left shows the direction in which the solidus slides as more water is added to the mantle; melting is made easier. From Rogers (1993).

preferentially move into the melt. This property of *partial melting* is not unique to melting rocks; water mixed with other materials (e.g., ammonia or organic solvents such as commercial antifreeze) undergo partial melting at a temperature well below 273 K.

What determines which elements move preferentially into the melt? One important determinant is discussed above: the size or ionic radius of the cations. Oversized cations (potassium, calcium, sodium) that have trouble fitting into the four-coordinate silicon-oxygen structure are energetically favored to be preferentially in the melt. This is not an either/or proposition; some fraction of the larger cations will remain in the solid, retaining their eight or larger coordination with oxygen, but overall they tend to concentrate in the melt.

Silicon and oxygen are so abundant that they remain the dominant constituents in both the melt and the remaining solid. Magnesium tends to reside preferentially in minerals with higher melting points, and hence favors the solid phase as the melt progresses. Iron and aluminum have a slight preference for the melt. Reference to table 11.1 in chapter 11 shows that these cannot be size effects, but are related instead to other properties of the bonding mechanisms between the cationic elements and the anionic oxygen.

As melt forms, being just slightly less dense than the solid, it moves upward more rapidly toward the surface. In effect, the lower-density melt is buoyant in the surrounding solid (just as a hot-air balloon is buoyant in the surrounding cooler air). Thus, unlike a chemistry experiment in which a material is melted in the flask and solid and liquid phases continue to interact chemically, the mantle melt leaves its solid residuum behind and moves upward through "fresh" mantle rock, some of which also melts. Pressure in the cracks through which the melt moves helps force it upward; this effect is particularly important at depths where the melt is just about the same density as the surrounding rock. By the time the melt reaches Earth's surface, it occupies a significant fraction of the total volume of the rock through which it moves.

Most of the mantle melt reaches the surface at midocean ridges where it oozes out, forming new oceanic crust. The composition of the new material is basaltic, corresponding roughly to that given in table 16.1. Sodium, potassium, iron, aluminum, and calcium are enhanced and magnesium is depleted in this melted derivative of the mantle. Not all basalts erupt at midocean ridges. Some come to the surface at isolated *hot spots* around the globe, such as the Hawaiian islands. Much of this hot-spot basalt appears to have been melted at somewhat deeper levels than the midocean-ridge basalts, with consequent interesting differences in element abundance.

The formation of basaltic crust represents a second stage of the chemical differentiation of Earth, the first stage being the primordial separation and fallout of iron into a core, with loss of some fraction of additional elements that tend to combine with iron, such as sulfur and oxygen. The partial melting of rising mantle material as basalts is probably a process common to Earth, Mars, and Venus. What differs about Earth is that basalt is not the sole crustal material. Most of the continental crust is not basaltic in composition, and its composition is not derivable from the simple partial melting process considered here. Instead, the origin of continental rocks appears to be intimately tied up with the special feature of Earth as a planet, namely its abundance of liquid water.

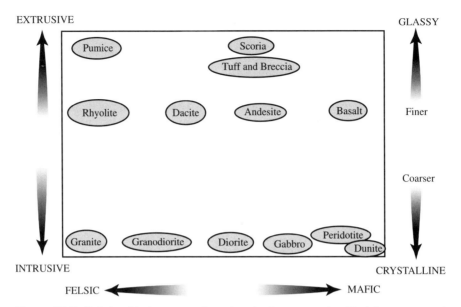

Figure 16.3. Relationship between selected major igneous rocks. Rock types are positioned according to the amount of silica they contain, and whether they are plutonic (intrusive) or volcanic (extrusive). Plutonic rocks do not breach the surface during eruptions, but cool and solidify buried in the crust. Because these rocks cool slowly, they are composed of large crystals. Volcanic rocks, which erupt to the surface, are cooled suddenly and have glassy textures, or very fine crystals.

16.5 FORMATION OF ANDESITES AND GRANITES

16.5.1 Rock Relationships

To gain insight into the formation of terrestrial rocks other than basalts requires understanding the chemical relationships between the major rock classes. Figure 16.3 illustrates these relationships, for the major igneous rocks only; recall from chapter 8 that metamorphic and sedimentary rocks begin their existence as igneous rocks. The primary distinguishing feature among the igneous rocks is their silica (or SiO_2) content. Felsic rocks (richer in SiO_2) are light colored, relatively low density; mafic rocks are poor in SiO_2, have proportionately more iron and magnesium, and are therefore higher density and darker color. There is a progression of rocks from felsic to mafic. This progression occurs along two tracks, corresponding to whether the rock erupted from a volcano, and hence is *extrusive*, or instead, the melt stopped rising within the relatively low-density continental crust (hence was no longer buoyant), and cooled to solidification as an *intrusive* rock.

The major extrusive/intrusive equivalent rock types, in progression of chemical composition from mafic to felsic, are basalt/gabbro, andesite/diorite, rhyolite/granite. The more common felsic rock class is the intrusive granites; the more common extrusive mafics are the basalts. In some of what follows, we rather loosely describe oceanic crust as basaltic, continental crust as granitic, and crust produced at continental margins above descending slabs as andesitic. Mantle rock is even more mafic than basalt; an example of this kind of rock is called *peridotite*.

16.5.2 Seismic Waves and Composition

As a brief aside, the seismic P-wave velocity (chapter 11) changes smoothly with composition; granites have a P-wave velocity of 6 km/s, gabbro and basalts 7 km/s, and peridotite 8 km/s. The seismic-wave explorations described in chapter 11 for Earth show that the oceanic crust is basaltic (along with some gabbro) with no granites, and the upper mantle is consistent with an ultramafic composition like that of peridotite. Interestingly, the thick continental crust is not entirely granitic; the P-wave velocities suggest that some gabbro is present in the lowermost continental crust. In this dual nature of continental crust lies further clues to continent formation.

16.5.3 Role of Water in Partial Melting

As table 16.1 shows, a granite is made from a basalt by reducing the iron and magnesium contents while increasing the sodium and potassium abundances. Andesites are, compositionally, an intermediate step. The challenge lies in identifying the geologic environment in which large amounts of such chemical alterations can occur. Simply cycling the basalt back into the mantle is not sufficient to produce andesites and granites; the melting of dry basalt occurs at a fairly high temperature, thus fairly deep, in such a fashion that the liquid retains a composition quite similar to that of the original rock. (Recall that, because basalt is the solidified product of partial melting of mantle rock, it will melt under different conditions than the original mantle rock.)

What is required to make more felsic rocks is the addition of a material that will alter the melting relationship of the basalt, and water is an excellent candidate. Figure 16.2 shows that, in the presence of sufficient amounts of water, the melting point of the basalt drops dramatically. Further, the melting is partial, with larger ions again partitioning preferentially into the melt. The resulting liquid has an andesitic composition, partway between basalt and granite.

The physical environment in which water plays a role is supplied by plate tectonics (figure 16.4). The basalt at the midocean ridges solidifies as it approaches the surface. Because the ridges are underwater, the basalts react with the water, which becomes incorporated in the crystal structure of the rocks, *hydrating* the minerals. The process can be very efficient because the midocean ridges are filled with cracks, through which water circulates in an intricate network of hydrothermal systems. Thus, much like the cells in our body receive nourishment through an intricate network of capillaries, oceanic basalts enjoy extensive and intimate contact with water.

As oceanic crust moves away from midocean ridges, it cools and thus becomes denser than the rock beneath. Sitting on less dense rock is an unstable situation, and the crust eventually founders and sinks into the mantle below. This sinking, or subduction, can take place at the boundary of a continent or purely within an ocean basin. Basalt that was not fully hydrated at the ridges now has a second chance as it is warmed up during its descent below the continental edge. However, temperatures continue to climb as the subducting *slab* descends, and eventually become too high for water to remain stably bound in the basaltic rock. The water becomes unbound from the minerals and, being buoyant, rises upward through the boundaries between mineral grains.

The release of water has two profound effects on the subducting slab and the adjoining mantle. First, the slab rock becomes denser, aiding subduction. Second, the water rises above the slab into the wedge of mantle above; as it contacts basaltic and mantle grains, the melting point of the grains plummets, and partial melting of the slab and mantle begins at temperatures and pressures much lower than that which originally formed the basalts. The partial-melt products are less dense than the mantle and hence rise, though more slowly than the water because the molten rock is closer in density to the mantle itself. The partial melt, andesitic in composition, erupts onto the surface in the form of volcanic lavas; some may come to rest near the surface as igneous intrusions of diorites.

Andesitic volcanism, or *arc volcanism*, is common along the margins of active subduction zones, for example, along the Andes, Japan, and the northwestern United States. Some of these margins are actually disconnected from the adjoining continent (as with Japan). Aside from an enrichment in sodium and potassium, the andesitic or dioritic rock is enriched in the large ions uranium and thorium. Because these elements, along with potassium, have relatively abundant, long-lived radioactive isotopes, the effect of the low-temperature melting of the mantle at plate margins is to concentrate the heat-producing elements in the andesitic crust.

The reader may notice an interesting correspondence between the cycling of water into the mantle and that of carbon, described in chapter 14. Indeed, in both cases, plate tectonic subduction provides a mechanism for recycling volatiles trapped chemically in oceanic sediments (carbonates) or basaltic crust (water) back onto the surface. In the case of carbon, the recycling is key to sustaining a warming atmospheric greenhouse; in the case of water, its release from rock provides the key step in low-temperature partial melting of basalts and mantle to form andesites and diorites.

16.5.4 The Puzzle of Granite Formation

The upper crust of the continents is not made predominantly of andesites or diorites. The granite of which it is composed is even further in composition from basalt than are the andesites, and hence must represent an additional cycle of differentiation. However, no obvious simple process exists by which such differentiation might occur. Although it generally is acknowledged that arc (subduction zone) volcanism – the production of andesites – is a principal means by which mantle material is converted to a buoyant state and accreted onto continents, the eventual conversion of that material to more grantic composition is a major puzzle.

The currently favored picture for granite formation, shown in figure 16.4, is that partial melting within the continental crust itself produces felsic rocks, such as granites and the granodiorites (intermediate between granites and diorites), which rise to the surface in the form of large

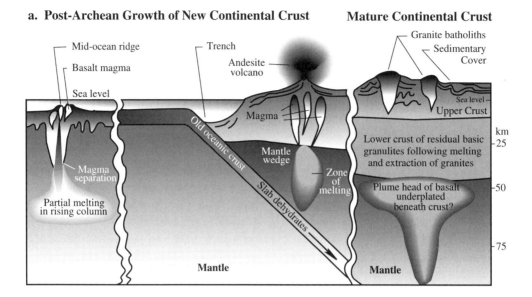

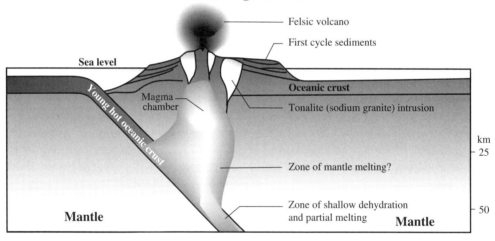

Figure 16.4. View of the formation of continental crust: (a) today and (b) in the Archean. The size of the continent is truncated in the top panel; only the edge and a portion of the interior continental shield are shown. At the continental margin, formation of andesite is fairly well understood. In the interior, it is suggested that the formation of granites results from melting of more mafic rocks in the lower continental crust, with the granites rising to the top as batholiths. In the bottom panel, it is speculated that Archean plate tectonics involve rapid recycling of ocean crust and small continental minishields which are the products of partial melting of subducting slabs. However, because hydration of the basalt (that is, suffusion with water) is not complete, and Archean temperatures were quite high even at shallow depths, the melt product is not andesite but a suite of rocks including unusual granites rich in sodium (*tonalites*). From Taylor and McLennan (1995).

intrusive masses called *batholiths*, leaving behind a residue in the lower continental crust. However, samples of the lower continental crust are not consistent with being simply a residue of such melting. These *xenoliths* contain trace elements, such as the rare-earths, whose composition is altered by the formation of granite. The abundance pattern seen in these elements in granites versus xenoliths requires that granite formation be more complicated than simple melting and differentiation of the lower continental crust.

One way to explain the pattern is to invoke basaltic magmas beneath the continental crust that rise and plate (or essentially stick to) the bottom of the continental crust. These hot plumes might trigger episodes of deep continental melting and consequent differentiation, while contaminating the lowermost continental crust so as to produce the observed composition of xenoliths. The apparent presence of gabbro in lower continental rock, based on seismic data, is consistent with this idea. Such a model must be tentative at least, because xenoliths may or may not be representative of lower continental rock. More widespread samples of metamorphic rocks called *granulites*, which appear to have formed under high pressure and temperature, may have originated in the lower crust as well; however, there is no consensus on whether these are truly rocks from the lower continental crust.

The difficulty in understanding formation of granitic rock is in large measure a result of the very complex nature of the continents. Unlike the ocean floor, with its simple geology that is erased on 100-million-year timescales, the continents are cumulates of geologic processes stretching over billions of years. To understand how this process began requires examining the nature of rocks from the Archean time of Earth's history.

16.6 FORMATION OF PROTOCONTINENTS IN THE ARCHEAN

There are significant differences between Archean rocks and younger continental materials. In Archean igneous and metamorphic rocks, sodium is more abundant than potassium, in contrast to modern granites, which are potassium-rich. Archean volcanic rocks have iron and magnesium content much closer to the mantle values than do Proterozoic and more recent volcanic rocks. They are much less abundant in the rare-earth elements and large-ion elements such as potassium, rubidium, uranium, and thorium than are modern andesitic volcanics. A characteristic of modern continental volcanics and sediments derived therefrom is a depletion of the element europium relative to the other rare earths; no such depletion is seen in comparable rocks from the Archean. Furthermore, Archean sediments show a large degree of variability in their rare-earth element abundances, unlike the more uniform sediments of later times. The general impression of Archean rocks, then, is of a more chaotic, less regular production process that yields a less extreme fractionation pattern than in the granites of today, and more variability. Further, the continental environment upon which these rocks eroded apparently was less well-developed in terms of extensive stream transport and sorting of sediments: Archean sediments seem poorly processed by water, usually being angular and not well sorted by size compared to modern sediments.

Although several models are offered for the formation of continental masses during the Archean, one particular story stands out that links Archean-type granites to high heat flow from Earth and small continent sizes. Because more accretional heat and undecayed radionuclides were present in the Archean Earth than is the case today, the average heat flow was perhaps double the present-day value. This is an interesting number that corresponds to the rate of heat flowing today from the region of Iceland. Iceland is a minicontinent atop the mid-Atlantic ridge, built up from basaltic magmas and lacking a granitic core. Although its origin in a special event of a hot plume intersecting a mid-ocean ridge is not directly analogous to Archean continental growth, it is a reminder that continents need not (and cannot) start out as granites.

This possible story, then, for the origin of continents goes this way: The higher heat flow of Earth may have organized the structure of the earliest Archean crust into small plates bounded by hot spots through which basaltic magmas rose. These basalts, building the crust up in selected places, perhaps provided the very first protocontinents poking above the ocean surface. Melting within the basaltic cores of these continents, facilitated by the hot spots and generally hotter nature of the Archean crust, allowed more felsic materials to build up while the residue sank into the lower part of the lithosphere (the rigid, nonconvecting part of the mantle), or flaked off the bottom of the plate into the asthenosphere (flake tectonics).

Eventually, the conditions evolved to the point at which subduction of crust could begin. How and why this happened remains unclear; the contrast in stiffness (viscosity) of lithosphere and asthenosphere may have become appropriate, or perhaps the presence of growing protocontinents on plates forced adjoining plates to sink underneath. The subducting basaltic slab would have been warmer than present-day subducting slabs at comparable depths. For this reason, melting of the subducting Archean slabs could have occurred at much shallower depths, and prior to complete dehydration of the slab. Such premature melting is actually inferred to take place today under the southern Andes mountains of South America, where the subduction zone is so close to the East Pacific Rise (a midocean ridge) that the subducting oceanic crust has not cooled as much as is typical elsewhere around the present Earth (see figure 9.9 of chapter 9).

Consideration of the composition of the resulting partial melt suggests that it could be sodium-rich granites and granodiorites consistent with Archean composition. These melts would intrude into the basaltic protocontinents, cool, and solidify. As the Archean continents grew in this way, erosion would eventually expose the granites and form sediments. The small size of the Archean continents limited the length and size of river systems, accounting for the poor sorting and shaping of sediments. As sodium-rich granites became more massive than the hot-spot basalts, and hot-spot injection of basalts became proportionately less important, melting within the continental mass lessened. Its effect on the sodium-rich granites became minor, in contrast to the modern continental granites whose origin in intracontinent melting leads to strong fractionation of elements.

In this picture, summarized in the bottom panel of figure 16.4, the Archean crust is derived from two sources: basalts possibly originating in numerous crustal hot spots, and sodium-rich granites and granodiorites whose formation is made possible by subduction of warm basaltic crust under marginally buoyant protocontinents. Proponents of this view argue that the rare-earth-element patterns in Archean rock are consistent with this model, and it has the advantage of at least one location on the present Earth where an analogous subduction episode is taking place. Other proposals for the origin of sodium-rich granites have been made, for example, through direct melting of mantle material that was previously altered through some earlier melting episode. What is of primary importance here is the notion that continents in effect have bootstrapped their way into the crust. An initial episode of the formation of basaltic protocontinents above hot spots provides the seed for further chemical differentiation within or beneath such continents. Once such differentiation produced felsic, low-density crust, the die was cast for the permanent existence of buoyant continents rising above a denser, surrounding basaltic crust. It is natural to ask whether the evidence for the earliest type of basaltic protocontinents might be found anywhere. On Earth, the remarkable transition in continental growth at the Archean-Proterozoic boundary erased that evidence forever. Beyond the modern terrestrial analog of Iceland lies the potentially remarkable record in the highlands of Venus, to which we return in section 16.9.

16.7 THE ARCHEAN-PROTEROZOIC TRANSITION

As time progressed through the Archean, heat flow from Earth declined, and the area of the surface occupied by the growing continents increased. The lower heat flow may have encouraged an evolution toward larger plates typical of those characterizing Earth's surface today. The change to large-plate tectonics is not recorded directly in the geologic record, but the buildup of continental-type crust is, defining the transition between the Archean and the Proterozoic. In the early Proterozoic, at different times in different locations, sediments rapidly transition from rare-earth patterns typical of Archean to those of the Proterozoic and more recent times. Potassium-rich granites begin to appear and dominate during this time. The widespread occurrence of uranium deposits in the earliest Proterozoic sediments reflects the increased abundance of large-ion elements typical of potassium-rich granites in the upper crust. The bacterial colonies of stromatolites become much more widespread at 2.5 billion years, reflecting perhaps a larger area of stable continental-shelf territory than available before.

What precipitated this dramatic change in the geochemistry of the continents? Most popular is the view that the growing area of the continents themselves was the cause. As thick, rigid lithospheric crust topped by buoyant continents grew in area, the fraction of mantle plumes under continental crust increased. The insulating effect of thick continental crust and the physical deflection of upwelling plumes by the continental base must have had an increasingly profound effect on the manner in which the crust responded to the heat flow from the interior. Collisions of growing continents on adjoining plates to form larger continental masses exacerbated these effects.

One might speculate that a critical point was reached in the system such that massive melting in the interiors of the continental crust was initiated, triggering large-scale differentiation of Archean continental material and basaltic or mantle material underplating the continents. The increased concentration of radioactive elements in the upper continental crust may have encouraged this process. At what point such massive melting would be triggered, and the details of how the growing continents actually precipitated such a crisis, are not yet understood. In this picture, the potassium-rich granites that are so dominant in the continents of today are secondary granites, a consequence of crustal buoyancy ensured by the Archean generation of sodium-rich granites and granodiorites.

Whether the Archean-Proterozoic transition reflects just a change in the composition of the existing continental crust, or a real increase in the volume of continental crust, is controversial. One interpretation of elemental and isotopic data in continental rocks would have the volume of continental material at 10 to 20% of the present value at the onset of the transition some 3.2 billion years ago increasing rapidly to 60% by the close of the Archean at 2.5 billion years (figure 16.5). However, others have argued that 40% of the present continental volume was already present early in the Archean. The idea of more continental area created earlier in the Archean is attractive in providing a more dramatic perturbing effect on mantle heat flow, so as to initiate large-scale continental melting at the Archean-Proterozoic boundary.

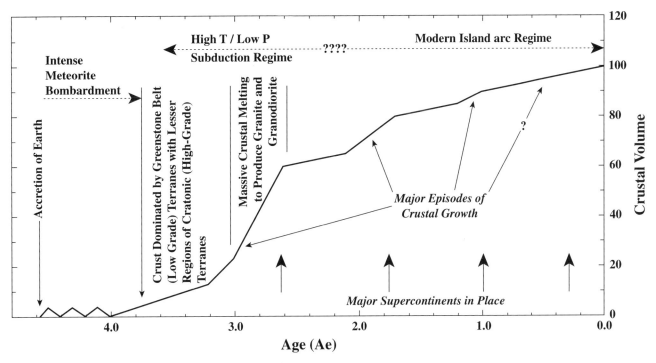

Figure 16.5. One view of the growth of continental volume, relative to its present value, over time. Some other major events in Earth's crustal history are shown, as well as a rough guess as to when modern arc volcanism came into play. Age is in billions of years. From Taylor and McLennan (1995).

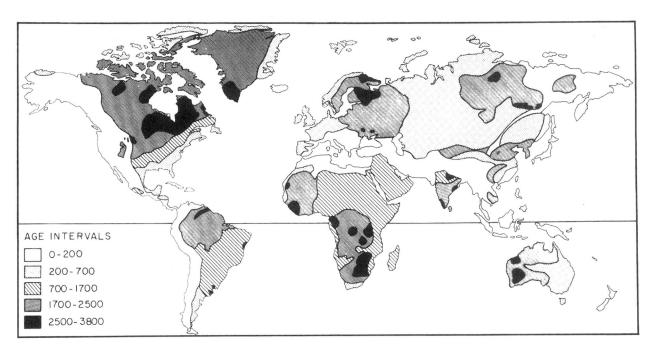

Figure 16.6. Map of Earth showing approximate ages of continental material, in millions of years. Reproduced from Broecker (1985) by permission of Eldigio Press.

Regardless of just how much continent was produced in the Archean, what is not in dispute is the tremendous change in the nature of the granitic rocks: Only 7% of the present continental area of Earth is composed of Archean-type rocks, either exposed or buried beneath younger sediments (figure 16.6). Although some Archean granites might be lurking very deep within continental crust, the consensus view is that much or most of Archean continental material was remelted and differentiated at the end of the Archean.

A final point is that, as the whole of the continental crust was transformed by major melting events, the composition of the lower part of the crust must have become increasingly distinct from the upper. Today, the upper 10 km of the continental crust is a distinct geochemical entity from what lies beneath. The lowermost 20 to 30 km of continental crust is the "other side of the coin" that holds important clues to how potassium-rich, modern granites were produced – clues that can be glimpsed only dimly through xenoliths and other bits of the lower crust that are by chance exposed.

16.8 AFTER THE PROTEROZOIC: MODERN PLATE TECTONICS

The large-scale buildup of continental crust was likely the last step in the development of modern plate tectonics. The continued growth of continents to the present is best explained as beginning at arcs, where buoyant andesite is produced. Motions of converging plates cause continents to ride up over subduction zones, and the andesite is accreted onto the edge of the continent, along with an amalgam of seafloor sediments and small amounts of seafloor crust. As collisions between continents on opposing plates proceed, coastlines disappear and the *melange* of andesitic and other materials is thrust up in mountain belts along the line of the collision. Metamorphism in these events, along with possible further episodes of crustal melting, continues the geochemical transformation of these rocks.

Geologic evidence suggests that three times in post-Archean Earth history, the continents have joined to form one or two supercontinents. (There is also evidence, albeit very weak, for such a merger at the close of the Archean.) As the continents merged and oceans closed up, the rate of subduction of old ocean crust would have been unusually high, and hence production of andesites at arcs would have increased. The continental collisions themselves could have encouraged further episodes of crustal melting and production of granites, though none as dramatic as that at the close of the Archean. There is weak and controversial evidence that continental crustal growth peaks around times of supercontinent formation.

16.9 VENUS: AN EARTH-SIZED PLANET WITHOUT PLATE TECTONICS

Imagine taking hold of Earth in the mid-Archean and stopping once and for all whatever primitive plate tectonics were occurring on the surface. Perhaps a few protocontinents have reached respectable size (close to that of Australia or Antarctica); sodium-rich granite or granodioritic material is beginning to accumulate within the continental cores. Other protocontinents are present, but haven't yet reached the state at which such felsic materials ensure buoyancy; instead, they are rather unsteady basaltic rafts floundering at near-neutral buoyancy in a basaltic crust. In the ocean basins, most of the area of the planet, small-scale subduction zones have been established at the edges of some of the protocontinents, carrying hydrated rocks down to shallow depths where partial melting occurs. As these cease to function (by our imagined interdiction), felsic products of basaltic partial melting stop being produced.

The interrupted planet must still rid itself of heat, and so, basaltic protocontinents continue to form; as they get bigger they eventually founder in the crust. Other hot spots pop up elsewhere and produce new sites of volcanism and growth of plateaus destined eventually to sink. The ocean floor never becomes part of a conveyor belt recycling crust and volatiles; it therefore ages with time along with the rest of the planet. Over billions of years, this ocean floor records the scars of hot-spot formation, foundering of plateaus, abortive attempts at subduction, and asteroid impacts. Over parts of this terrain, large basaltic flows associated with hot spots spread across the surface, renewing portions of it geologically and hiding some of the evidence of past episodes of volcanism.

To find a planet whose surface seems to record such a story, we need look no farther than Venus. Venus illustrates what happens to an Earth-size and Earth-composition planet on which plate tectonics fails to take hold beyond the early formation of protocontinents. Figure 16.7 compares the distribution of ages of crust on Earth, Venus, Mars, and the Moon. Venus lacks the bimodal ages of continent and ocean floor that Earth possesses. It also lacks the accompanying bimodal height distribution of Earth discussed in chapter 9, in which the mean elevation of continental crust is well separated from that of oceanic. Instead, a broad range of heights exists on Venus, consistent with: (i) no continuously-renewing ocean floor; and (ii) no large-scale production of buoyant granites from the basaltic crust, hence no mechanism to create large continental shields. The apparent youth of the entire Venusian crust, attributed by some to recent global volcanic outpourings, is somewhat controversial. Others argue that the Venusian crater record is consistent with a smooth distribution of crustal ages. Regardless of the interpretation, Venus lacks the sharp distinction between old continental and young oceanic crust seen on Earth.

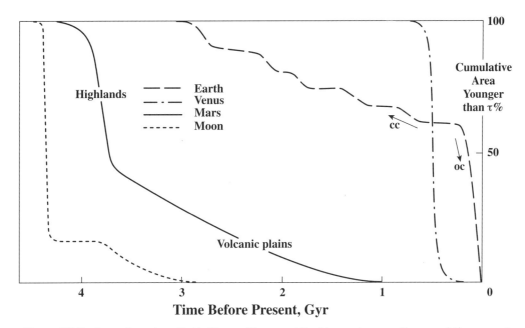

Figure 16.7. Ages of crust on Earth, Venus, Mars, and the Moon, shown as the cumulative amount of area younger than a particular age. The horizontal axis is the age in billions of years before present. Mars is divided roughly into ancient highlands and younger volcanic plains; Earth is even more sharply delineated in terms of age into continental (cc) and oceanic (oc) crust. Adapted from Turcotte (1996).

The geology of the Venusian surface is as complex as that of Earth, and an adequate description would fill a book. Careful study of the *Magellan* images and other data returned by that spacecraft strongly suggests that Venus does not currently have plate tectonics and instead rids itself of internal heat through vertical tectonics. An analogy provided by planetary geologists R.J. Phillips (Washington University) and V.L. Hansen (Southern Methodist University) is that "the crust is not recycled, but instead acts like a rug that locally rips and crumples as a result of relative displacements of domains within the mantle below. The local rips and rumples in the crustal rug of Venus are never far from where the rock that comprises it differentiated at depth. The surface is replenished or repainted with volcanism..." (Phillips and Hansen, 1994, p. 648).

Few scientists who study Venus would argue with this view; the disputes are in the details of individual types of surface features. Figure 16.8 provides one view of the origin of major types of Venusian terrain, by Phillips and Hansen. Although others have alternative mechanisms for forming, for example, Ishtar Terra, the general picture that there is little or no horizontal movement is fairly widely accepted.

The Venusian mesolands are the closest analog to midocean ridges on Earth. They are places where plateaus are created because of upwelling of warm rock in the mantle. Here, new crust may well be generated. However, the crust does not spread horizontally but piles up vertically. The accumulating material may be slightly buoyant as a result of partial melting but, in the absence of water, the production of any kind of granites is not expected.

It is unresolved whether subduction of crust occurs anywhere on Venus; various features have been argued to be the surface expression of subduction on a planet without well-defined plates. Such subduction is likely to be less important today than *delamination* or flake tectonics, in which material comes off the bottom of the crust and falls into the mantle. On Earth, seafloor spreading and subduction of plates accounts for about 60% of the heat flow from the interior; on Venus, delamination could be as important. Where such delamination is occurring is unclear, but both the plains and the lower parts of Ishtar have been invoked as candidates.

Ishtar is a particular enigma. It contains some of the highest terrain on Venus, and displays well-formed mountain belts. It is tempting say that Ishtar is the most developed of the continental masses on the planet, and indeed is the size of Antarctica. However, the origin of Ishtar is controversial, and it has been argued that this unusual part of Venus formed as a result of mantle downwelling beneath a buoyant part of the crust. The buoyant zone corresponds to Lakshmi Planum, a broad plateau at the center of the Ishtar mass framed by two mountain ranges. Thus, Ishtar may represent the response of the planet's crust to the presence of a buoyant pseudocontinent, Lakshmi Planum. This model, like all others, will undergo

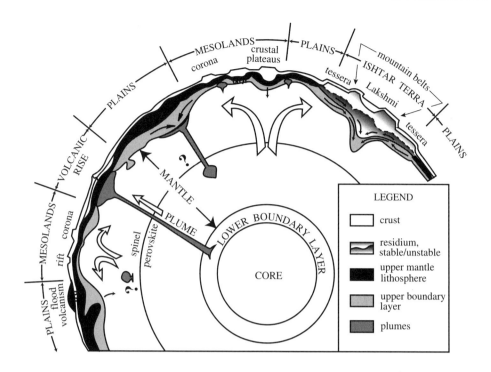

Figure 16.8. One view of the geology of Venus. Mantle plumes rise beneath volcanic regions, and may cause some plains to sag and force crust downward. Other plains may lose crust by delamination or flaking of crust into the mantle. Upwellings are topped by mesoland plateaus, which may be buoyant in crude analogy to Earth's continents. Ishtar is a special region where a particularly buoyant plateau, Lakshmi Planum, is dominating the tectonics. Compare these processes with those for the present and Archean Earths in figure 16.4. Adapted from Phillips and Hansen (1994) by permission of Annual Review, Inc.

criticism and modification, but it is interesting that even in this picture a part of Ishtar is, or once was, the Venus equivalent of a continent.

Finally, volcanism powered by hot spots from the deep interior is responsible for volcanic rises on Venus today. Evidence for substantial amounts of volcanic flows is seen in *Magellan* data. The widespread occurrence of such flows is a caution to those who would use the surface of Venus as an analog to the early Archean Earth: Since the close of the Earth's Archean, Venus has had 2.5 billion years of geologic evolution, during which features have been obliterated and new ones formed in their place. The analogy of Venus as an Archean Earth frozen in time is an imperfect one, insofar as Venus has remained active through most or all of its history.

16.10 WATER AND PLATE TECTONICS

An examination of Venus is inconclusive with regard to when it diverged from Earth in terms of geologic styles. Nonetheless, it is tempting to ascribe the difference to the lack of water on Venus. On Earth, modern plate tectonics became possible in the Archean when felsic magmas were produced, allowing protocontinents to be buoyant in the basaltic crust. The precise means by which such magmas form is not important here, because most proposals require that water be carried into the mantle to lower the melting point and alter the nature of the melt products of mantle and basalt. In the absence of such water, buoyant felsic magmas cannot be produced in abundance.

Geochemists I.E. Campbell of Canada and S.R. Taylor of Australia go further. Over a decade ago, they pointed out that large amounts of granite must have been produced in the Archean, and this is true regardless of whether by mid-Archean there was 10% or 40% of the present continental volume. Hence, large amounts of water must have moved through hydrothermal regions in the basaltic ocean crust to hydrate it effectively. It is not sufficient to have had a few lakes here and there on the surface of Earth; a deep ocean is required over much of the Archean planet to ensure large-scale production of granites.

Campbell and Taylor state it eloquently in their 1983 paper: "*Water is essential for the formation of granites and granite, in turn, is essential for the formation of stable continents. The Earth is the only planet with granite and continents because it is the only planet with abundant water.*"

Venus, having lost its ocean early on, was stopped at the protocontinent stage when, at best, only small amounts of felsic rock could be produced, and the protocontinents could not become permanent buoyant fixtures of the crust.

Even if Venus did not have continent formation, could it have sustained plate tectonics for a long period of geologic time? The extensive resurfacing recorded in *Magellan* radar images makes answering this question difficult. However, subduction may work on Earth because water weakens faults in the lithosphere, allowing this relatively rigid layer of the crust and upper mantle to slide over itself. The lack of water on Venus might have shut down plate tectonics by creating a much stronger lithosphere. Mantle convection would have continued (and probably does today), but underneath a strongly rigid lid that prevents subduction and encourages fixed sites of prodigious volcanism that covers the surface with basaltic lavas.

To fully comprehend the history of tectonics on Venus requires a technologically-challenging program to return to the torrid surface of that planet, to chemically-sample the rocks over large areas of the highland regions and plains. The high temperatures of the Venusian surface are very hard on present-day electronics and machinery, but to drill beneath the surface of Lakshmi Planum in search of granites is worthy of new technologies that will allow such explorations.

One final, speculative point about plate tectonics and Earth's oceans should be made. The mean elevation of the ocean relative to the continents has fluctuated over time, but not by very much. Most of the fluctuation comes from pulses in plate motion: When the continents are merged together, new ocean floor is not produced at many sites, and the ocean crust is unusually cool in the absence of active ridges. A cooler crust is contracted relative to a hot one, and hence the sea level falls relative to the continents. Conversely, when the continents are dispersed and many active ocean ridges exist, the average ocean crust is hotter and expanded, and hence sea level is higher. These sea-level changes have created and destroyed ecological environments on continents, and perhaps have been an important stimulus for the evolution of life.

More controversial and less well understood is whether the sea level has fallen progressively over time from the Archean to the present, as the heat flow from Earth has decreased and plate activity has fallen. The post-Archean geologic record provides no evidence of changes in sea level greater than a few hundred meters in either direction. Either geologic processes have not been such as to force a larger variation, or the volume of water in the oceans adjusts in some way to large changes in the volume of the seafloor crust. This latter possibility is intriguing because the mantle can hold significant amounts of water. Models have been proposed in which the dynamic exchange of water between ocean and mantle, via plate tectonic processes, regulates the volume of ocean water through various feedback mechanisms. As yet these seem speculative.

16.11 CONTINENTS, THE MOON, AND THE LENGTH OF EARTH'S DAY

The growth of continental landmasses created obstacles around which the ocean waters flow; at continental edges, ocean waves eroded rocky material and created shallow shelves and beaches. In such regions the paired gravitational pull of the Sun and the Moon produce the high- and low-tide patterns with which we are familiar. As the Moon orbits Earth, the oceans respond to its pull much more than does the solid crust of the Earth. However, the effect of the rising and falling of the oceans is to dissipate some of the energy associated with the tides, in the form of ocean waves and friction along the sea bottom. The net result of this loss of energy is the continuous transfer of angular momentum (chapter 10) from the rotation of Earth to the orbit of the Moon: the Moon is spiralling outward and Earth's rotation is slowing.

Evidence for this gravitational game of tug-of-war should exist; the length of the day must have been progressively shorter further in the past, and indeed a variety of indicators show this to be the case. The most ancient reliable records stretch back to the Proterozoic in the form of mudstones and sandstones that are stacked in sequences of thicker and thinner layers. These layers are created by the daily changes in the velocity of currents in regions sensitive to tides: the fronts of river deltas, tidal channels, tidal flats, and estuaries. Variations in the daily ebb and flow have a definite relationship to (a) the monthly modulations of high and low tides, as the Moon swings around Earth once every 28 modern days, and (b) the annual cycle of the Earth-Moon system's motion about the Sun. Both the number of days in a lunar orbit and in the year are clearly seen in recent mudstones and sandstones.

Analysis of 900 million year old mudstones and sandstones by University of Arizona planetary scientist Charles Sonett and colleagues shows layering implying a different number of days per lunar orbit and per year than at present: The data require the days' length in the late Proterozoic to be only 85% of the length of modern days. Furthermore, by comparing with later mudstones and sandstones, as well as with the modern rate of retreat of the Moon determined by precise measurement of the Earth–Moon distance by laser over 25 years, they conclude that the Moon is moving away from Earth more quickly today than 900 million years ago. The rate of lengthening of the day must also be higher today than during the late Proterozoic. Thus the particular configuration of the modern continents may be especially favorable for the dissipation of tidal energy.

The significance of the existence of continents in the context of the lengthening day is that these landmasses create the environments within which tidal effects are amplified (estuaries, tidal flats, etc.). Absent the continents, tides would still occur, but the dissipation of tidal energy and transfer of Earth's rotational angular momentum to the lunar orbit would be much smaller. Therefore, the outward spiral of the Moon and the lengthening of the day was likely a much more gradual affair before the Archean-Proterozoic growth of continents. What the day length was 3 billion years ago is not known, but much of the lengthening may have occurred in the last half of Earth's history. For those of us who find the length of the day much too short, there is at least some comfort in the notion that, absent tides, it could be much shorter.

16.12 ENTREE TO THE MODERN WORLD

From a planetary perspective, the shift to a fully modern plate tectonic mode of crustal heat loss by 2.5 billion years ago represents a key departure of Earth's history from that of Mars and Venus. No more significant geologic change has happened to Earth up to the present. From the standpoint of life, the growth of continents opened up whole new places to live, but it would require another 2 billion years for life to take full advantage of the vast spaces of exposed land.

As the Proterozoic eon began, increasing amounts of photosynthesis, reflecting the growing abundance of life, began to alter the composition of the atmosphere toward an oxygen-rich state. This in turn allowed a profound alteration in the nature of cellular life that was the prerequisite for the kinds of continental ecosystems that we see today. How the oxygen revolution came about, and its implications for life, are the subject of chapter 17.

16.13 QUESTIONS

a. Suppose Earth had remained a waterworld with few continents. How would this have affected the evolution of life, recycling of carbon dioxide, and Earth–Moon orbital evolution?

b. What definitive chemical tests are required on Venus to determine that plate tectonics have not operated there for billions of years?

16.14 READINGS

16.14.1 General Reading

Broecker, W. 1985. *How to Build a Habitable Planet*. Eldigio Press, New York.

Rogers, J.J.W. 1993. *A History of the Earth*. Cambridge University Press, Cambridge, UK.

16.14.2 References

Broecker, W. 1985. *How to Build a Habitable Planet*. Eldigio Press, New York.

Campbell, I.H., and Taylor, S.R. 1983. No water, no granites – no oceans, no continents. *Geophysical Research Letters* **10**, 1061–1064.

Kasting, J.F., and Holm, N.G. 1992. What determines the volume of the oceans? *Earth and Planetary Science Letters* **109**, 507–515.

Kröner, A. 1985. Evolution of the Archean continental crust. *Annual Review of Earth and Planetary Sciences* **13**, 49–74.

Kröner, A., and Layer, P.W. 1992. Crust formation and plate motion in the early Archean. *Science* **256**, 1405–1411.

Mason, S.F. 1991. *Chemical Evolution*. Clarendon Press, Oxford.

Phillips, R.J., and Hansen, V.L. 1994. Tectonic and magmatic evolution of Venus. *Annual Review of Earth and Planetary Sciences* **22**, 597–654.

Press, F., and Siever, R. 1978. *Earth*. W.H. Freeman and Company, San Francisco.

Rogers, J.J.W. 1993. *A History of the Earth*. Cambridge University Press, Cambridge, UK.

Sonnett, C.P., Kvale, E.P., Zakharan, A., Chan, M.A., and Demko, T.M. 1996. Late proterozoic and paleozoic tides, retreat of the moon and rotation of the Earth. *Science* **273**, 100–104. Corrigenda *Science* **273**, 1325 and *Science* **274**, 1065.

Taylor, S.R., and McLennan, S.M. 1995. The geochemical evolution of the continental crust. *Reviews of Geophysics* **33**, 241–265.

Turcotte, D.L. 1995. How does Venus lose heat? *Journal of Geophysical Research* **100**, 16,931–16,940.

Turcotte, D.L. 1996. Magellan and comparative planetology. *Journal of Geophysical Research* **101**, 4765–4773.

17 THE OXYGEN REVOLUTION

17.1 INTRODUCTION

Perhaps the most fundamental shift in the evolution of Earth's surface and atmosphere was the oxygen "revolution," an event stretching over the Proterozoic eon when molecular oxygen levels in the atmosphere rose and carbon dioxide levels decreased. (Hereinafter, for brevity, we refer to molecular oxygen, which is O_2, simply as oxygen.) In consequence, the fundamental chemical nature of the atmosphere and its interactions with life changed drastically. Life helped precipitate the drastic increase in oxygen levels and, as a result, was set on a radical new course. Earth's atmosphere today is not the sedate carbon dioxide atmosphere in chemical equilibrium with the surface, as on Mars and Venus. Instead, it is an atmosphere far from equilibrium, held in a precarious chemical state by the biosphere. As Margulis and Sagan (1986) express it, the modern biosphere hums "with the thrill and danger of free oxygen."

In this chapter we explore how this change came about on the Proterozoic Earth, by first examining the present-day oxygen cycle and the evidence in the rock record for an oxygen-poor Archean and early Proterozoic Earth. We then consider a model that, although not necessarily fully correct, illustrates very well how the change might have taken place. Such models often have critical utility in science, in that they point the way toward new observations and investigations that will yield deeper insight into a particular process (even while proving the model itself to be incomplete or incorrect).

17.2 MODERN OXYGEN CYCLE

Figure 17.1 shows the sources and losses (sinks) of oxygen on Earth today. The total oxygen in the atmosphere today is roughly 6×10^{17} kilograms and is held in balance by production (gain) and loss processes, the importance of which may have varied on geologic timescales. (Readers who wish to review scientific notation should turn to chapter 1.) Here we outline the most important gain and loss processes. We give rates only to the nearest order of magnitude; this is good enough for our purposes, and in many cases the uncertainties do not justify any higher accuracy.

(i) *Photochemistry and escape of hydrogen to space*. The absorption of ultraviolet photons from the Sun by water (H_2O) causes the molecule to break up, forming hydrogen and oxygen. The hydrogen can escape from the atmosphere, preventing recombination. The oxygen left behind makes molecular oxygen (O_2) and ozone (O_3). The rate of oxygen production is 10^8 kilograms per year (abbreviated as kg/yr).

(ii) *Weathering of rock*. Oxygen and carbon dioxide in the atmosphere, with the help of water, attack minerals in the rock to make new compounds which precipitate out as sediments (figure 17.2). In the case of oxygen, which attacks the iron in the rock, the process is akin to rusting. Estimating the rate of this process is not easy because it depends on how rapidly the weathered products are transported to the ocean by river systems, but is approximately -10^{11} kg/yr. The negative sign indicates that this is a loss process.

(iii) *Volcanism*. Volcanoes on land and the ocean floor emit reduced gases, such as carbon monoxide and sulfur compounds, that strongly tend to combine with oxygen in the atmosphere. The resulting rate of oxygen loss is about one-third the rate caused by weathering. Volcanoes also emit water vapor which, through photochemistry and loss of hydrogen, produces oxygen as described above.

(iv) *Photosynthesis*. Carbon dioxide is removed from the atmosphere by plants and bacteria and molecular

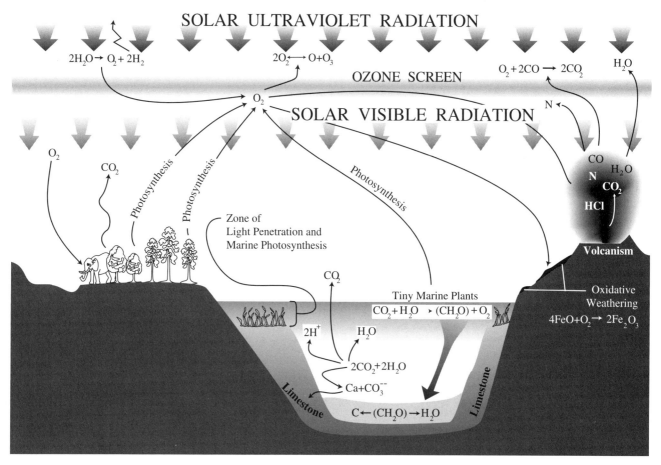

Figure 17.1. Oxygen cycle on Earth today, showing processes that are significant in producing or destroying oxygen. Chemical reactions involving oxygen are summarized; the actual chemistry involves many more steps than the equations on the figure show. Based on Cloud (1988).

oxygen is produced. The rate of oxygen production from photosynthesis is 10^{14} kg/yr.

(v) *Respiration and decay.* These two processes are, with respect to oxygen, the reverse of photosynthesis. Oxygen is taken up from the air by animals, plants, and certain bacteria and combined with sugars or other organic compounds to generate energy (along with carbon dioxide, water, and other products). Decay refers to organic matter that is no longer living, but that is attacked by bacteria through the consumption of oxygen to generate energy as described in chapter 12. The observation that the present level of molecular oxygen is approximately constant and the fact that respiration would deplete the atmosphere of oxygen in 6,000 years imply that respiration/decay is in balance with photosynthesis. If the number of plants were to suddenly increase, enhancing the level of oxygen, surface decay processes would speed up as the bacterial population grew to take advantage of the additional oxygen. Note that only three-fourths of the surface organic reservoir in contact with the atmosphere today is living. The sum rate of oxygen loss from these processes is -10^{14} kg/yr.

(vi) *Burial of carbon from organisms.* Computations show that, on average, the remains of dead organisms lie on the surface, in contact with the atmosphere, for several decades or more. We refer to this carbon, which is relatively rich in hydrogen and tends to soak up oxygen, as *reduced carbon*. (Some workers in the field refer to this material as organic carbon, but we have previously used the term organic in other ways.) The primary means of burial of the reduced carbon is deposition in continental and oceanic sediments, which breaks the contact with the atmosphere and allows the carbon to be preserved. Because the buried carbon is no longer available to soak up oxygen, the net result is that oxygen is added to the atmosphere over time. The effective rate of oxygen production is 10^{11} kg/yr.

(vii) *Recycling of buried sediments.* As discussed in chapter 14, ocean-floor sediments containing trapped

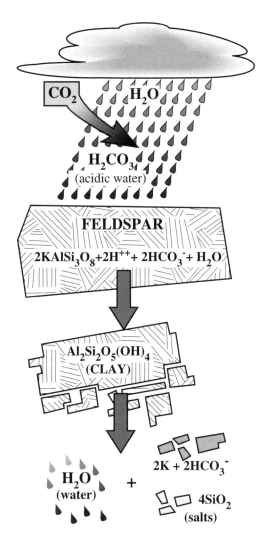

Figure 17.2. Example of the weathering of rock, in this particular case through the action of water and carbon dioxide.

carbon are recycled through the upper mantle by plate tectonics. The cycling time is roughly 100 million to 200 million years. The result is the re-emergence of reduced carbon at the surface, a net source of carbon dioxide and sink of atmospheric oxygen. The amount of oxygen loss is somewhat less than the production rate associated with the sedimentary burial of reduced carbon given above.

(viii) *Fossil fuel combustion.* This is an artificial form of weathering, caused by human burning of oil, coal, natural gas, and other fossil fuels extracted from deep sedimentary layers. The rate of oxygen loss, -10^{12} kg/yr, is much larger than for natural weathering. It will be short-lived on geologic timescales because such burning began in earnest during the seventeenth century Industrial Revolution and will cease as we deplete these resources within the next century or so (chapter 23).

17.3 OXYGEN BALANCE WITH AND WITHOUT LIFE

A look at the numbers given above shows that photosynthesis is the most important source of oxygen. Respiration/decay must be the primary balancing mechanism for losing oxygen because none of the geologic processes are speedy enough to balance photosynthesis. What was the situation before life became abundant? We can compare the most important nonbiological processes for gaining and losing oxygen, which are photochemistry at 10^8 kg/yr and weathering at -10^{11} kg/yr.

Clearly, photochemistry cannot generate oxygen quickly enough to keep pace with the destruction by weathering and volcanism. Because there are currently 6×10^{17} kg of oxygen in the atmosphere, weathering and volcanism could destroy almost all of the oxygen in the atmosphere in 6 million years at its present rate [6×10^{17} kg/10^{11} kg/yr = 6×10^6 years]. So, if we consider the roughly one-billion-year period before the emergence of photosynthesizing life-forms, it becomes a sensible notion that free molecular oxygen must have been very scarce in that atmosphere in the presence of weathering and volcanism and in the absence of photosynthesizing life.

17.4 LIMITS ON THE OXYGEN LEVELS ON EARLY EARTH

Photosynthesis in the time of the first stromatolites that have been found in the fossil record (3.5 billion to 3 billion years ago) was not widespread; consequently, the rate of oxygen production was less than it is today. The early oxygen would have been "soaked up" by weathering and by vigorous volcanic activity. Evidence for the early oxygen abundance and its increase through time is to be found in a number of parts of the rock record, the most important of which are outlined below.

17.4.1 Minerals Unstable in the Presence of Oxygen

The early continental rock record, up to about 2.7 billion years ago, shows broken-up material, that is, small debris, containing the minerals pyrite and uraninite. Pyrite is FeS_2 and, in the presence of oxygen, would react such that the iron combines with some of the oxygen to form iron oxides. Uraninite is UO_2 and uranium tends also to form other oxides. Note that the presence of significant amounts of uranium in the crust was, as discussed in chapter 16, a consequence of the partial melting process that led to continent formation. Here, the particular chemical form in which uranium exists in ancient rock deposits tells us something about the amount of free oxygen that could have existed in the atmosphere at the time that rock was first exposed at the surface. Had there been significant amounts of oxygen in the atmosphere 2.7 billion years ago, the uraninite and pyrite fragments would have been chemically altered through exposure to the air. (Subsequent burial of the rock, until more recent extraction, ensured that the more modern oxygen-rich atmosphere had no effect; undoubtedly other uraninite deposits have been destroyed over time.) Pyrite and particularly uraninite suggest that the Archean and early Proterozoic atmospheres had very little molecular oxygen.

17.4.2 Banded Iron Formation

The banded iron formations (BIFs) occur commonly among sedimentary rocks dated in the 2-billion- to 3-billion-year-old range, with a few older examples. They are extremely rare or nonexistent in younger rocks. They consist of alternating dark bands containing up to 30% iron, and light bands made of silica (chert) (figure 17.3). These bands retain their distinctiveness over vast horizontal lengths of hundreds of kilometers. To form such bands required that iron be dissolved in ocean water, then deposited repeatedly on top of layers of accumulating chert on the seabed. The sediment then was compressed, forming over time a hard rock. BIFs are found essentially on all continents, and make up more than 90% of the world's commercial iron supply.

The curiosity about BIFs lies in the need to dissolve iron in water during their formation – it cannot happen under today's atmospheric composition. The form of iron that dissolves in water (FeO) is called ferrous iron. Oxygen in the atmosphere today is partly dissolved in the ocean, and can then combine with the ferrous iron to make ferric iron, Fe_2O_3. Ferric iron is more oxidized than ferrous – that is, the element iron has bonded with more oxygen atoms than in the ferrous state: three oxygen atoms for every two iron atoms, instead of 1 to 1. The ferric iron immediately precipitates out of the water and falls to the seafloor as iron-rich particles.

To maintain iron in the ferrous form, and hence soluble in the ocean, required an atmosphere that was relatively oxygen-free. This sets limits on the amount of oxygen in the late Archean and early Proterozoic, 2 billion to 3 billion years ago, at a few percent of the present-day value or less. However, a mystery still remains: Given a mechanism for dissolving the iron in the water, the production of BIFs then requires periodic precipitation of the iron out of the water.

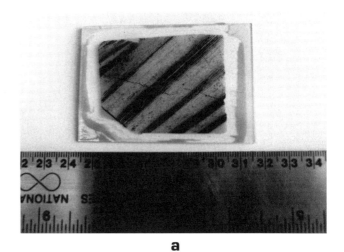

a

b

Figure 17.3. BIF rocks from (a) the Proterozoic and (b) the late Archean. Panel (a) is a thin section of the Proterozoic rock mounted on a glass plate. Scale below is in centimeters.

The problem is unsolved, but one idea goes as follows: Dissolved iron was contained in deep-ocean water near active vent sites. These iron-rich waters would spread by mixing over large areas of the ocean. Upwelling of this water to the near-surface brought it into contact with regions in which cyanobacteria existed, and hence photosynthesis took place. At certain seasons of the year, or perhaps stimulated by sufficiently large amounts of dissolved iron, the bacteria would increase their photosynthetic output of oxygen. Beyond a certain point, the oxygen produced by the bacteria would combine with ferrous iron to make ferric iron, which would precipitate out. In shallow ocean areas, the iron would precipitate out onto chert layers, forming one set of alternating bands. As the photosynthesis slowed again, oxygen levels in the water would decrease, iron could be stable again in the dissolved ferrous form, and the cycle would repeat. These layered sediments, over time, eventually would be compacted and lithified.

The cyanobacteria were not the only life-forms participating in this process. Certain other kinds of "rusting" bacteria take oxygen from the surrounding environment and combine it with iron, creating stored energy usable for their life processes. This would have assisted the process of iron precipitation. In the summer, when cyanobacteria were active in producing oxygen, the rusting bacteria would have been more abundant and extracted more iron from seawater. In the winter, with less oxygen produced by cyanobacteria, biological deposition of iron-bearing sediments would have slowed or stopped. The variation from place to place in the width of the iron bands – from micrometers to meters – suggests that oxygen levels fluctuated on seasonal and longer (perhaps decades or more) timescales in different places at different times.

An explosion in the production rate of BIFs in the 2.2 billion- to 1.8-billion-year time frame suggests that oxygen levels worldwide had reached a threshold at which variations in photosynthetic activity modulated the precipitation of iron from oceans. However, some rare cases of BIFs occur prior to 3 billion years ago (perhaps as early as 3.86 billion years), when worldwide oxygen levels were very low. These oldest BIFs hint that, in localized areas, some form of photosynthesis intensive enough to produce significant quantities of oxygen might have occurred. Alternatively, mechanisms have been suggested by which molecular oxygen produced by atmospheric photochemistry might periodically have been concentrated in localized environments, but they remain speculative.

17.4.3 Redbeds

Beginning about 2 billion years ago and extending to recent times, sediments appear in the rock record that require oxygen for their formation. These *redbeds* form when iron is weathered out of rock in the presence of oxygen. The threshold amounts of oxygen that are required to make redbeds are significant but still small enough to permit BIFs to exist; the two overlap in the geologic record by several hundred million years.

17.4.4 Fossils of Aerobic Organisms

Before the advent of free oxygen, organisms produced energy for biochemical processes in a number of ways. The most familiar process, one still in operation in oxygen-poor (anaerobic) environments, is fermentation (chapter 12). Here, sugar is converted to ethanol and other molecules, with release of energy. The energy is stored in a biological molecule containing phosphate bonds, called adenosine triphosphate (ATP). One molecule of sugar makes enough energy to be stored as two molecules of ATP.

Respiration, as discussed in chapter 12, uses oxygen to convert sugars to carbon dioxide, water, other products, and a great deal of energy. Respiration can produce up to 36 ATP molecules from one sugar molecule. This tremendous boost in bioenergy efficiency allowed explosive growth in the number of forms of cyanobacteria in the Proterozoic eon, and later enabled complex cellular life (eukaryotes) and multicellular eukaryotic life (plants and animals).

The times at which these biological events appear in the fossil record in rocks and the known biochemical requirements for oxygen among such species today allow the increase in oxygen in the atmosphere to be tracked. The late and relatively rapid appearance of large, complex, multicellular animals only 550 million years ago suggests that oxygen levels may have remained well below the present value (perhaps 10 times less) until then. The evidence for charcoal in the fossil record of the past 100 million–200 million years implies forests capable of undergoing combustion (burning); this requires oxygen levels close to those at present (13% compared to the present value of 21%).

17.5 HISTORY OF THE RISE OF OXYGEN

With the evidence for an early time of little or no atmospheric oxygen and a significant increase beginning in the Proterozoic eon, we can put together a chart (figure 17.4) of the amount of oxygen in the atmosphere over Earth's history. The chart is rough, showing much uncertainty in the actual levels, but the general nature of the conclusion is clear: Before the start of the Proterozoic, Earth's atmosphere was, for all intents and purposes, devoid of oxygen.

How did the change come about? The clues are present in the evidence described here, but must be assembled carefully into a working hypothesis. Such a hypothesis ought to explain the physical evidence in terms of the processes that occurred over time to generate the oxygen-rich atmosphere. We next consider one possible model for the growth of oxygen.

17.6 BALANCE BETWEEN OXYGEN LOSS AND GAIN

Earlier in the chapter we considered present-day rates of oxygen production and loss. These rates were different during the Archean epoch compared to the present. Biological

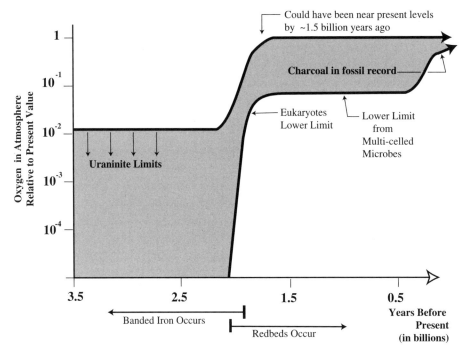

Figure 17.4. History of oxygen abundance in the atmosphere of Earth, assembled from diverse pieces of evidence described in the text. In most cases, the constraints are weak, or provide only upper and lower limits. Hence, oxygen levels at any given time could have been within the shaded area of the chart. After Kasting (1991).

processes were not nearly as important then, and respiration in the absence of significant amounts of oxygen must have been negligible or nonexistent. Recycling of crust and mantle may have been much more rapid in the Archean than at present, leading to a greater rate of volcanism at that time. This led to a higher flux of reduced gases into the atmosphere, which could soak up oxygen at a higher rate.

The change in amount of oxygen in the atmosphere per unit time is simply the rate of production minus the rate of loss. In figuring out how production and loss work to produce a particular amount of oxygen, an important fact is the following: The loss processes depend on how much oxygen is available, but the production processes usually do not. If we start with zero oxygen, there can be no loss of oxygen from weathering or volcanic gases. As more oxygen is produced by photochemistry, more can be lost by weathering and volcanic gases. A graph of the amount of oxygen as a function of time will then look something like figure 17.5.

As any new source of oxygen arises, loss rates (proportional to the oxygen abundance) increase, until a new steady state is reached, characterized by a constant or only slowly varying oxygen abundance. Alternatively, some loss processes might saturate as the oxygen abundance rises; this would effectively increase the rate at which the oxygen abundance grows. The sinks of oxygen on the early Earth must eventually have been overwhelmed by increasing rates of oxygen production.

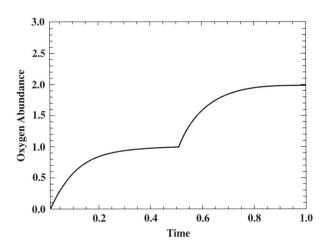

Figure 17.5. Example curve of oxygen abundance in the face of changes in production or loss processes. The horizontal axis is time, and the vertical axis is oxygen abundance (both in arbitrary units). As the oxygen abundance approaches a constant value, reflecting a balance between production and loss, a suddenly increased production rate causes a jump in abundance followed by leveling-off at a new, higher value. The increased production rate might be due to a novel source, or simply an increase in production from an established source of oxygen.

17.7 RESERVOIRS OF OXYGEN AND REDUCED GASES

The situation on the early Earth is best summarized by considering reservoirs of oxygen and the substances that can soak up oxygen, via weathering, volcanism, or organic matter from life-forms that have died but have not been deeply buried in sediments. We will simply call these substances *reducing compounds*, meaning elements or molecules that like to combine with oxygen. A simplified model of Earth as just atmosphere and ocean is sketched in figure 17.6.

We ignore the continents because, even though volcanoes may be on land or sea, and weathering processes start out on land, the "action" ends up being in the ocean or atmosphere. Most of the continental weathering products end up in the ocean, and the volcanic gases are present in the atmosphere or dissolved in the ocean. We must distinguish between the deep ocean, which has slow, limited contact with the atmosphere, and the shallow upper part of the ocean, where photosynthesis takes place (because some sunlight is present) and gases are exchanged with the atmosphere. Included in the shallow part of the ocean are rivers and lakes.

Furthermore, we do not consider the variation from place to place in oxygen content, only the difference between the three environments – atmosphere (top), shallow ocean (middle), and deep ocean (bottom). This is called a *one-dimensional model*; it is useful in understanding many physical situations because of its simplicity. Obviously, such models cannot explain fine details, and may miss important processes that occur or vary from one place to another, but our information on oxygen abundance on the early Earth is so limited that this simple model has great utility. A reminder of its limitations is the presence of BIFs in the Archean, which indicates oxygen variations from one location to another on Earth.

The model of oxygen production on Earth that we consider was developed by James Kasting of Pennsylvania State. Its general outline can be sketched as follows:

Over time, we distinguish between three states of each of the three reservoirs:

(i) *Reducing.* This means that the reservoir has so little oxygen that minerals such as uraninite will be stable, and iron can remain in solution in the ocean water which is required to produce BIFs.

(ii) *Oxidizing.* Here, the reservoir has enough oxygen to make minerals such as uraninite unstable, and to prevent iron from staying dissolved in seawater. However, not enough oxygen is available to sustain aerobic respiration.

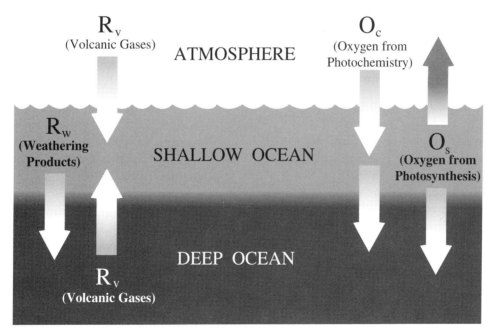

Figure 17.6. Box model of Earth used to understand the growth of oxygen. Three components of Earth are atmosphere, shallow ocean, and deep ocean. Sources of oxygen are labeled "O," and sinks are labeled "R," with subscripts to distinguish among them. Weathering products include not only sediments but also reduced carbon from dead organisms which, left exposed to the atmosphere, can soak up oxygen. Based on the model of Kasting (1991).

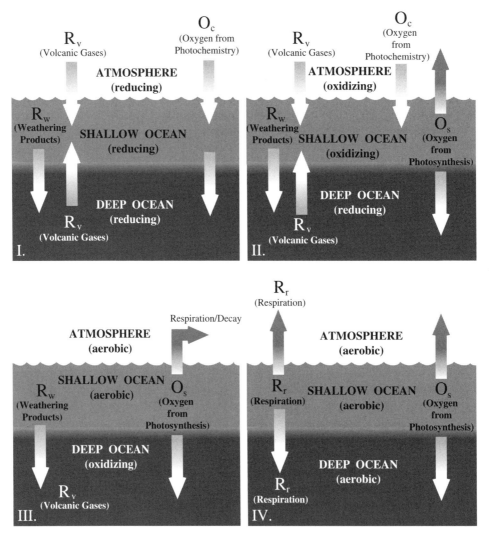

Figure 17.7. Four stages in the history of oxygen on Earth, distinguished by the oxidation state of the three major oxygen reservoirs considered in the model. Important production and loss processes in each stage are shown. Based on the model of Kasting (1991).

(iii) *Aerobic.* Enough oxygen is present to allow aerobic respiration to occur.

With this model we can map the history of oxygen in the four stages illustrated in figure 17.7. All oxygen abundances are listed in fractions of the present atmospheric level (PAL).

17.8 HISTORY OF OXYGEN ON EARTH

17.8.1 Stage 1

Once water is established on Earth (Hadean-Archean boundary), photochemistry begins to produce oxygen. The oxygen levels off as production is balanced by weathering and volcanism. Oxygen in the atmosphere ranges between 10^{-8} and 10^{-14} PAL, based on detailed calculations. This reducing environment would preserve uraninite. BIFs from this time occur, and were formed either as solar ultraviolet radiation (which reached Earth's surface in the absence of an ozone shield), oxidized iron in water, or in localized environments where primitive photosynthesizing bacteria were concentrated.

17.8.2 Stage 2

The spread of photosynthesizing organisms around the planet initiated a new source of oxygen. At first, photosynthesis would have been limited in extent, but would have increased as numbers of organisms expanded with time. The geologic data suggest that oxygen in the atmosphere jumped to 10^{-2} PAL around 2.2 billion to 2 billion years ago, enough to be considered oxidizing for most minerals. The geologic record for this time shows an overlap between the occurrence of BIFs and redbeds. The oxygen abundance in stage 2 is small enough that the deep ocean could have remained oxygen-poor (reducing), whereas the upper ocean, where photosynthesis took place, would have been oxidizing. Under such circumstances, deep-ocean water containing dissolved iron may have slowly circulated up to the surface, where it encountered oxygen-rich conditions and precipitated out iron, forming BIFs.

The steep rise in oxygen during this time period has prompted speculations on mechanisms beyond increased photosynthesis to pump up the atmospheric oxygen level. Geologic evidence suggests that around this time a number of small continents collided to form the first *supercontinent*, a process to be repeated again and again in more recent history (chapter 19). NASA Ames scientist David Des Marais suggests that the assemblage of continental fragments into larger masses had as a side effect the increasing rate of burial of dead organisms (what we have called reduced carbon). The heightened burial rate occurred both directly in the extensive continental interiors and on the seafloor; rates on the seafloor were enhanced as large mountain ranges, built up on the colliding continents, sped the delivery of sediments to the sea. With much smaller amounts of reduced carbon exposed to the atmosphere, less absorption of oxygen by these compounds could occur; in effect an important sink of oxygen was eliminated. Alternatively, Kasting argues that, by this point in Earth's history, large amounts of ocean water were mixed into the mantle by plate tectonics (equivalent perhaps to half the volume of the present oceans). This process would gradually have turned the mantle from a reducing to an oxidizing chemical state, such that volcanic gases emanating from the mantle became progressively less effective in soaking up atmospheric oxygen. The decreased importance of volcanism as a sink combined with increasing rates of oxygen production from photosynthesis led, in this picture, to a steep increase in abundance of atmospheric oxygen.

17.8.3 Stage 3

As photosynthesizing organisms proliferated, the oxygen content of the atmosphere increased. However, primitive anaerobes (organisms that do not use oxygen in metabolism) cannot tolerate oxygen beyond a certain level.

Oxygen levels in the Proterozoic may have been limited by this intolerance: Too high a production rate would have poisoned the photosynthesizers. Eventually, organisms had to evolve to tolerate large amounts of oxygen in their environment. At this point, perhaps 1.7 billion years ago, increased photosynthetic production of oxygen and higher net abundance could be safely sustained. Then, the atmosphere and surface ocean reservoirs became aerobic, with the deep ocean oxidizing. BIFs could no longer be produced because iron dissolved in seawater was always unstable. Redbed formations became more widespread.

17.8.4 Stage 4

Eventually, the deep ocean received enough flux of oxygen to become aerobic as well. The advent of oxygen respiration (aerobic metabolism) was initiated among living forms, and the number and vigor of photosynthesizing species increased. The new balance in oxygen production and loss was between photosynthesis and respiration/decay, with photochemistry, weathering, and volcanism now insignificant in their effect on oxygen levels. The balance was such as to permit a gradual increase in oxygen to the current abundance within the past billion years.

17.9 SHIELD AGAINST ULTRAVIOLET RADIATION

The damaging short-wavelength ultraviolet (uv) photons from the Sun are today shielded by O_3 in Earth's stratospheric layer (chapter 14). Ozone is produced photochemically from molecular oxygen (O_2) by absorption of uv photons.

Based on chemical models, to maintain an ozone shield requires 10^{-2} PAL or higher of oxygen. Clearly, then, an ozone shield was not available up to about 2 billion years ago. Because uv radiation is absorbed more effectively by water than is visible radiation, photosynthesizing organisms in the oceans could have been protected from uv radiation, even at shallow depths. Other atmospheric gases and aerosols, such as sulfur-bearing molecules, also might have afforded protection from some of the uv radiation, making life on land surfaces possible. Because these shields likely were not as effective as the current stratospheric ozone layer, organisms had to develop protection themselves; the common tendency of bacterial colonies to form mats, the evidence of which is the stromatolites, would have shielded such colonies from the uv flux. In spite of various survival strategies, incomplete shielding of continents and the ocean's surface from uv radiation probably restricted severely the number of viable life-forms and viable habitats on the Archean and early Proterozoic Earth.

17.10 ONSET OF EUKARYOTIC LIFE

The dramatic rise in oxygen levels around 2 billion years before present resulted in two events that enabled a large increase in the forms and number of living organisms, and the ecological niches that they could occupy. These were (i) the enabling of aerobic respiration, which dramatically increased the energy that life could generate and use from the environment, and (ii) the development of an ozone shield.

All aerobic cells, be they prokaryotic or eukaryotic, contain enzymes that are required to detoxify the molecular fragments, or *radicals*, that contain oxygen. Without such enzymes, these free radicals would react with and destroy cellular structures. Anaerobes must avoid oxygen by existing in oxygen-poor environments or mounting defenses against oxygen similar to those of the aerobes. Even more intriguing is that, with just a few exceptions, oxygen is not used in the chemical pathways synthesizing proteins and other biological molecules – it is just used as an energy source. Had abundant free molecular oxygen been available at the origin of life, one might expect it to have been used to full advantage in cellular structures.

Although some prokaryotes evolved to take advantage of oxygen and employ it in their metabolism, the advent of oxygen apparently led to the successful spread and diversification of a new kind of cell. Around the 2-billion-year mark, in the mid-Proterozoic, fossil evidence appears for eukaryotes, in which cellular function is divided among individual areas (the organelles described in chapter 12) separated by membranes and, in some cases, containing their own separate DNA and RNA. The cell's central genetic code is isolated in a nucleus, and organized into *chromosomes*; there is far more genetic material wrapped in the chromosomes than in the single strand of DNA contained in prokaryotes. (However, bacteria are far more genetically flexible than eukaryotes in that they readily pick up mobile packages of genes from other bacteria, allowing drastic changes in structure and function. Such package transfers to eukaryotes; the viruses, for example, almost always disrupt cell function.)

Essential to the workings of the eukaryotes in the current biosphere are the plastids (for example, the green chloroplasts) and the mitochondria, defined in chapter 12. The plastids convert sunlight, carbon dioxide and water into sugars. The mitochondria take alchohols and lactic acid – products of fermentation of food products that takes place in the cytoplasm of the cell – and conduct a set of chemical reactions involving oxygen and the fermentation products to create the enormous phosphate-bond storehouse of energy characteristic of aerobic metabolism.

The mitochondria and plastids are important also for providing a clue to the origin of the complex eukaryotes: both resemble bacteria. Mitochondria have their own DNA, messenger RNA, transfer RNA, and *ribosomes* (the sites of protein synthesis in prokaryotic and eukaryotic cells) within the mitochondrial membrane. The DNA floats within the mitochondria as strands, and is not bound in chromosomes. The ribosomes look like bacterial ribosomes, and are sensitive to the same antibiotics. Mitochondria divide at times different from the rest of the cell, by simple pinching and division, as do bacteria. Plastids resemble bacteria even more than do mitochondria in the appearance and arrangement of their internal structures.

Boston University biochemist Lynn Margulis proposed some years ago that the eukaryotic cell is the result of symbiotic (cooperative and dependant) relationships between bacteria of various types. Sometime in the past, presumably in the mid-Proterozoic as aerobic metabolism became possible, various symbiotic relationships between aerobic bacteria, cyanobacteria, and larger host bacteria created combined organisms that survived and prospered, eventually becoming fully internally dependent such that the resulting composite cells were the eukaryotes that we are made of today.

Although mitochondria and plastids cannot exist outside of their own cells, there are plenty of examples of symbiosis among bacteria, and between bacteria and eukaryotes, in both the natural world and in laboratory experiments conducted over the past few decades. Some eukaryotes are actually anaerobic, lacking mitochondria but, in some cases, containing organelles specialized for fermentation; examples include the protozoan *giardia intestinalis*, responsible for severe diarrhea in humans. Some anaerobic eukaryotes exist in a tightly dependent relationship with other organisms, including bacteria, or actually harbor bacteria within their cells in a symbiotic relationship. Removal of the bacteria usually leads to death of the host eukaryote. In one case the symbiotic bacteria belong to a group generally thought to be good candidates for the ancestors of mitochondria. Laboratory experiments have successfully forced symbiosis between bacteria and amoebas that do not normally engage in such processes; by then selecting the amoebas that best accommodated the invaders, a colony of healthy amoebas was created. The bacteria lived off the amoebas and, curiously, the amoebas became dependent on the bacteria as well.

A further clue to the origin of eukaryotes lies in the predatory nature of some bacteria that will invade the cell walls of other bacteria. Although most such encounters eventually result in the death of the host, and hence of the invaders, in some cases the prey have evolved a tolerance for the predatory bacteria.

Margulis and others have proposed that several extant aerobic (and predatory) bacteria are descendants of bacteria that evolved into mitochondria. A large photosynthesizing bacterium, *prochloron*, with unusual plant-like properties and a taste for symbiosis in sea animals, is perhaps descended from a similar bacterium that infected certain cells and evolved into plastids. Similarly, the large host cellular mass of eukaryotes is echoed in the large bacterium, *thermoplasm*, that is modestly oxygen tolerant. Other candidates for the cell nucleus and additional eukaryotic cellular structures have been proposed.

The notion that complex plants and animals are the result of symbiotic relationships between bacteria may be shocking to some, but it is increasingly accepted by biologists. The structural similarities between organelles and some bacteria, the symbiosis between anaerobic eukaryotes and bacteria, between different types of bacteria, and the tolerance of some eukaryotes for forced laboratory symbiosis all suggest that such dependencies have arisen throughout the history of Earth.

In fact, it is not even clear that the first eukaryotes waited until 2 billion years ago to appear, though earlier fossil evidence is not present. (Eukaryotic cells can be distinguished from prokaryotes as fossil imprints because of the much larger size of the former.) Comparison of the genetic structures of eukaryotes and prokaryotes have led some molecular biologists to propose an earlier origin for eukaryotes, perhaps prior to 3 billion years ago. This interpretation of the data is controversial, and most biologists argue that eukaryotes most probably first appeared around 2 billion years ago. Nonetheless, one could imagine bacterial experiments in symbiosis, at some earlier times, leading to the production of anaerobic eukaryotes without mitochondria prior to the onset of abundant atmospheric oxygen.

Regardless of the original appearance of eukaryotes, their success was in the utilization of atmospheric oxygen. As cyanobacterial photosynthesis polluted the atmosphere and shallow ocean with oxygen, some bacteria evolved to be oxygen-tolerant and then oxygen-dependent. The ancestors of mitochondria were efficient enough at using oxygen that their symbiotic relationship with host bacteria created an energy-efficient, adaptable cell that would become the basis for animals and fungi. Further symbiosis brought photosynthesis into the eukaryotic realm. Eukaryotes spread into a variety of ecological niches – some even adapted to survive in anaerobic environments, in contrast to those eukaryotes that lack mitochondria and were anaerobic from the start.

Interestingly, the advent of aerobic eukaryotes continued the trend toward increased morphological diversity and decreasing chemical variety that was discussed at the end of chapter 13. Bacteria exhibit a wide range of metabolisms used to derive energy from the environment, including fermentation, sulfur metabolism, nitrogen consumption, and aerobic combustion of hydrogen. On the other hand, aerobic eukaryotes threw their lot in with the mitochondria, so that, in spite of the enormous diversity of shapes and types of multicelled animals and fungi, the power source is almost entirely aerobic respiration through the mitochondria. Most nucleated cells have essentially the same kind of metabolism. This metabolism, coupled with plant and bacterial photosynthesis to create oxygen, largely determines the atmospheric composition that we see today.

The advent of aerobic eukaryotes enabled predator-prey food chains to come into existence. Although predator-prey relationships exist among bacteria, the food chain is one level: Predator bacteria invade prey bacteria, and the food chain ends there. Anaerobic organisms are not efficient enough at producing energy from fermentation and other mechanisms to create enough food for a multilevel food chain, and aerobic bacteria do not come in enough morphological varieties to create such a chain themselves. Eukaryotes have the high energy efficiency and morphological diversity to sustain the multilevel food chains that every student learns about in biology classes. In this regard, biologists T. Fenchel and B.J. Finlay point out that evolution toward large size and complexity is a tremendous evolutionary advantage – one can swallow smaller organisms or avoid (through size, speed, and smarts) being swallowed in turn. The high-efficiency oxygen metabolism and complex versatility of eukaryotes were prerequisites to innovations such as ourselves.

Following on the heels of symbiotic creation of aerobic eukaryotes, cooperative colonies of eukaryotic cells developed into the first multicellular organisms – plants and animals. By the end of the Proterozoic, a bit more than a half billion years ago, the O_2-rich, CO_2-poor atmosphere of Earth supported – and was sustained by – a wide variety of eukaryotic, multicellular species. This was perhaps the last step in the departure of Earth from the history of its neighboring planets; our atmosphere would never be the same. Nonetheless, many of the same external influences affected Earth as well as the other planets – variations in the Sun's brightness, shifts in orbits, occasional impacts of large objects to form craters. It is the response of Earth's atmosphere, oceans, and biological systems to such events that make up much of the story of the last half-billion years of Earth history – a story to which the remainder of this book is devoted.

17.11 QUESTIONS

a. How might the history of oxygen on Earth have been altered if the oceans were very shallow – only 100 meters in depth, for example?

b. How in turn might this have altered the evolution of life and the sustained habitability of Earth?

17.12 READINGS

Cloud, P. 1988. *Oasis in Space: Earth History from the Beginning*. W.W. Norton, New York.

Fenchel, T., and Finlay, B.J. 1994. The evolution of life without oxygen. *American Scientist* **82(1)**, 22–29.

Kasting, J.F. 1991. Box models for the evolution of atmospheric oxygen: An update. *Paleogeography, Paleoclimatology, Paleoecology* **97**, 125–131.

Margulis, L., and Sagan, D. 1986. *Microcosmos: Four Billion Years of Microbial Evolution*. Summit Books, New York.

Press, F., and Siever, R. 1978. *Earth*. W.H. Freeman and Company, San Francisco.

18 THE PHANEROZOIC: FLOWERING AND EXTINCTION OF COMPLEX LIFE

18.1 INTRODUCTION TO THE PHANEROZOIC

The Phanerozoic eon is a major division in the fossil record that dates radioisotopically at a bit younger than 600 million years before present. Its geologic marker is the appearance of numerous complex multicellular organisms in the fossil record. This eon has no counterpart on any other planet, even if Mars harbored simple life-forms within the first billion years of its history. On Phanerozoic Earth, life began to occupy just about every conceivable niche on land, sea, and air. Geologically, Earth was more or less modern in form as the eon opened: The total continental mass was comparable to that today, modern-style plate tectonics were operating, and oxygen levels in the atmosphere were approaching present-day values.

The Phanerozoic eon is divided into eras, eras into periods, and periods into epochs. The boundaries between most of the periods are defined by extinction episodes in which a number (sometimes very large) of species disappear and are replaced in the sedimentary fossil record above that point by new species. Although the resulting story of complex multicellular organisms is too large to tell in detail in this book, some of the highlights are shown in figure 18.1.

The presence of multicellular organisms per se was not new. Multicellular bacterial colonies had existed since the Archean; multicellular algae (for example, green seaweed) made their appearance shortly after the first unicellular eukaryotes in the fossil record. In each of these, and many other cases, there is little or no specialization among cells, and only limited communication. The Phanerozoic biological revolution was about organisms composed of cells, the forms of which were altered to conduct specific functions, and which were wholly dependent on one another. Animals are the extreme expression of this intricate symbiosis; plants exhibit this to a lesser extent.

What precipitated the rapid flowering of life into the diverse and complex multicellular entities of the modern era remains a mystery. Biologists, particularly those whose specialty is the understanding of the genetic code, tend to look for answers in the nuclei of eukaryotic cells, to see whether the vastly increased complexity of the genes precipitated this new way of living. At the heart of the mystery, though, is why it took a billion years after the appearance of eukaryotes for such an innovation to become widespread. Some very recent fossils, collected in China, suggest that multicellular plants existed at the 1.7-billion-year mark, but the rapid proliferation of complex multicellular life-forms did not occur for another billion years – 20% of Earth's history.

Geologists and planetary scientists tend to look for external causes behind the great flowering of complex life at the dawn of the Phanerozoic. Indeed, over the past several decades, the importance of environmental changes as a stimulus for major evolutionary changes and extinctions has been recognized. Even more recently, the possibility that impacts of asteroidal or cometary fragments could be the cause of episodes of large-scale extinctions has gained respectability. As environments change or whole ecosystems are wiped out, new forms of life develop from old to occupy the now-empty niches.

In fact, and as is realized by biologists and physical scientists alike, the flowering of whole new phyla of life-forms and their elaboration through the Phanerozoic require both internal genetic change and the external stimuli to make them happen. Coupling these leads to a view of evolution that is much more complex than Darwin's eternal set of battles for survival of the fittest member of a species played out against the background of a gradually

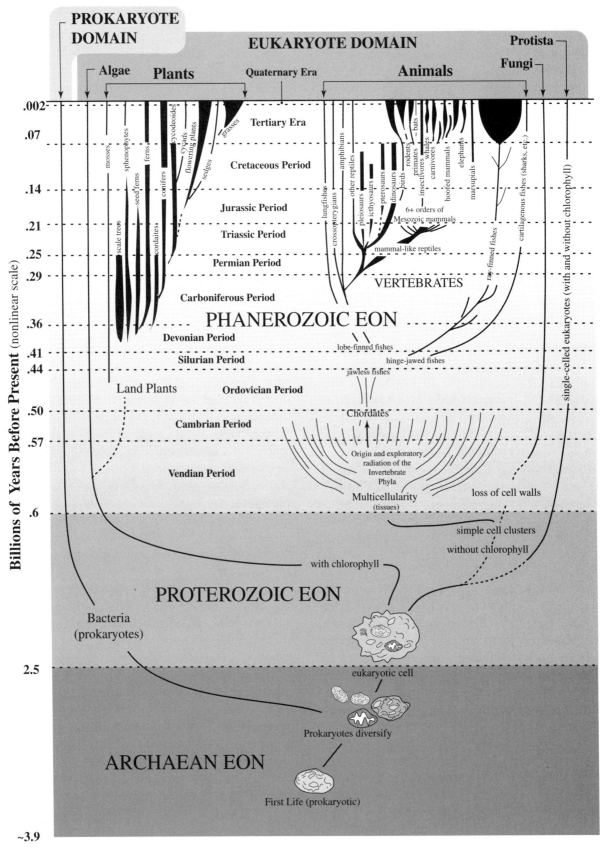

Figure 18.1. Major biological events in the Phanerozoic. The chart is divided into prokaryotes and eukaryotes, and among the eukaryotes into the major kingdoms. Eukaryotic lines of descent shown at earliest times are phyla, then in later times some (but not all) major classes and orders of vertebrates and plants are shown. Of note is the very large number of different animal phyla in the early Phanerozoic, some of which are no longer represented by species. An attempt to show the range of diversity of forms in the plants and vertebrates is given by the thickness of the lines. Particularly uncertain aspects of the histories are indicated by dashed lines. Redrawn and modified from Cloud (1988).

changing Earth. The fossil record seems to demand a more complicated view.

In this chapter we examine the internal and external mechanisms behind the evolution of species. We focus on two major biological events in Earth's history: the flowering of multicelled life at the start of the Phanerozoic and the extinction of many families of species at the close of the Cretaceous period some 65 million years ago. These are by no means the only watershed moments in the biological evolution of the past 15% of Earth's history, but they are illustrative examples of two different kinds of evolutionary events – one perhaps more internal than external, and the other quite the opposite.

18.2 EVOLUTION

Much controversy still surrounds the concept of the evolution of species. The controversy is not so much scientific as it is political, centered on the question of the special and simultaneous creation of all extant species as a literal interpretation of the Judeo-Christian bible would require. It is not the place of this book to argue the merits of this point of view on spiritual grounds, but the following are offered in support of the notion that species evolve with time:

(i) The fossil record shows a wide range of extinct lifeforms that bear an increasing relationship to current living organisms in progressively younger sediments.
(ii) The source of the instructions by which the form and function of organisms are defined, DNA and RNA, was identified almost a half-century ago. Since that time, the ability to manipulate and transfer genetic code to effect a change in form and function of organisms has been repeatedly demonstrated; this *recombinant DNA* technique is now routinely used to make agricultural and pharmaceutical products.
(iii) Natural selection and consequent evolution have been directly observed for a few species with short reproductive cycles that have experienced sudden changes in their environments.

Broadly defined, *evolution* is the formation of new species from old. Viewed in a slightly more specific way, it is the change of form and function of living organisms sufficiently profound to create a new, self-contained breeding population. The two definitions are tied through the concept of *species*, a taxonomic label whose precise characterization has proved difficult. We recognize intuitively the human species – all humans alive on Earth today are the same species and can produce offspring with each other.

Our ability to distinguish between members of different species versus more general taxonomic orders (see section 18.3) decreases as the relation to us becomes more distant. A chimpanzee is a chimpanzee – but the bonobo "chimps," it turns out, are a different species. It is much harder to decide what constitutes a species when staring at different kinds of fungi – unless one is a professional biologist.

The best and simplest definition of species, at least for animals, is that it is a cohort of living organisms that can produce viable and fertile offspring that interact sexually and hence are isolated from other organisms in the reproductive sense. The qualification "fertile" is required because different species – tigers and lions, or horses and donkeys – can produce common offspring that are, however, sterile. This definition does not work as well for many kinds of sexually reproducing plants; and among forms of life that do not reproduce sexually, or do so in the "bacterial fashion" through frequent exchange of DNA fragments, the concept of species is almost irrelevant.

Evolution would not be possible if the reproduction of the genetic code were error-free from generation to generation. However, alteration of the DNA code, or *mutation*, occurs by random copying errors and by errors induced by ultraviolet light (particularly at wavelengths of 2,600-2,800 Angstroms) and impacts of cosmic rays (high-energy protons streaming through Earth's atmosphere) on the DNA molecule.

Mutations induced in the sexual cells – eggs and sperm – may result in a change of form or function of the offspring. Most mutations are harmful, many are fatal, but a large number – those that alter certain codons such that no new amino acid is specified for that position – are neutral. However, others may confer some advantage to the offspring and to future generations in which the new genetic code is perpetuated.

Although the genetic code is voluminous, recent discoveries of key gene sequences, or "trigger" genes, suggest that some single mutations may have dramatic effects. Trigger genes are those that control the activation of large sets of genes, which in the aggregate control a major structural or functional characteristic of a species. In fruit flies, for example, the activation of a single or small number of trigger genes means the difference between development of a leg versus an antenna.

18.2.1 Traditional, Darwinian, Model of Evolution

Given a mechanism for change in species, how does evolution actually work? The traditional mechanism, still taught in high school biology classes, was first clearly articulated by Darwin, and is popularly called "survival of the fittest." In this view, individuals compete within a more-or-less stable or slowly changing environment; individuals with the characteristics that make them most competitive survive to produce multiple offspring, often with multiple reproductive partners, offspring who carry those characteristics.

The implications of this view for evolution are that the transformation of ancestral populations into a new species is slow and fairly constant in time, and a large fraction of the population, over most of its geographic range, is transformed. Evolution is a gradual process, in this view, with two important implications for the fossil record:

(i) An ideal sedimentary (fossil) record of the origin of new species will contain a sequence of forms intermediate between the ancestral and descendant species.
(ii) Breaks or gaps in the forms intermediate between the old and the new species are caused by the intrinsic incompleteness or imperfections in the sedimentary record.

Examination of the fossil record shows few or no intermediate forms between the new species. One objective interpretation is that species do not evolve, but arise out of whole cloth. The other interpretation is based on mechanisms of fossilization (chapter 8), namely that fossilization is so rare and the record so imperfect that the chance of catching a species in the act of changing is vanishingly small. This latter argument has been used by geologists for roughly a century, and it is not unreasonable: For a body form to be well preserved in the sedimentary record, it must be allowed to remain intact against bacterial decay and physical damage. Once fossilized, the sediments themselves must remain relatively unaltered through enormous spans of time. As noted in chapter 8, only a very small fraction of the world's organisms have had the "privilege" of becoming fossils.

The lack of transitional forms is a problem in spite of the difficulty of fossilization, and it provided fertile ground for antievolutionists to argue that the whole concept of natural changes in and formation of new species was wrong. Even Charles Darwin regarded the fossil record as an embarrassment for his theory. But the problem was rationalized away and new generations of paleontologists (those who study fossils), raised on the textbooks and canons of their mentors, carried on the tradition of arguing that the absence of transitional forms reflected the imperfections of the geologic record. As more and more sedimentary layers were analyzed and revealed more detail in the progression of life-forms – without solving the transition problem – paleontology reached something of a crisis by the 1960s.

Science is a self-correcting process, and flawed hypotheses find themselves defeated by the sword of new data. But often a whole picture or paradigm of the way things work, maintained over decades or more of work in a particular field, finally succumbs to a mountain of evidence. In this case, a wholly new paradigm usually comes to the fore, and often we view this in retrospect as a revolution in science. The development of quantum mechanics to supplant classical physics was one such revolution. The development of a new model for the way evolution works arguably could be called a minirevolution, one that introduced the concept of *punctuated evolution*.

18.2.2 Punctuated Equilibrium Approach to Evolution

On the island of Bermuda lies an excellent fossil record of the history of a particular snail species, over the last 300,000 years of Earth history (a period we cover beginning in chapter 20). The sedimentary record is well preserved, a large number of fossils occur in a fairly small area, and one subspecies is currently in existence. (A subspecies, even more difficult to define than a species, is a population with distinctive characteristics but still able to produce fertile offspring with other members of the larger species. That is, different subspecies of the same species interact genetically with each other, whereas different species do not. Breeds of dogs, or horses, constitute examples of subspecies.)

Harvard paleontologist Stephen Jay Gould studied this high-fidelity fossil record over a quarter of a century ago and came to a surprising conclusion: He could understand the formation and disappearance of new subspecies, not in terms of gradual evolution, but in terms of isolation and subsequent rapid changes in the characteristics of a given population of snails. Characteristics studied included variations in the color banding of the shell, in the general form of the shell's spire and lip, and in its thickness.

The details of Gould's analysis of the Bermuda fossil record are not repeated here, but the interested reader is referred to the article by Eldredge and Gould (1972). Figure 18.2 illustrates the relationships in time among the various subspecies of snails, based on the sedimentary sequence. In terms of timescales, the production of new subspecies was very rapid and occurred in populations at the edges of the main species itself. New subspecies of the snails appeared on the geographic periphery of a population of established snails with similar body characteristics. Body characteristics of a western subspecies did not appear on the periphery of the eastern subspecies, and vice versa. Eventually, the original subspecies became

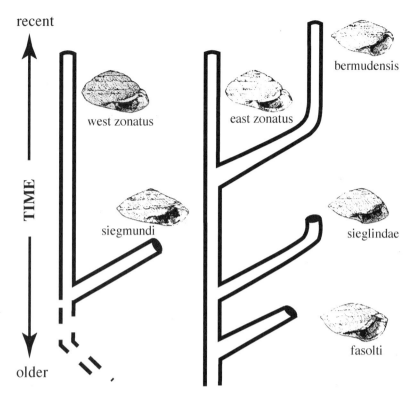

Figure 18.2. Results of Gould's analysis of populations of snails in Bermuda. The vertical axis is time, increasing upward. The branches are used to show the named subspecies branching off from the main species. Such branchings occurred on short timescales compared to that covered by the entire sequence. Although not indicated on the graph, these branchings occurred in geographically isolated populations. The two main species, west zonatus and east zonatus, presumably diverged from a common ancestor at a time well before that represented on the graph. Redrawn from Eldredge and Gould (1972).

extinct, leaving the newer ones to continue in the fossil record.

Other examples are cited by Eldredge and Gould in the fossil record, wherein new species are seen appearing suddenly in the fossil record at the geographic periphery of existing species, with body types clearly most closely related to that proximate population. At issue is how such new species could have evolved out of the old in a relatively short time span. Eldredge and Gould suggest that the new population, being on the periphery of the old, becomes isolated geographically, perhaps through migration, climate change, sea level rise or fall, or other factors. The members of this population then interbreed among themselves, so that particular characteristics (including recessive ones) of that genetic pool are distributed and enhanced through that population. In cases where new features are neutral or advantageous with respect to survival, the isolated population as a new subspecies or species might well outlast and replace the old.

That such isolated populations occur is amply illustrated by the cheetah, which today is remarkable in the enormous genetic similarity among different members of the species. Typical species, including humans, exhibit much greater genetic variety among individuals than does the cheetah. Sometime in the past, most of the cheetah population was destroyed by environmental changes, or perhaps by human hunting. Only a small, isolated, interbreeding subgroup remained. Inherent in this genetic identity lies the danger of a single disease wiping out the entire population, but such identity also implies that the particular special characteristics of this group are now part of what defines "a cheetah."

The bottom line of the punctuated equilibrium evolutionary model is that the fossil record accurately but incompletely reflects the history of the species contained within it. Species characteristics are stable over long periods of time, and then isolated populations of a given species may change rapidly as they breed among their limited group. The potential for change lies within the genetic code, expressed as an existing characteristic that gets amplified or more prevalent among certain members through interbreeding, or a random DNA mutation leading to an advantageous new characteristic or function. Evolution does *not*, in this view, proceed by gradual competition and elimination of the less fit members of a large breeding population; the action occurs among the few, in isolation, perhaps at times of great environmental stress.

Eldredge and Gould's idea was and still is controversial. It clearly explains some aspects of the fossil record, particularly its inability to show gradual morphological (shape) changes between one species and a clearly related one in younger sediments. But some changes seem so abrupt that punctuated equilibria cannot be the entire explanation for the nature of the fossil record. As sedimentary layers of rock are subjected to new tectonic forces in a region, they may be overturned, partially altered, or destroyed, taking key pieces of the fossil record with them. The vast majority of organisms that have lived on Earth did not have the honor of becoming fossils; their body parts were consumed by other organisms, including bacteria, to a complete or nearly complete extent. Finally, major disasters in Earth's history, one of which we document below, killed off species (often large numbers of species) very quickly, creating a break in the paleontological record.

For these reasons, the abrupt transitions in the fossil record must be a combination of imperfect preservation, breaks due to geologic processes and major extinction events, and the tendency of species to undergo rapid changes in isolated populations. Punctuated equilibria may well be the whole story in evolution, but it is only part of the story in explaining the appearance of the fossil record itself.

A lingering question that must be asked is whether the nature of the genetic code allows for rapid and ultimately profound changes in species characteristics. At one level, the punctuated equilibrium model doesn't make any demands on the genetic code: It simply requires that a small breeding population tend to enhance the particular special characteristics of group members, and we know this to be the case from breeding dogs or royalty. However, the genetic code must have within it the capacity to enable the remarkable development of intricate structures such as eyes, heart, brain, and hands, regardless of how evolution proceeds (that is, by punctuated equilibrium or otherwise).

The discovery of classes of genes that trigger, or turn on and off, whole groups of other genes, leading to drastic differences in form and function, provides a mechanism for dramatic changes in the morphology and functionality of eukaryotic cells and multi-celled organisms. These and other features of the genetic code seem to be capable of priming organisms for significant changes in their appearance and complexity. Over the enormous spans of geologic time, specialized cells and organs bootstrapped new functions out of old. But not in a gradual fashion; punctuated equilibrium implies long periods of stability in a given species followed by rapid, and perhaps dramatic, change.

Whether triggering genes or other components of the complex genetic code provide the key to the enormous variations seen in biology remains an open question, because we exist still in the youth of the genetic revolution. But that such a revolution exists at all is one of the wonders of twentieth century science. As recently as a century ago, the human egg was regarded as bearing homunculi: fully formed but miniature human beings awaiting the sperm to make them grow. The discovery of DNA followed by modern techniques of mapping the genetic code and manipulating genes have revealed a richness of malleable information at the heart of every cell that provides the enormous potential both for stability and change. Eukaryotic cells contain far more genetic material than is actually used to code for body type and function; the purpose of the rest remains unclear.

The rapid production of new forms and hence new species is not the purview of sexual reproduction alone. Many creatures clone themselves; some (like aphids) do so at particular times of the year and then later conduct sexual reproduction. The acquisition of extra chromosome pairs, crucial to development of commercial cotton, wheat, and other crops, also may play a role in development of new species. The splitting of chromosomes at their tie points occurs occasionally, and there is evidence in chromosome maps that some mammalian species are derived from others by this process (dogs from wolves, for example). Although the usual result of extra chromosomes is harmful (Down syndrome, for example), the genetic evidence tells us that there are occasional cases where something advantageous happens.

We need not understand the detailed working of the genetic code to understand that species are, in fact, complex dynamical systems. They draw energy through their cells, conduct regulated chemical processes to sustain and reproduce those cells, and they produce copies of themselves. Complex dynamical systems have the property of remaining in one sort of stable state or mode for some time and then suddenly, through a relatively small change in the external conditions, undergoing drastic change to a different state. Species are like that: They exist in stable form for an enormity of generations (tens of thousands to millions) and then, perhaps stimulated by a changing environment, an isolated fringe of that species will shift in relatively few generations to a different form or mode of operation. Nothing supernatural need happen: just the availability and activation of a complex but malleable sequence of genes with a range of regulatory and control functions.

In chapter 20, we examine the evidence for such rapid changes in the history of the precursors of modern humans, a story that is rich in long-lived hominid species punctuated by rapid changes. But long before those events came an extraordinary change in the nature of eukaryotic organisms, near the beginning of the Phanerozoic. It is this, the Cambrian revolution, that provides the most dramatic example of an explosion of novel body types in new species, one that defined the body plans for much of what was to follow in life's history.

18.3 VENDIAN-CAMBRIAN REVOLUTION

The time periods immediately preceding and following the start of the Phanerozoic eon, the Vendian and Cambrian periods, represent a biological fomenting unprecedented before and since that time. The fossil record of this time is seen in sediments that were originally seafloor environments and were later pushed onto the continents by plate tectonic processes to become part of the melange of preserved sediments. Most famous among these Cambrian sediments is the Burgess Shale, layers thrust upward in geologically recent times to form part of the British Columbian Cascade Mountains. However, many other, originally marine sediments from this time are found around the world. They all tell the same story of dramatic biological change. To appreciate the change that took place requires understanding a little of the way biologists classify organisms, a digression we turn to before examining the fossil record.

18.3.1 Taxonomy for the Restless

A necessary evil is at least a cursory acquaintance with the system for classifying life, established in its earliest form some 250 years ago. This *taxonomic* system classifies species according to morphological (form-based) relationships into progressively higher and broader categories, ending with the major eukaryotic *Kingdoms* of algae, plants, animals, fungi, and single-celled protists. (A higher level, *domain*, separates the two prokaryotic types from each other and from eukaryotes.) The order of the major taxonomic categories, from most to least specific, are species, genus, family, order, class, phylum, kingdom, domain.

A eukaryotic life-form can be identified in this scheme by its genus (capitalized) followed by the specific name – hence the domestic dog is *Canis familiaris*. The mountain coyote, which right now is hunting somewhere near the author's house, is *Canis latrans*; the grey wolf, *Canis lupus*. These and all other members of the genus *Canis* belong to the family Canidae, or true canines, which lies within the order Carnivora – flesh-eating mammals including cats, racoons, bears, hyaena, and many others. The class Mammalia is broader still, embracing the Carnivora, kangaroos, rats, aardvarks, dolphins, whales, humans, and others. The principal commonality is that all are vertebrates and all female mammals secrete milk for feeding their young. Mammals are part of the phylum Chordata, meaning animals with internal skeletons (hence excluding insects, crabs, etc.), central nervous systems, and with a few exceptions, possessing a vertebrate-type backbone and skeletal arrangement. (Strictly speaking, vertebrates are a subphylum of Chordata.) Finally, chordates are members of the animal kingdom, a multicelled eukaryotic organism, nonphotosynthesizing (hence not plant or algae), with distinct cell walls and a single nucleus per cell (hence not fungi).

The organization of creatures into this framework is an enormous undertaking, one that has been at the core of botany and other types of biological field endeavors. Although form was the original criterion, allowing extinct creatures to be classified on the basis of fossil morphology, sequencing of the genetic code to determine species relationships at the blueprint level is now a standard technique used on living species. Such an exercise might seem to some the ultimate in butterfly collecting, but it has revealed over decades of work the pattern of biological forms and has allowed inferences as to the genesis of particular species from each other. In the animal kingdom, 30 phyla exist, comprised of perhaps 10 million species.

18.3.2 Establishment of the Basic Plans

To appreciate the revolution at the start of the Phanerozoic requires recognizing that the taxonomic level *phylum* refers to basic, fundamental body plans. We instinctively think of insects and spiders (members of the phylum Arthropoda) as wholly different from us vertebrates because their body plan is based around a fundamentally different architecture – that of the exoskeleton. What is remarkable about the start of the Phanerozoic is that, within about 10 million years, all but perhaps one of the animal phyla in existence today appears in the fossil record. (This revolution is restricted to animals – higher plants began to appear in abundance as land-colonizing descendants of algae roughly 460 million years ago, after the Cambrian.) Included in the Cambrian shales is a fossil animal no more than 2-inches long with a clearly defined spinal rod, the 525-million-year-old ancestor of the Chordata – the mother of us vertebrates.

Understanding of the importance of the Cambrian explosion has grown, and especially in recent years. Examination of the Burgess shale early in this century revealed the appearance of many invertebrate phyla, but only in the past 30 years has re-examination of the shales revealed the forms of many soft-bodied animals previously missed. In the past decade, examples of chordate ancestors have been found, and better radioisotopic dating, using volcanic ash with uranium-bearing minerals, has shrunk the whole revolution down in time to a narrow 10-million-year window.

Not all phyla made it. Roughly a third again as many extinct phyla are seen in the Cambrian record as exist today. Moreover, each phylum is represented by far fewer species than are contained in extant phyla today. Clearly, the remarkable radiation of body forms involved large leaps in structures, without the more minor elaborations and variations at the family-through-species levels yet to come.

Equally important in this biological revolution is what didn't happen later: At no other time in the subsequent history of animal life did such abundant and new innovations occur. Once formed, the many phyla of the Cambrian became all there is. When we look at the remarkable diversity of mammals, reptiles, and birds, for example, we see that all have a remarkably similar body plan, on which small changes in form and function are elaborated. A human-sized, twelve-legged, exoskeletal merry-go-round with an extendable mouth at the bottom and a crop of bushy hair at the top, foraging across the grasslands, is not in the cards because the basics of such a body plan did not appear in the Cambrian. Nonetheless, many of the Cambrian pioneers of the dozens of new phyla were monstrous in themselves, such as those seen in figure 18.3.

18.3.3 Clues from the Vendian

What happened in the Cambrian need not invoke invention out of whole cloth. Recent careful work on fossil sediments from the period immediately preceding the Cambrian – the Vendian – has revealed increasing numbers of interesting, organized, multicellular animals. These sea creatures, existing over a time of several tens of millions of years, look very much like palm fronds arranged in linear, radial, or bilateral schemes. Some bear enough resemblance to worms, seaweed, and jellyfish to be argued as their ancestors, but the jury is still out. Are the Vendian creatures another failed experiment, on the eve of the Cambrian, in multicelled organized animals? Or are they the seedcorn from which the Cambrian phyla exploded, a kind of biological bootstrap that allowed modern body plans to be achieved?

18.3.4 Causes of the Vendian-Cambrian Revolution

Whether the Vendian biota were related to those in the Cambrian is secondary to the important innovation that was taking place: In a short span of geological time, complex and mobile creatures, organized of assemblages of eukaryotic cells, were appearing on a number of different body plans. Many of the innovations were relatively simple, animal hard parts being little different than the calcareous wastes or shells of more primitive unicellular creatures. More dramatic was the differentiation of cells into interdependent entities with distinct forms and functions. No long, gradual set of changes from multicelled, undifferentiated colonies to animals is seen – the eukaryotic menagerie remains relatively uninventive for the better part of a billion years prior to the Cambrian.

A number of possible factors, working together or separately, may have contributed to the Vendian-Cambrian revolution:

(i) *Increased oxygen levels.* By 600 million years ago, oxygen levels were not far from the present values and were increasing. A decrease in the abundance of biologically preferred carbon isotope ^{12}C relative to the rarer and less-palatable ^{13}C, suggests that just prior to the Cambrian a large amount of carbon was being removed from continents and shallow oceans and deposited on the deep ocean floor. Circumstantial evidence for less carbon in the system (hence less carbon dioxide) is the onset of an ice age at this time (chapter 19). Removal of large amounts of carbon from the biologically active shallow marine environments could have boosted the levels of molecular oxygen, which otherwise would have combined with the decaying carbon to make carbon dioxide.

The origin of the large carbon sequestration might lie in the expansive amount of continental area now available for erosion and rapid deposition into the sea (burying carbon detritus with it), along with changes in ocean circulation associated with plate tectonic cycles. Enhanced oxygen levels meant more energy for complex life-forms that require large amounts of aeration of tissues.

(ii) *Genetic complexity.* To take advantage of a new kind of environment requires that the genetic mechanisms be sufficiently complex to allow drastic changes in form and function. Eukaryotic cells have much more genetic

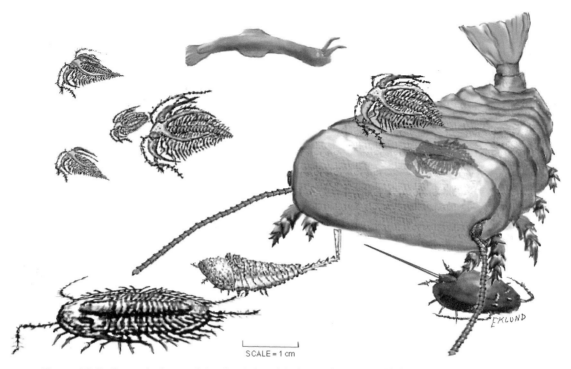

Figure 18.3. Some denizens of the Cambrian: (clockwise from top left) five marrellae, an amiskwa (swimming rightward), a habelia (with turned-up tail), the very large sidneyia under which is a burgessia, and a trilobite. Figure courtesy of Philip Eklund, Sierra Madre Games.

material than prokaryotes, and in fact only a very small portion of it is actively used to control protein production and other functions. The latency present in underutilized genetic information is a potentially powerful force for change. If multicelled, complex animals could not have survived below certain atmospheric oxygen thresholds, that would not have prevented repeated experiments in this direction, all of which failed before becoming abundant enough to show up in the fossil record. When the oxygen levels rose above a requisite threshold, the next set of organizational attempts worked, and animals began to multiply and diversify.

Some biologists have argued that the threshold trigger was not oxygen but genetics itself. More primitive multicelled animals have fewer regulatory, or trigger genes, than do more complex forms. Presumably there is a threshold number to achieve any sort of multicelled organization at all. In this view, the Vendian-Cambrian opened with a chance production of additional regulatory genes, presumably co-opted from the existing genetic codes available in eukaryotic cells. Possibly the Vendian innovation was oxygen-related, followed quickly and coincidentally by the innovation of more trigger genes and consequent explosion of body types. Or perhaps the trigger genes were available first, followed by the rapid rise in oxygen.

(iii) *Absence of predators.* Another contributing factor to the rapid appearance of different body types was the absence of other creatures in the same ecological niches and, specifically, the absence of predation. Simpler unicellular or colonial eukaryotes could not take advantage of the environment in the way that complex multicellular animals could; hence, once the latter appeared, they had no competition. Many different experimental body plans would all have flourished in this environment, until competition became intense enough, because of crowding, that predation took place. Some paleontologists see evidence for predation beginning at the end of the Vendian.

(iv) *Artifact of the geologic record.* Until recently it was thought that the sudden appearance of Cambrian shelled organisms was an artifact of their descent from earlier, soft-bodied creatures. More detailed analysis of sediments revealed large numbers of soft-bodied imprints in both the Cambrian and the Vendian, indicating that the sudden flowering of life was not an artifact of the data record. Nonetheless, one cannot discount entirely the possibility of future discoveries of animals in the fossil record *prior* to the Vendian.

Regardless of the specific causes, the Vendian-Cambrian revolution seems, in retrospect, inevitable. At issue is only the timing. It is quite possible, and intriguing to consider, that over the prior billion years similar experiments did happen. Absent enough oxygen, these creatures were ill-fated; alternatively, they might have flourished in environments temporarily rich in oxygen. As oxygen levels dropped again, such grand experiments disappeared, never abundant enough or sustained enough to show up in fossils recovered to date. Understood as a complex, self-organizing system (chapter 13), life's giant step across the multicelled-body threshold is an inevitable behavior of a dynamical system. Change the conditions and the system changes its mode of operating. Provide plenty of energy flow and more complexity and new examples of organization will appear spontaneously.

18.3.5 Why Has It Not Happened Again?

More difficult than why is "why not again." It is possible that Cambrian and post-Cambrian levels of biological complexity fully exploit the genetic capabilities of our cells, at least within the current environment. Little has changed on Earth in the past half-billion years. Oxygen levels have stabilized, plate tectonics moves continents around, and those as well as orbital cycles cause the climate of Earth to oscillate in ways that we describe in chapter 19. But fundamental changes in atmospheric or oceanic chemistry have not occurred during this time. Life has not had access to new energy-producing systems allowing new capabilities. Until such time as this happens (perhaps accelerated by human invention), we might expect that the Vendian-Cambrian revolution will continue to be played out in relatively minor innovations in body forms and functions.

Stimulating these minor innovations are the modest environmental changes hinted at in the preceding paragraph. These environmental changes, cyclic and stochastic, along with life's response to them, constitute the post-Cambrian history of Earth. In chapter 19, we examine some of the cyclical changes. In the remainder of this chapter we focus on catastrophic change as a vehicle for "emptying ecosystems," allowing creatures with new innovations on the old body plans to gain ascendancy.

18.4 MASS EXTINCTION EVENTS IN THE PHANEROZOIC

Five mass extinction episodes occurred in the Phanerozoic eon. These events are identified in the fossil record by the apparently sudden disappearance of large numbers of families of organisms. The severity of extinction events is classified in terms of the fraction killed of families that leave reliable fossil remains. For example, shelly marine organisms are often a good marker of extinctions: A major late-Permian event killed off half of the shallow-water shelled families, and 80% of the species contained in total in those families. Figure 18.4 shows the number of families in certain marine and nonmarine phyla during the Phanerozoic; major extinction events are marked, but many smaller extinction events are reflected in dips in the number of families. The graphs show that such events can have a long-lasting effect on the diversity of living forms. The graph for marine organisms also shows that the initial production rate of different families at the start of the Cambrian, extrapolated to today, would have yielded 10^{27} families at present, instead of the roughly 2,000 actual living marine families. Clearly, both catastrophies and the saturation of ecological niches in between extinctions work to limit the variety of life.

The causes invoked for mass extinctions have been many and varied: Drastic changes in climate such as major ice ages, large-scale volcanic eruptions, and impacts of large asteroids or comets are the most plausible candidates. The underlying causes of climate changes are themselves many and varied. We defer a discussion of cyclic or periodic climate-change processes to chapter 19. Here we consider in detail the evidence for, and effects of, impacts. Many of the effects of volcanic eruptions are similar to those of impacts, and the focus on impacts is not intended to minimize the role of large-scale eruptions. However, only recently has the importance of impacts in terrestrial processes been recognized, a result of the reconnaissance of the solar system and its accompanying images of crater-pocked surfaces (figures 7.3, 7.4). In particular, a compelling case has been made that the great extinction at the close of the Cretaceous, some 65 million years ago, was precipitated by a large impact.

18.5 CRETACEOUS-TERTIARY EXTINCTION

In the time from the Cambrian to the Cretaceous period, a span of nearly half a billion years, animal phyla spread from the sea to continental environments, starting just shortly after the higher plants came onto the land. The vertebrate subphylum was represented first, primarily by amphibians and then by reptiles about 300 million years ago. Subsequent to the major Permian mass extinction, a subclass of reptiles called dinosaurs (Archosauria) diversified and occupied a large number of different niches on land and sea, equivalent roughly to those occupied today by mammals and birds. Mammals had developed by this time, about 250 million years ago in the Triassic, but were less successful than the dinosaurs and were restricted to species of small rodent-like creatures. Famous for the enormity of their size and diversity of forms and habits,

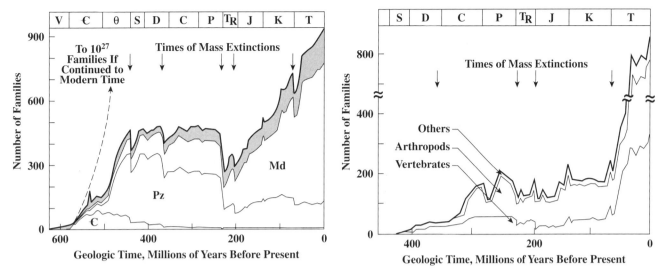

Figure 18.4. Number of biological families of (left) shelly marine fauna and (right) nonmarine animals over time. Dashed line extrapolates to later times the increase in number of families in the early Cambrian in the absence of extinctions. Arrows show times of mass extinctions. On the nonmarine graph, vertebrates and arthropods are broken out as well. Note that animals did not appear in nonmarine sediments, and hence were not on land, until almost 200 million years after the start of the Phanerozoic. The top portion of the graph labels the geologic periods using traditional symbols. "C, Pz, Md" refer to the three major faunal types that dominated the Earth's seas in succession: Cambrian, Paleozoic, and Modern. From Milne et al. (1985).

dinosaurs were the dominant large animal on land and sea up to 65 million years ago.

At the end of the Cretaceous, a dramatic demarcation in the fossil record occurs, wherein 15% of the shallow-water marine families become extinct – including 80% of shallow-water invertebrate *species*. The dinosaurs disappear too, though the massive size of their fossilized skeletons make it difficult to conclude definitively that their disappearance was sudden (i.e., dinosaur fossils are rare and the youngest occur a few meters below the upper end of the Cretaceous sediments). Above the Cretaceous sediments lies the Tertiary, a time when mammals diversified and occupied the niches left empty by the demise of the dinosaurs.

The apparently contemporaneous disappearance of the dinosaurs, other land species, and large numbers of marine organisms as recorded in the rock record is called the *Cretaceous-Tertiary boundary extinction*, or K/T boundary event. "K" is the common geologists' symbol for the Cretaceous period of Earth history, based on the German word *Kreide* for Cretaceous. The K/T boundary in the geologic record worldwide is dated isotopically at 65 million years ago.

18.5.1 Boundary Sediments

The dividing line at the K/T boundary is a thin (inches in extent) layer of clay. It has been identified in sediments worldwide. In addition to the sudden disappearance of many small marine organisms just below the boundary, replaced in sediments above by more modern forms, the clay contains peculiar abundance anomalies. The platinum-group elements – iridium, osmium, gold, platinum, and others – more closely resemble abundances in meteorites than in the crust of the Earth. As we have discussed, the crust of the Earth is a differentiated material relative to the bulk Earth abundances. In particular, at the K/T boundary, iridium is more abundant than in normal crustal rocks. Other properties of the thin boundary clay layer or adjacent layers include the following items:

(i) *Shocked quartz*. The impact of a large asteroidal or cometary fragment with the Earth's crust produces shocked grains of quartz with distinctive structure that allows their ready identification. These are seen in K/T boundary sediments around the world.

(ii) *Melt spherules*. Large impacts eject droplets of molten rock from the forming crater. These cool rapidly and solidify during their flight to form distinctive spherules. Their specific characteristics and occurrence over a large geographic area are diagnostic of an impact origin.

(iii) *Graphite*. Graphite, a form of carbon, and other evidence for burning is found at the K/T boundary in some parts of the world. Such burning would be expected from the large amount of debris lofted high into and even above the atmosphere: During re-entry this material would heat by friction (as does the Space Shuttle during its re-entry) and radiate this heat down to the surface. Distributed over a geographically large area by the impact, the molten debris would heat the air and the ground enough to ignite forests.

(iv) *Tidal wave action*. Impact into the ocean would generate large waves, with destructive effects upon reaching shore. There is evidence in some K/T boundary sediments adjacent to the Gulf of Mexico of patterns interpreted to be sudden wave action at the time the sediments were deposited.

18.5.2 Interpretation of the K/T Boundary as an Impact Event

An asteroidal or cometary fragment roughly 10 km in size, striking Earth at high velocity can explain the characteristics of the K/T boundary layer and adjacent material. Such an impactor would gouge a crater in excess of 100 km in diameter, blow a temporary hole of that size through the atmosphere, and fling dust into the upper atmosphere. The molten flying debris and pressure wave would burn and knock down trees across thousands of kilometers of land. Rock near the impact site would be "shock heated," changing its character to that seen, for example, in the shocked quartz of the K/T boundary. If the impact were into water, the resulting tidal wave would inundate adjacent land areas for hundreds of miles around. The material blown into the stratosphere (20 km or more above the surface) would circle the Earth; enough dust would be available to shroud the Earth in darkness for months. The dust, a mixture of impactor and crustal material, would fall onto continental and seafloor surfaces, carrying with it the chemical signature of the asteroid.

The general properties of the K/T boundary appear to be best explained by an impact. Volcanism involving magma from the deep interior would produce an iridium enhancement, but the ratios of iridium to the elements gold and osmium are different from those in meteorites and the K/T sediments. Fire fountaining in volcanoes can produce spherules, but usually on a local scale, and in basaltic eruptions; by contrast the K/T spherules are andesitic or dacitic. The arguments against volcanism as a cause of the K/T boundary phenomena do not rule out episodes of widespread volcanism as causing mass extinctions at other times in Earth's history; they pertain only to this particular event.

The energy released from such an impact is extraordinary, as shown in figure 18.5. Over a million times more powerful than the Mount St. Helens eruption or the largest nuclear test explosion, such an impact has no rival with regard to anything experienced by humankind. However, serendipity allowed humans to view a similar impact into

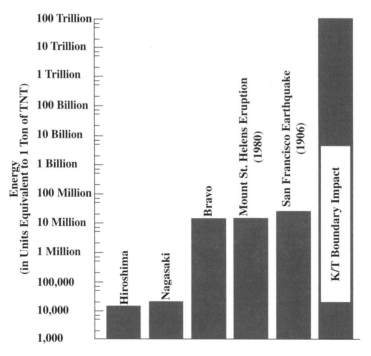

Figure 18.5. Comparison of the energy released in a putative K/T impact event with other energetic processes. Bravo is the largest U.S. nuclear explosion. From Kring (1993).

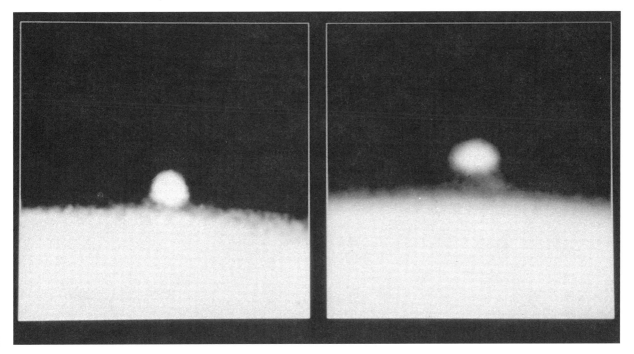

Figure 18.6. Hubble Space Telescope views of the 1994 impact of the Comet Shoemaker-Levy fragments into Jupiter. (a) Mushroom cloud rising over the Jovian limb.

another planetary atmosphere, that of Jupiter, in 1994. In 1992, the orbit of Comet Shoemaker-Levy 9 was perturbed by a close pass to Jupiter and the comet itself was torn into two dozen fragments. Two years later these fragments collided with Jupiter's atmosphere. Although the largest pieces were a kilometer or less in size, the great gravity of Jupiter resulted in a much larger entry speed into the atmosphere of Jupiter than would have been typical at Earth. Since the energy of motion (kinetic energy) scales as the speed squared, this higher velocity was such as to make the energy of the biggest impacts similar to that of the K/T impactor. Clouds of dust the size of Earth, created by the comet fragments and by chemistry in the hot plume of rising gas after impact, were imaged by telescopes around the world and in space. Hubble Space Telescope in particular caught a spectacular glimpse of the mushroom clouds raised by the impact (figure 18.6 and color plate VII).

18.5.3 Biological Effects of the Impact

In addition to the direct effects of the impact, such as widespread forest fires and tidal waves, the plume of debris and smoke injected into Earth's upper troposphere and stratosphere would have a devastating effect on life through alteration of the climate. By physically blocking the rays of the Sun, the dust would cause the lower atmosphere and the surface of Earth to cool suddenly, and remain this way for weeks or months. Models suggest much of the continental area of Earth would have average daytime temperatures of only $10°C$, much lower than the present average. In addition to the direct effects of cold on large animals and plants, the reduced sunlight would slow or shut down photosynthesis for up to a year, killing off large numbers of species dependent on various marine and continental food chains.

Analysis of the K/T boundary sediments suggests that the impact site might have been a seafloor carbonate platform, where the carbonate was deposited over long periods of time by shell-forming organisms. The amount of carbon dioxide released from the carbonate target on impact could have been very large, forcing the atmosphere to heat strongly after the cooling dust settled. Thus months of global winter might have been followed by years, decades, or more of global warming.

Another important effect of the impactor was on the chemistry of the atmosphere. As the impactor, or bolide, streaked through the atmosphere, it heated the air around it and caused chemical reactions to take place between the nitrogen and oxygen to produce nitrous oxides. Lightning

Figure 18.6. (Continued) Hubble Space Telescope views of the 1994 impact of the Comet Shoemaker-Levy fragments into Jupiter. (b) Clouds of dark material from multiple impacts, each of which is larger in area than the Earth. Courtesy NASA/Space Telescope Science Institute.

does the same thing when it strikes, but the bolide could have produced enough that, when dissolved in water, the nitrous oxides and other synthesized compounds would have killed off large numbers of marine organisms. Calcium-shelled species could have had their shells dissolved by a change in the acidity of lake waters; such shelled organisms seemed to have suffered the most in terms of the fossil record of species extinctions. Sulfur oxides released from the seafloor would have converted into sulfuric acid in the stratosphere, creating acid rain that may have defoliated vegetation and altered water acidities over large areas.

Other effects of the impact may have been longer lasting. Tremendous amounts of water could have been injected into the stratosphere by the impact, reducing the ozone abundance which then would have increased the incidence of cellular damage in surface organisms through elevated levels of uv radiation. Hydrogen sulfide produced

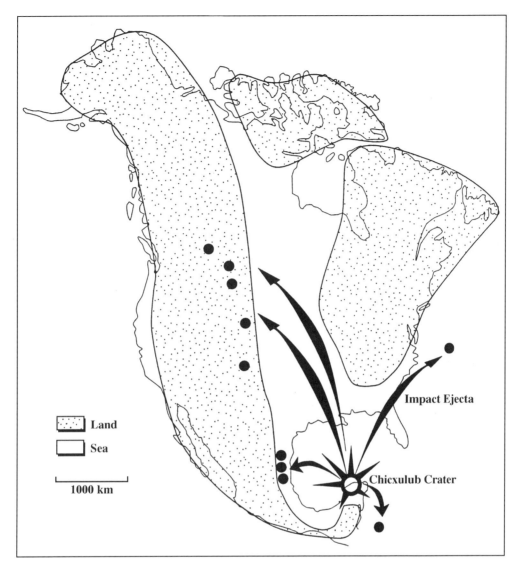

Figure 18.7. Map showing the location, off the Yucatan Peninsula of Mexico, of the Chicxulub crater. The present outline of the North American continent is shown. Shading indicates the approximate positions of North American continental areas that were above sea level during the late Cretaceous. Areas that are not shaded were under water, in part because of the very high sea levels of the late Cretaceous (see chapter 19). Locations are labeled where impact ejecta associated with the crater have been collected. Courtesy of David Kring, University of Arizona.

in large quantities by worldwide decaying vegetation may have caused secondary food chain disruptions.

18.5.4 Where Is the Crater?

The inability to identify a large eroded crater remnant 65 million years old was for many years a puzzle. If the impact was in the ocean, the part of the crust containing the crater may have traveled to a subduction zone by now and been destroyed by subduction. However, careful detective work by a number of U.S., Canadian, and Mexican geologists has unveiled a host of evidence from rocks in North and Central America that pinpoints the location of the crater. If the interpretation is correct, the crater is not subducted, but instead is buried in the Gulf of Mexico right off the tip of the Yucatan Peninsula, at a site called *Chicxulub* (figure 18.7). Evidence supporting this site for the crater includes:

(i) The thickness of deposited debris at the K/T boundary (*ejecta*) decreases with distance from Chicxulub, consistent with it being the source (figure 18.8). Sites relatively close to the postulated crater show evidence for two layers: the low-energy debris thrown out and

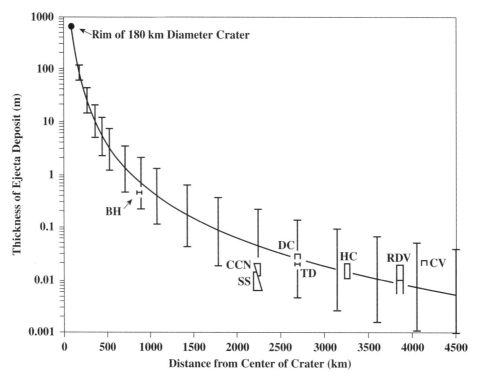

Figure 18.8. Thickness of K/T deposits as a function of distance from the Chicxulub (Yucatan, Mexico) site as analyzed by Vickery et al. (1992). Abbreviations label different geographical locations where the thickness of K/T boundary sediments, interpreted to be ejecta from an impact, has been measured. At most of the locations, the ejecta thickness varies over some range, and this is indicated by the irregularly shaped boxes. The curved line is from a computer model predicting the thickness of ejecta from an impact event at Chicxulub, assuming the crater to be 180 km in diameter. Uncertainties in the predicted thickness are expressed by the vertical error bars spaced along the curve. Figure courtesy of Ann Vickery, University of Arizona.

deposited locally and the high-energy material projected into a globe-circling layer.

(ii) The size of spherules and amount of shocked quartz is much larger in the Caribbean region than elsewhere. The chemistry of some of the spherules close to Chicxulub is high in calcium, consistent with impact into the calcium carbonate deposits associated with the Chicxulub site.

(iii) Studies of the geology of the Chicxulub site, including mapping of the density of the crust by measuring gravity variations, and mapping of iron concentration using magnetic measurements, are all consistent with a buried impact crater being located there.

Detailed studies of the site, many of which were conducted years ago for oil exploration, reveal a circular structure roughly 180 kilometers in diameter, with several other rings interior and exterior to the main one. Shocked material in and around the site dates isotopically to a single time of 65 million years. Field studies are ongoing, and evidence that the Chicxulub structure is the remains of a huge impact crater continues to mount.

18.5.5 Impacts and Other Extinction Events

Although several other major extinction events, which mark boundaries in the geologic time of the rock record, have been tentatively proposed as being associated with impacts, no records with the clarity of the K/T boundary exist for them. Hints of iridium enrichment, shocked material, or other evidence appear sporadically in other sedimentary layers, but never as abundantly as at the K/T boundary. Either the quality of the impact evidence at K/T is an anomaly, or impacts are not the primary cause of the bulk of Phanerozoic mass extinction events.

The characteristics of the Chicxulub impact site, with its carbon- and sulfur-laden marine sediments, probably ensured an unusually severe reaction on the part of Earth's atmosphere and biosphere to that impact event. It may well be that other impacts of similar magnitude at other times in the Phanerozoic did not have such a profound effect on life. The other major extinction events in the Phanerozoic may have had other causes, such as massive volcanism, or the more cyclic types of climate change to be discussed in chapter 19.

Although this line of argument separates extinction events from the requirement that they be impact-related, it does not explain why the sedimentary record does not show other impact events with the clarity of the K/T boundary event. Deep-ocean impacts, far from land, might well have lofted less dust worldwide – and the resulting ocean-floor crater might have been subducted and hence lost forever to discovery. Other impact events are preserved (usually poorly) on continental shields, but only a few rival Chicxulub in size. Evidently, the low frequency of large impacts in the Phanerozoic and the very active processes of erosion and subduction on Earth conspire to make the record of large impacts sparse indeed.

Many of the severe and sudden changes in climate and atmospheric chemistry induced by the bolide at the K/T boundary have their analogs in today's human activities, though on a less massive scale. Acid rain, enhancements to the greenhouse effect, and ozone depletion are with us today. A nuclear war could release enough dust through burning of cities to cool the surface of Earth and destroy agricultural food production for months or years. Direct human destruction of habitat, occurring now, may lead to loss of one-fourth of all species present on Earth over the coming decades. Such an extinction event, seen from the perspective of the fossil record, would be classified as intermediate or major along with the K/T event, the Permian event, and other great catastrophes of the Phanerozoic eon.

18.6 QUESTIONS

a. Can you conceive of several alternative explanations for the lack of transitional forms in the fossil record?

b. Is the Vendian-Cambrian revolution an inevitable result of increasing genetic complexity? If so, what might you imagine could happen in a putative future revolution? Is such a revolution prohibited by external environmental conditions?

18.7 READINGS

18.7.1 General Reading

Margulis, L., and Sagan, D. 1986. *Microcosmos*. Summit Books, New York.

Sagan, C., and Druyan, A. 1992. *Shadows of Forgotten Ancestors: A Search for Who We Are*. Random House, New York.

Taylor, G.R. 1983. *The Great Evolution Mystery*. Harper and Row, New York.

18.7.2 References

Cloud, P. 1988. *Oasis in Space: Earth History from the Beginning*. W.W. Norton, New York.

Eldredge, N., and Gould, S.J. 1972. Punctuated equilibria: An alternative to phyletic gradualism. In *Models in Paleontology* (T.J.M. Schopf, ed.). W.H. Freeman and Company, San Francisco, pp. 82–115.

Gould, S.J. 1985. *The Flamingo's Smile: Reflections in Natural History*. W.W. Norton, New York.

Gould, S.J. 1969. An evolutionary microcosm: Pleistocene and recent history of the land snail P. (*Poecilozonites*) in Bermuda. *Bulletin of the Museum of Comparative Zoology* 138, 407–531.

Kring, D.A. 1993. The Chicxulub impact event and possible causes of K/T boundary extinctions. In *Proceedings of the First Annual Symposium of Fossils of Arizona* (D. Boaz and M. Dornan, eds.). Mesa Southwest Museum and Southwest Paleontological Society, Mesa, Arizona, pp. 63–79.

Milne, D., Raup, D., Billingham, J., Niklaus, K., and Padian, K. (eds.) 1985. *The Evolution of Complex and Higher Organisms*, NASA SP-478. U.S. Government Printing Office, Washington, DC.

Vickery, A.C., Kring, D.A., and Melosh, H.J. 1992. Ejecta associated with large terrestrial impacts: Implications for the Chicxulub impact and K/T boundary stratigraphy. *Lunar and Planetary Science* XXIII, 1473–1474.

19 CLIMATE CHANGE ACROSS THE PHANEROZOIC

19.1 INTRODUCTION

The preceding chapter focused on singular events in the later history of the Earth—the flowering of multicellular complex organisms at the start of the Phanerozoic eon and the widespread extinction of species some 65 million years ago at the close of the Cretaceous period. Although these events stand out in their drama and the mystery of their causes, any understanding of the interactive history of life and Earth's environment cannot rest on their study. Throughout the Phanerozoic, and before, the relatively steady rhythms of plate tectonics brought continental masses together and then moved them apart, creating new seafloor and destroying old. The process of moving great landmasses around the planet must have had profound effects on the environment, and indeed this is seen to be the case in the geologic record.

This chapter begins by reconsidering plate tectonics with an eye to understanding the apparently cyclical creation and breakup of multicontinent landmasses, or *supercontinents*. We consider the effects of such supercontinent cycles on the amount of volcanic activity, and hence atmospheric chemistry, on the ocean circulation patterns, on mountain building, and hence on the available area for storage of continental snow and ice deposits. Such considerations touch on a major theme of the latter portion of Earth history, the comings and goings of great ice ages. Finally, we draw our attention in detail to a particularly warm time in recent Earth history, the Cretaceous period. Ice-free and showing much less drop in temperature from equator to pole than Earth experiences today, the Cretaceous has become a proving ground for climate modelers who seek to predict the amount and nature of global warming in humankind's future.

19.2 THE SUPERCONTINENT CYCLE

The ultimate causative agent of plate tectonics must be the release of heat from Earth's interior through mantle convection, but the details of continental movement and seafloor subduction cannot be tied directly to the interior convective patterns, at least based on computer models simulating those deep motions. Instead, the surface patterns of plate motion depend upon several things visualized in figure 19.1: the age and density of the oceanic crust, collisions between continents, and the deflection of mantle heat sources by piled-up supercontinental masses.

Oceanic crust newly created at midocean ridges is hot, and hence relatively buoyant. As this crust is displaced by yet younger crust, it rolls laterally away from the ridge, cooling as it does. Cooler crust contracts, and becomes denser. If the older oceanic crust does not encounter a pre-existing subduction zone, forcing it under, it eventually will cool and densify enough to sink spontaneously, creating a new subduction zone. Evidence from magnetic reversals on the seafloor (chapter 9) that no portion of oceanic crust is older than 200 million years is buttressed by computer models suggesting that beyond that age the ocean crust is indeed too dense to be supported by the asthenospheric part of the mantle.

Continental collisions are self-explanatory: Because continental crust is buoyant at any age, collisions between continental landmasses on adjacent plates force the directions of plate motions to shift. Strong compression during such collisions raises mountain ranges, such as the Tibetan Plateau (with Mt. Everest), raised by the current collision of India with Asia. Furthermore, as bigger aggregations of continents build, heat flow from the mantle is inhibited by the thick crusts and insulating properties of these buoyant

a. Seafloor Subduction

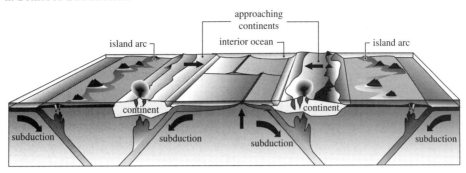

b. Continental Collisions

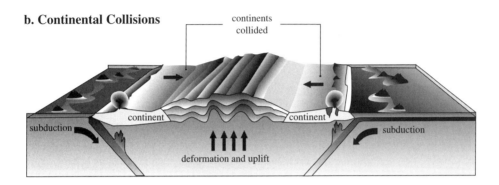

c. Mantle Plumes and Supercontinent Breakup

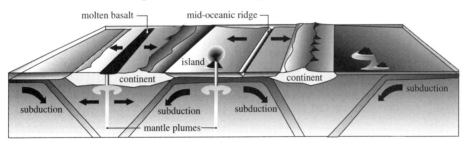

Figure 19.1. Three processes important in the determination of plate motions: (a) subduction of cold, dense ocean crust; (b) collisions between buoyant continental masses; (c) effect of thick continental crust on heat flow from the mantle. In panel (c), a mantle plume has developed beneath the supercontinent on the left, encouraging breakup.

masses. As a result, heat flow elsewhere may increase, precipitating new oceanic ridges, or may eventually rift the continents apart again. The idea that plate motions on long timescales have a cyclical character defined by continental collisions is suggested by the tracking of plate motions as far into the past as feasible, perhaps a billion years or more. Originally proposed by Toronto geophysicist J. Tuzo Wilson and refined by others, the supercontinent cycle goes as follows:

(i) The continents are collected together in a single amalgamated mass (a supercontinent), surrounded by a global ocean (a universal ocean).

(ii) Mantle heat is trapped beneath the supercontinental crust; temperatures rise within the crust, causing expansion, arching, and fracturing of the supercontinent. Additionally, the spin of Earth puts a small additional stress on the supercontinent, which sits like a raised pimple above the ocean floor and hence is subject to a higher centrifugal force than the seafloor crust.

(iii) Rifting of the supercontinental mass occurs along one or several lines. Mantle material rising up in the space between the newly-fragmented continents partially melts, forming oceanic crust along a new mid-ocean ridge in a growing "inland" ocean. As new seafloor is created, the continents spread apart, the boundary between continent and seafloor being a tectonically quiescent *passive margin*. In the universal ocean surrounding the exterior margins of the continents, subduction zones at continental margins and elsewhere consume seafloor, shrinking the universal ocean.

(iv) Seafloor at the continental margin of the new ocean becomes older and colder until finally buoyancy is lost and subduction begins. Subduction halts or redirects the growth of the new ocean. Continental masses no longer spread outward but may begin to converge again until collisions recreate a single supercontinent.

Figure 9.10, showing the motion of the continents over the past 200 million years, illustrates the first half of the supercontinent cycle. The breakup of the last supercontinent, Pangaea, initiated the opening of the Atlantic Ocean and the shrinking of the Pacific. The margins of the Atlantic do not contain subduction zones, but instead are passive boundaries with the surrounding continents. In contrast, the continental margins in the Pacific are sites of active subduction zones or, where lateral motion is taking place, transform faults. The earthquakes and volcanic eruptions along the Pacific ring of fire stand in stark contrast to the quiet of the Atlantic region. The supercontinent breaks up not once, but several times, until the current number of separate landmasses is reached. Eventually, perhaps in a few tens of millions of years or less, the Atlantic will develop subduction zones as cooling ocean crust loses buoyancy. Tectonic activity will develop along the Eastern seaboard of the United States, Western Europe, and West Africa. The expansion of the Atlantic will end and the continents will eventually collide back together to form a single landmass.

Reconstructions of early plate tectonic cycles support the notion that previous supercontinents existed, the one prior to Pangaea rifting apart perhaps 700 million to 800 million years ago. Tenuous evidence for an earlier episode also exists in rocks a half-billion years older still. So, supercontinents break up and reform on Earth every half-billion years or so, perhaps as far back as the end of the Archean when enough continental mass existed to influence the motion of the crustal plates.

19.3 EFFECTS OF CONTINENTAL BREAKUPS AND COLLISIONS

The separation and collision of continents does more than just alter the geographic map of the world over time; changes in continental positions and possible accompanying pulses of geologic activity play roles in altering climate. These effects continue to be an active area of research, and a detailed correspondence between plate positions and possible ancillary events in the geologic record remains elusive. However, several potential effects can be identified.

Mountain building is associated both with the expansion of continental masses away from the supercontinents and with subsequent collisions. Interior mountain chains and highland plateaus result from continents colliding with each other; mountain chains along the exterior of a continent are built up by volcanism associated with the subduction of ocean floor beneath the edge of the continent. In either case, the buildup of new continental highlands produces more land area for ice accumulation, with effects that we discuss in section 19.5.

Volcanism associated with the formation of new subduction zones along continental margins as well as in the seafloor exterior to the diverging continents puts large amounts of ash, aerosols, and greenhouse gases into the atmosphere. Like large impacts, the initial effect is a cooling as atmospheric aerosols reflect or absorb some sunlight. Eventually, the aerosols drop out, but the carbon dioxide and other greenhouse gases added to the atmosphere remain for much longer and contribute to a hotter climate. Volcanic gases and ash added to lakes and seas change the acidity of the waters, altering their suitability for adapted organisms.

Volcanic episodes are not restricted to the continental margins; a surge of eruptive activity associated with the initial rifting of a supercontinent may have dramatic climate effects as well. A massive extrusion of lava over a 200,000-square-mile region, the so-called Deccan Trap lava flood in India, occurred some 65 million years ago, associated with rifting away of part of the continent. Ancillary effects of the eruptions might have played a role in climate change and possibly extinctions near the K/T boundary, additional to (or, some argue, in place of) a large asteroid impact.

Changing continental positions have two primary effects on climate. First, the drift of continental fragments toward higher latitudes than those occupied by the supercontinents, which seem to have had their geographic centers at low latitudes, allows more snow and ice accumulation to take place. High-latitude continents are better accumulators of snow and ice than are high-latitude seas, primarily because continental areas have elevated terrains. Second, as continents drift, ocean currents, which transport warm and cold ocean water over vast distances, shift in their strength and direction. The so-called North Atlantic deep water, an area of sinking salty water that strongly moderates Europe's climate, is shaped in large measure by the North American continental margin. (The

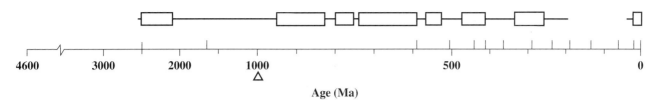

Figure 19.2. Timeline of ice ages. Boxes correspond to times in Earth's history during which widespread glaciation occurred, based on geologic data from a number of locations around the globe. Other times when at least some year-round ice was present on Earth are marked with the horizontal lines. Blanks characterize episodes when Earth had little or no year-round ice anywhere. Times are marked in millions of years before present: Thus, "1,000 Ma" is a billion years ago. From Barron (1992).

role of this major ocean feature in climate is discussed in chapters 21 and 22.)

If indeed the motion of tectonic plates plays a role in determining Earth's climate, such modulation should be present in the geologic record. And, so it is, in the form of long epochs of ice ages that have occurred several times over the past billion years of Earth history, and even earlier.

19.4 EVIDENCE OF ICE AGES ON EARTH

Ice ages is a loose term for times in Earth's history when glaciers covered large areas of the continents, down to mid-latitude regions and hence much farther equatorward than today. Glaciers, year-round sheets of ice and entrained rocks of all sizes from grains to huge boulders, leave characteristic signatures as they advance across the landscape and then break up. (Few glaciers recede large distances intact.) These features are distinct from the erosive effects of liquid water because of the very different mechanical properties of liquid water and ice.

Glaciers carve out U-shaped valleys and bowl-shaped *cirque* basins in mountainous terrain. On a continent-wide scale, the advance of glaciers with their embedded rocks scratches and striates the surface. Debris pushed ahead of glaciers and abandoned when the glaciers vanish creates the undulating *moraine* terrains. The shear weight of ice sheets that rise 3,000 meters above the surrounding terrain depresses the upper continental crust; as the glaciers disappear and the land rebounds, lines of stress called *strandlines* appear over large areas. On a small scale, glaciers do not sort and round rocks the way streams do—poorly sorted angular material is more characteristic of glacial debris. In some rare cases, freezing muds may capture the imprints of ice crystals at the base of the glaciers.

Such geologic signatures (and others) of glacial activity are present at sites where glaciers still exist—or did in historical times—and amply over the broad northern continental ranges affected by the glaciations of the past million years. To adduce the existence of much earlier ice ages, upwards of a billion years ago or more, is a much more difficult proposition. Perhaps the extreme case of this is the attempt to infer glacial epochs on Mars, as described in chapter 15, where geologic processes have been dominated by impacts and some volcanism, with ancient episodes of water erosion. Since only a tiny part of Mars has been examined by landed instruments, the search for glacial features is limited to orbital surveys, and hence only large-scale features serve for now as the (rather controversial) evidence for sheets of ice sometime in Mars' past.

On Earth, at least, the rocks can be examined at close range. Ancient rock strata preserved in the older shields of the continents must be examined for the small-scale evidence of glacial action; large-scale glacial terrains from ancient ice ages have been largely erased by subsequent tectonic and erosive processes. The most common and diagnostic indicators of the existence of ancient glaciers are rock surfaces that are polished and striated, pebbles with a characteristic shape associated with glacial scouring, and agglomerations of large angular rock fragments in a fine-grained matrix. These and other signatures in the sedimentary rock record have been used to infer several major episodes of glaciation over the past billion years; faint signatures of earlier events exist in the rock record up to 2.8 billion years ago (figure 19.2).

19.5 CAUSES OF THE ICE AGES

19.5.1 Positive Feedbacks in the Basic Climate System

Widespread continental glaciation represents a distinct state of the complicated physical system comprising Earth's atmosphere, oceans, and continents. As with many complicated, nonlinear physical systems, a series of small changes may push the system into an entirely different state, as positive feedbacks amplify the small perturbations. Continental ice cover is a good example. Adding ice sheets to a continent, for whatever reason, raises the *albedo* or reflectivity of the surface, ensuring that less sunlight is absorbed by the ground, and hence less energy is reradiated as infrared photons back into the atmosphere. The contribution to the annual mean temperature and the atmospheric heat budget of Earth is less from regions that become ice-covered, and global temperatures drop. This encourages more ice to form at even lower latitudes (on both land and oceans) and the system is driven toward a state in which large areas of Earth are covered in ice.

The triggers for such ice ages remain somewhat controversial. Clearly one trigger is the movement of continents, split off from a single supercontinent, toward higher latitudes. This drift puts more landmass in regions where cold climate allows ice accumulation. The production of mountain ranges associated with high-latitude continental collisions, collisions of continents with island arcs, or subduction zones pushes continental landmass to higher altitudes, encouraging further ice accumulation. The evidence for Proterozoic and Phanerozoic plate tectonic cycles of continental assemblage into supercontinents, followed by breakup, is strong. Although correspondence between past ice ages and dispersal of continents cannot be made confidently because of uncertainties in the ages of both and in the timing of the onset of glaciations relative to continental positions, it is a plausible connection.

19.5.2 Negative Feedbacks in the Climate System

In practice, negative feedbacks prevent Earth from going to a completely, permanently, ice-covered state. During ice ages, less precipitation occurs in the form of rainfall, and hence less erosion and removal of atmospheric carbon dioxide to the seafloor (as carbonates) occurs. This effect is accentuated if continental masses are at high latitudes, where precipitation is almost all snow and hence erosion is less effective (figure 19.3). The volcanic outgassing of carbon dioxide previously subducted as carbonates continues regardless of the carbonate production rate so that, during the ice age, there is a net tendency of carbon dioxide to increase. This in turn increases the infrared opacity of the atmosphere, enhancing the greenhouse warming and eventually offsetting or ending an ice age.

The negative feedback associated with the resupply of carbon dioxide and other volatiles to the atmosphere distinguishes Earth from Mars. Mars has been in a perpetual ice age since early in its history, punctuated perhaps by only the briefest of episodes of running water. As described in chapter 15, the absence of plate tectonics, the relative ease with which the atmosphere could escape to space, and the more distant Sun all played important roles in shunt-

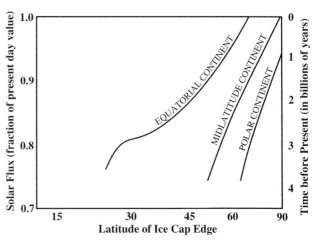

Figure 19.3. Example of the possible effect of continental positions on the severity of past ice ages. A model was constructed by University of Michigan scientists H. Marshall, J.C.G. Walker, and W.R. Kuhn of the balance between carbon dioxide consumption by weathering and release by volcanism and metamorphic heating. The weathering rate was varied depending on the latitude of an assumed single supercontinent. An equatorial supercontinent, receiving essentially all of its precipitation as rain, allows carbon dioxide to be consumed more quickly than does a near-polar supercontinent that receives its precipitation in the form of snowfall. The graph shows the lowest-latitude limit of ice sheets for three different model supercontinents at various times in Earth's history, corresponding to different values of the solar luminosity. At no time does the ice reach completely to the equator, but ice ages, once begun, are less severe when continents are confined to high latitudes. Adapted from Marshall et al. (1988).

ing the Martian climate to this state. Important here is the recognition that, although Earth's climate is not constant, but instead oscillates between warm and cold extremes, the feedbacks afforded by tectonic and other processes have kept these oscillations small enough that the basic state of stable liquid water is retained.

19.5.3 Additional Influences on Global Glaciation

Other effects act on the extent and duration of glaciation but the direction and magnitude of each are harder to quantify. The positions of the continents determine in part the pattern of ocean currents that transport warm equatorial seawater to higher latitudes. The presence of high-latitude continents and high-altitude ice sheets alters the patterns of storm systems, hence affecting timing and amounts of rainfall and snowfall. Buildup of mountain ranges and high plateaus also might increase the rate of weathering and subsequent loss of atmospheric carbon dioxide. Causes external to Earth may trigger ice ages as well. Early in Earth's history, the Sun's lower luminosity would have made it easier for Earth to slip into ice ages.

In fact, absent the enhanced carbon dioxide abundance postulated for the Archean and Proterozoic atmospheres (chapter 14), Earth would have been in a continuous ice age that could have thwarted the establishment and development of widespread life. Temporary dips in the Sun's luminosity later in Earth's history, or passage of the solar system through dusty molecular clouds, attenuating the sunlight reaching Earth, cannot be ruled out either as sources of cold episodes.

19.5.4 Snowball Earth?

Continued collection of geologic evidence is essential to understanding the extent of past ice ages. For a long time, the severity of Proterozoic ice ages posed a puzzle for climatologists: Continental fragments that should have been at very low latitudes during that time showed evidence of glaciation. Could Earth have been completely ice covered? Climate models failed to produce such global glaciation without plunging Earth into a permanent and irretrievable deep freeze. Recent new data showing the direction of magnetization of Proterozoic rocks have forced a reexamination of the geologic evidence. Michigan geologists J. Meert and R. Van der Voo, as well as others, have reconstructed continental positions and increased the latitude of many continental fragments showing evidence of ancient glaciation. Proterozoic ice sheets, it now appears, did not extend below 25° latitude. This still represents a very deep glaciation compared to even the coldest part of our current ice age epoch, and might well indicate the influence of a slightly less luminous Sun.

19.6 CRETACEOUS CLIMATE

The mid-Cretaceous, from roughly 100 million years ago to its conclusion 65 million years ago, appears to have been characterized by an Earth with no permanent ice caps, equatorial mean annual temperatures slightly higher than today, and polar-cap mean annual temperatures 40° to 60°C higher than today. Such a world would look from space much different than our present Earth, with the Arctic and Antarctic ice caps not present. It also would have been a far different place to live, with little variation in climate from the equator to high latitudes. It represents an extreme in climate, opposite to that of the deep global glaciations, and which can be studied in detail because it occurred recently in Earth history. Understanding this last warm time in Earth history is therefore a priority among climatologists, who also see in the Cretaceous a guide to the possible effects of human-induced global warming.

19.6.1 Evidence for the Cretaceous Climate Pattern

The following constraints exist on the Cretaceous climate:

(i) *Isotopic data.* Stable isotope ratios, primarily ^{18}O to ^{16}O, are available for a number of sediments from Cretaceous times that were formed in equatorial and midlatitude seas. By choosing sediments characteristic of both deep-sea and shallow-sea environments, it is possible to get a profile of ocean temperatures with depth, as well as latitude (chapter 6).

(ii) *Fossil organisms.* Plate tectonic motions have carried continents far in the 100 million years since the mid-Cretaceous. It is possible to reconstruct the pattern of continents, which then permits the location of Cretaceous fossils according to latitude to be determined. A number of fossils indicate equable climate to the poles at that time. Coral reef and carbonate formations extended 5° to 15° of latitude poleward of their current limits, because of warmer conditions. Fossil alligator and crocodile remains indicate that these tropical creatures lived at latitudes up to 60° north and south in Cretaceous times. Other fauna support this pattern; fossils of cold-water species are absent from the Cretaceous sedimentary record, and diverse numbers of warm-water species are present at high latitudes.

(iii) *Geology.* Glaciers are a primary force for erosion at high latitudes and high altitudes on Earth today. Yet the key patterns revealing glacial erosion are missing from Cretaceous rock formations that were at high latitudes. Some temporary ice may have formed during parts of the year during the Cretaceous, but year-round ice is largely ruled out by such findings.

The results of the range of evidence presented here are summarized in figure 19.4 as annual mean temperature as a function of latitude during the Cretaceous. Two estimates based on the data—lower and upper limits—are compared with the present annual mean temperature at each latitude. There are several interesting effects: The global annual mean temperature in the Cretaceous was 6° to 14°C higher than today. The annual mean temperature at the equator was 0° to 5°C higher; the polar mean temperature was as much as 50°C higher. Instead of the 41°C equator-to-pole contrast that we see today, the contrast in the Cretaceous was only 17° to 26°C. Permanent ice and widespread seasonal ice were absent from Earth at that time.

Such warmth exceeds by a large amount the visions of the human-induced global warming predicted by computer models discussed in chapter 22. To be able to

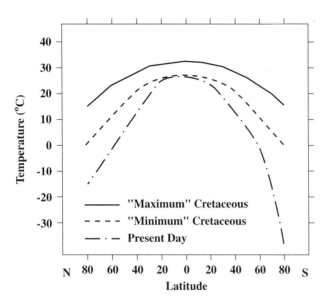

Figure 19.4. Estimated limits on temperature in the Cretaceous as a function of latitude. The plausible range (maximum and minimum) of annually averaged temperature at each latitude is shown, along with the value for the present-day Earth. From Barron et al. (1995).

reproduce such a different climate with computer models developed to predict weather today is clearly of keen scientific interest because such an exercise stretches the physical regimes under which such models have been fine-tuned.

19.6.2 Plate Tectonic Effects on Cretaceous Climate Change

Although the breakup of the supercontinent Pangaea began in the Jurassic, the Cretaceous Earth still had most of its continental landmass at low and mid-latitudes. With little land available near the poles, ice accumulation was difficult. The overall reflectivity of Earth was therefore lower than at present, allowing more sunlight to be absorbed and encouraging warmer conditions.

But tectonic effects on the Cretaceous climate were more complex than simple land distribution implies. As the supercontinental bottleneck was broken, plate spreading rates were probably fairly high. Very active seafloor spreading brought relatively hot, puffed-up crust rapidly away from midocean ridges. This, along with the absence of ice on the continents, implied a very high sea level, and water flowed over the continental lowlands to form vast inland seas. In consequence, the area of exposed land in the Cretaceous may have been only 60 to 70% that at present. These inland seas absorbed more sunlight than did the dry land, and may have been more important than the absence of ice in heating Earth's surface. Further, the inland seas were, on average, warmer than the ocean and probably helped to maintain mild sea-surface conditions at high latitudes through exchange of water with the ocean.

The spreading apart of Pangaea was a time of less mountain building, because continental collisions were minimal. Less mountain building meant less land area at high altitudes. The lower mean altitude may have implied less snow on the midlatitude continents, buttressing the effect of having little landmass near the polar regions. With fewer massive mountain ranges, as well as a higher sea level, the amount of continental surface area available for weathering may have been less than at present, leading to a lower rate of removal of carbon dioxide from the atmosphere. Also, faster plate tectonic recycling of the crust could have accelerated the rate of production of carbon dioxide from subducted carbonates, and injection of the gas into the atmosphere through greater volcanic activity.

19.6.3 Additional Important Effects on Cretaceous Climate

Ocean currents. The broad universal ocean undoubtedly had a different pattern of ocean currents than today. Less continental land area was affected by such currents than today because a single landmass has less coastline than the same mass fragmented, which could have led to more severe latitudinal variations in continental weather. As the continents broke up in the Cretaceous, currents of water in the new Atlantic Ocean changed this pattern substantially.

Water vapor and clouds. Increased temperature of the oceans increases abundance of water vapor in the atmosphere, which increases the greenhouse heating. It might also increase the cloud cover of Earth, which can add to or subtract from the heating, depending upon the thickness and geographic distribution of clouds. The effect of increased temperature on cloudiness, however, is very uncertain—for example, in the tropics, increased heating might lead to a greater preponderance of convective clouds (cumulus and thunderstorms), which create areas of locally heavy cloud but leave some of the sky cloud-free.

19.6.4 Causes for Climate Change That Probably Are Not Important in the Cretaceous

There are other possible causes of climate change that cannot be directly ruled out but are either less likely to be relevant, or somewhat arbitrary in the way they must be invoked.

Solar output. Because the Sun has been heating up over time, we do not expect this trend to explain the relative difference between the Cretaceous and the present climate,

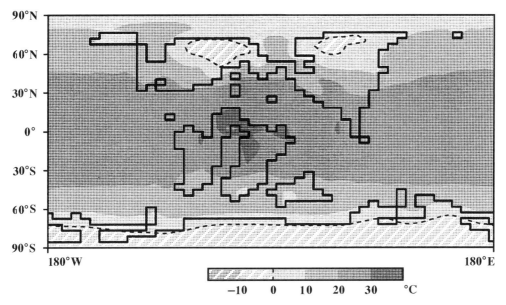

Figure 19.5. Predictions for annually averaged temperatures in the Cretaceous from a computer climate model. The results are displayed on a map of the world with the rough outlines of the continents as they would have appeared in the Cretaceous. Shading shows the annually averaged surface temperature, with a key at the bottom, in Celsius. The model has four times the present atmospheric carbon dioxide value and four times the present-day oceanic transport of heat from equator to poles; it satisfies the temperature constraints shown in figure 19.4. From Barron et al. (1995).

because the effect works the wrong way. Astrophysicists have suggested ways the Sun might brighten temporarily, and such a brightening could have triggered a warming, but it is impossible to determine whether the brightening timescale is commensurate with the duration of the warm period (some tens of millions of years).

Orbital variation. The variations in the orbit and tilt of Earth, described in section 19.8, occur on timescales much shorter than those required to explain the Cretaceous warmth.

Galactic effects. Passage of Earth through dusty clouds in the galaxy cools Earth rather than warms it. Perhaps we are in such a cloud now, and were not 100 million years ago. However, the magnitude of the cooling from the Cretaceous to present, and its gradual long-term nature, are hard to explain given what we know of the properties of such clouds.

19.6.5 Model for the Warm Cretaceous

Scientists have used computer models to predict changes in the present Earth's climate on timescales of days, weeks, months, years, and decades. Such a computer model was adapted by E. Barron (Pennsylvania State University) and colleagues at the National Center for Atmospheric Research in Boulder to simulate the Cretaceous climate. The model simulates the atmospheric greenhouse effect discussed in chapter 14, along with the transportation of heat in the oceans. We discuss this and models like it in much more detail in chapter 22, where we consider concerns about present-day global warming.

The first test for the model was to change the positions of the continents to correspond to Cretaceous times without changing anything else. This produced only a fraction of the temperature increase over the present-day climate required to explain the Cretaceous warm period. Adding four times the present carbon dioxide abundance to the atmosphere enhanced the atmospheric temperature at the poles to close to the values inferred from the data, but then the equatorial temperatures were too high.

It appears that to explain the temperature pattern shown in figure 19.4, there must have been enhanced transport of heat from the equator to the poles in the Cretaceous compared to the present. It is hard to make the atmosphere in the model transport the heat required, because a smaller temperature contrast from equator to pole actually means less efficient heat transport: The oceans must do the job (figure 19.5). It is possible that the ocean circulation at the Cretaceous time was organized in such a way as to promote very efficient transport of heat from equator to pole; computer models only recently have gained the sophistication to explore this possibility in detail.

It appears from the computations done to date that the most important differences between today's world and the

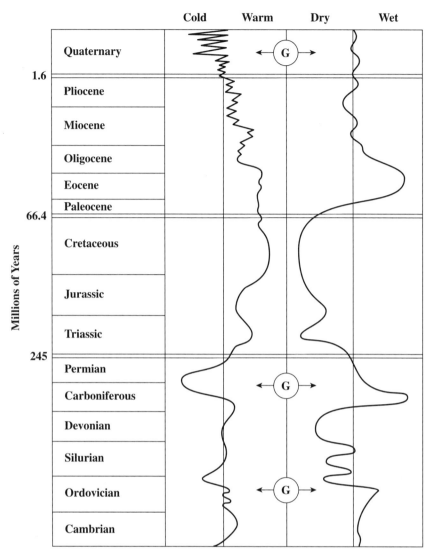

Figure 19.6. Temperature and precipitation for the past half-billion years of Earth history, based on isotopic and other indicators. Times before present and geologic periods are given on the left side of the graph. General times of very low temperatures, and possible ice ages, are shown with a "G." Adapted from Milne et al. (1985).

Cretaceous that determined the warmer climate are (i) the pattern of continents, which was more consolidated toward equatorial latitudes in the Cretaceous; (ii) enhanced Cretaceous ocean circulation from equator to pole; (iii) enhanced Cretaceous carbon dioxide levels. In today's world, human activities have an effect only on (iii). Until more accurate representations of the roles of clouds, precipitation, and other effects can be included in the models (chapter 22), these conclusions must be regarded as tentative.

19.7 THE GREAT TERTIARY COOLDOWN

The impact event that destroyed much of the Cretaceous fauna apparently did not have a long-term effect on climate, because the early Tertiary was similar to the late Cretaceous with respect to climate. As figure 19.6 shows, however, by the mid-Tertiary the climate was cooling down. Isotopic data on climate are excellent for this time, as the level of detail in the figure indicates.

Excursions in the Tertiary climate may be associated with increased volcanism, rapid changes in plate tectonic patterns (for example, India collides with the Asian continent in the Eocene), large variations in the Sun's luminosity, or even additional impact events such as the Cretaceous-Tertiary event (but smaller). Lesser extinction events occur in the late Eocene and in the Pliocene; one or both could be associated with climate shifts triggered by volcanic, tectonic, or impact events. The causes behind various shifts and overall cooling in the Tertiary part of the climate record remain poorly understood. Perhaps the climate record simply reflects the progressive departure of the tectonic and atmospheric states away from the Cretaceous condition of ice-free continents and high carbon dioxide content. The increasing ice coverage of the high-latitude continental areas, enabled by plate motion, reinforced the slow cooling, punctuated by occasional warmings of unknown origin.

It is unsatisfactory not to have a specific mechanism for the cooling and decline of carbon dioxide, and a dramatic one has been offered from the observation that, roughly 40 million years ago, the crustal plate carrying the Indian subcontinent collided with the massive Asian continent. Since that time, India has continued to plow into Asia to build up the enormous Tibetan Plateau, location of the world's highest mountains. M. Raymo of MIT and colleagues W. Ruddiman (U. Virginia) and P. Froelich (Georgia Institute of Technology) have proposed that the continued buildup of the Tibetan Plateau to the present has increased weathering and loss of carbon dioxide from the atmosphere. The presence of the plateau forces moist winds from the Indian ocean to rise and produce prodigious amounts of rain, which enhance weathering rates as well as feed eight of the Earth's large rivers, which in turn carry bicarbonates and other weathering products to the sea. Additionally, the presence of a large plateau with steep slopes increases the surface area of rock available for weathering, relative to a low flat plain. Computer simulations suggest that these results of the rise of the plateau have significantly increased the rate of removal of carbon dioxide from the atmosphere in the post-Cretaceous world. Indeed, the plateau may be so effective that only the negative feedback of an increasingly-colder climate has prevented an essentially total and catastrophic removal of the carbon dioxide.

Figure 19.6 shows that the progressive decrease in temperature toward the present shifts suddenly to very dramatic oscillations in temperature beginning early in the penultimate geologic epoch, the Pleistocene. These climate oscillations are characteristic of the ice age that continues to the present. Readers may be surprised that our time is identified as such; however, the ice age epoch in which we live is characterized by long stretches of glacial conditions punctuated by shorter intervals, only one-tenth as long as the glacials, of warm interglacial climate such as the current Holocene. Prior to the onset of glacial oscillations, the slowly cooling climate led to conditions in which mammals flourished, having filled most of the ecological niches vacated by the dinosaurs. Only the air remained the domain of dinosaurs or, more precisely, their close descendants, the birds. The Eocene, Oligocene, and Miocene epochs, stretching from 56 million to 5 million years ago, are really the golden age of mammals, with many more wonderful species of large mammals than humans have ever been privileged to see.

19.8 CAUSES OF THE PLEISTOCENE ICE AGE AND ITS OSCILLATIONS

The onset of the Pleistocene Ice Age was the end result of progressive cooling over the 50 million years prior to it. With enough continental landmass at high latitudes, low carbon dioxide in the atmosphere, and other properties that may not show in the geological record (for example, enhanced ocean mixing leading to more marine planktonic mass and higher consequent uptake of carbon dioxide), the climate system shifted abruptly to a state of widespread continental glaciation. There is more to the Pleistocene story than the glaciation itself, and that is the oscillatory nature of the climate. Long spans of continental glaciation, ranging from 40,000 years early in the Pleistocene to 100,000 years in the later cycles, are punctuated by interglacials of roughly 10,000–20,000 years characterized by warmer (or in some cases, unstable and rapidly shifting) climate (figure 19.7).

The origin of these oscillations most likely lies in the nature of Earth's spin and its orbit around the Sun. Currently, Earth's axis is tilted some 23 degree from a line perpendicular to the plane of its orbit around the Sun, and the orbit itself is slightly elliptical, or noncircular. The closest approach of Earth to the Sun happens to occur when the southern hemisphere is tilted toward the Sun, that is, during southern summer and northern winter. Because most of the continental mass lies in the northern hemisphere, this orbital state is one in which most of our planet's continental area does not experience the most summertime heating possible, because Earth is slightly farther from the Sun in July than in January. The difference in received sunlight, 8% from the closest to most distant point of the orbit, is significant in affecting climate.

Over a time of some 26,000 years, Earth's axis precesses around a fixed point in space, much like a toy gyroscope can be made to do by pushing its axis once it is set in motion. Viewed from the northern hemisphere, the current star closest to the north celestial pole, Polaris, will not forever be the north star: In 12,000 years Vega will be the pole star. The effect of this precession is to reorient the northern and southern hemisphere summers relative to the close and far points of the Earth in its orbit. Roughly 11,000 years from now, the northern hemisphere summer will occur when the Earth is closest to the Sun, opposite to the current state. Since plate tectonic motions are too slow to have shifted continental positions more than a few kilometers during that time, geography will be the same,

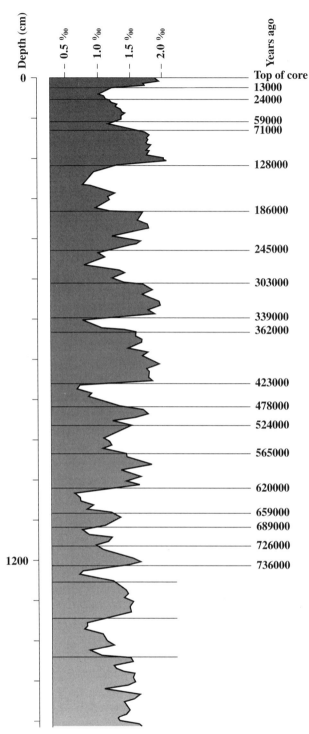

Figure 19.7. Data from a deep-sea drill core showing $^{18}O/^{16}O$ ratios over the past million years. The ^{18}O to ^{16}O value is an indirect measure of temperature: higher ^{18}O values (toward the left of the graph) mean more of Earth's water is locked up in continental glaciers, lower means more water is in the oceans (chapter 6). On this graph, then, values toward the left indicate cold times, and toward the right warm times. The dates on the right are assigned to particular depths in the core sample by examining the magnetic orientation of sediment grains layer by layer and comparing this to the known pattern of reversals of Earth's magnetic field from a million years ago to the present. For further details on magnetic reversals, see chapter 9. Adapted from Stringer and Gamble (1993).

and the heating of the northern hemisphere continental masses will be more severe 11,000 years from now than today.

Other oscillations in the motions of Earth are known to occur. The other planets of the solar system, in exerting very slight tugs on Earth, not only cause the axial precession but also slightly alter the magnitude of the tilt (from 21 degrees to 25 degrees) on 41,000-year cycles. The ellipse that is Earth's orbit drifts or rotates as well, which shortens the precessional period from 26,000 to 22,000 years. Furthermore, these planetary tidal pulls also modulate the eccentricity of Earth's orbit in a complicated way that approximates two cycles at 100,000- and 450,000-year periods (figure 19.8). The idea that all these cycles might affect climate goes back to the Scottish geologist J. Croll in the mid-nineteenth century. Serbian physicist M. Milankovitch, in the first part of the twentieth century, showed how these cycles affect the annual distribution of solar energy received by the Earth and hence the climate.

The relationship between the changing pattern of solar heating and glaciation is illustrated by the comparison in figure 19.9. The lower graph shows the amount of sunlight received at 65° north latitude in July; this is one measure of the effect of the shifting orbital and tilt parameters on the distribution of sunlight falling on Earth. The upper chart, keyed to the same timescale, shows oxygen isotope data as a proxy for sea level and hence temperature (cold temperatures, lower sea level, higher ^{18}O). Careful examination shows that major decreases in the extent of glaciation occur when northern summer has its greatest solar heating.

The correspondence is not perfect but other effects come into play as well. In particular, the change in climate alters the carbon-cycle balance and hence the carbon dioxide abundance. Detailed models that account for the carbon dioxide effect show a much better correlation, especially at later times. The progression of glaciation episodes from 40,000 to 100,000 years as the Pleistocene progresses is not fully accounted for by orbital cycles; other effects not tied to the orbits (modulations in volcanic activity, minor plate movements changing ocean currents) might be involved. We defer a more detailed discussion of these effects to chapters 21 and 22, in the contexts of the most recent glacial episode and concerns about the future of our planet's climate.

Why have the orbital cycles of the Earth amplified climate changes only in the last few million years? Earth's climate up to a few million years ago was sufficiently warm that the orbital changes in the pattern of sunlight on Earth (that is, in the strength and timing of the seasonal variations) were not enough to trigger formation of sufficient ice at high latitudes to force the onset of widespread

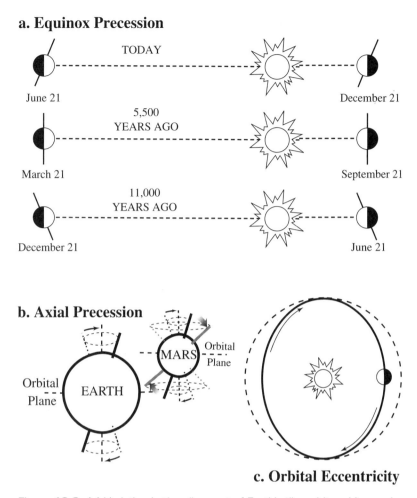

Figure 19.8. (a) Variation in the alignment of Earth's tilt and its orbit, causing a phase shift in the seasons. About 11,000 years ago, summer came to the northern hemisphere at a time when Earth was at perihelion; today northern summer occurs at aphelion. (b) Swings in the axial tilts of Earth and Mars over time; note that Mars suffers more extreme swings because it has no large moon to dampen the pull of the other planets. (c) Change in the shape of Earth's orbit on 100,000- and 450,000-year cycles and rotation or drift of the ellipse (indicated by the arrows).

glacial episodes. As global average temperature continued to decline in step with the removal of carbon dioxide, the amount of high-latitude ice formed during colder times in the Milankovitch cycles increased. Eventually, a point was reached several million years ago at which the amount of ice formed in the colder part of the cycle was sufficient to drive further cooling, and hence buildup of massive ice sheets across large areas of the planet's continental masses. In effect, the gradually cooling conditions brought Earth's climate to the threshold of instability, and the colder portions of the orbital cycles forced the climate across that brink into glacial episodes.

We do not know whether the Pleistocene Ice Age epoch is unique in its oscillatory behavior. Earth's orbit and spin likely have undergone cyclical variations throughout the solar system's history. Oscillations in ancient ice ages caused by such cycles may show only faintly in the geologic record, which becomes more imprecise with age. In the longest glacial period, some 340 million to 250 million years ago, there is evidence in sediments of cyclical changes in sea level, and some attempts have been made to ascribe these to climate oscillations induced by orbital and tilt cycles. Alternatively, at least some of the more ancient ice ages may have been "deeper" in the sense that conditions were such as to plunge Earth into a climate much colder than that of the Pleistocene. The geologic evidence of ancient glaciations suggests that glaciers occurred much closer to the equator than in the Pleistocene. Whether due to lower solar luminosity or other factors, a more severe ice age than the Pleistocene's would be less sensitive to orbital

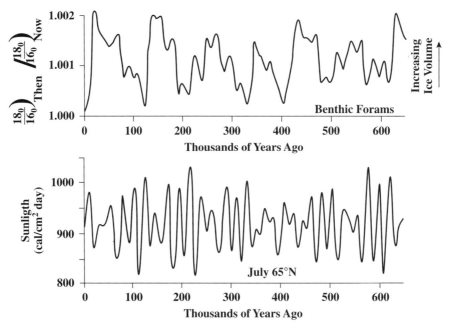

Figure 19.9. An examination of the effect of Earth's orbital and axial tilt variations on climate. Lower graph shows the amount of solar energy falling on the 65° latitude belt on Earth in July. This is affected both by tilt (more sunlight is received when the Sun is more directly overhead, i.e., when the northern hemisphere is tipped toward the Sun) and the varying distance from the Sun. The interplay between the various orbital and tilt cycles yields a complex pattern, much as musical instruments produce overtones and beats to achieve their rich spectrum of sound. The upper graph is a measure of the global temperature from oxygen isotope data. The climate is cooler when the relative oxygen-18 abundance is higher; on this graph the ratio $^{18}O/^{16}O$ is plotted *relative* to the present-day value (a ratio of ratios!). Reproduced with permission from Broecker (1985).

fluctuations, which could come and go without drastically altering the climate. If this is correct, then the environmental and biological effects of the severe climate oscillations between glacials and interglacials in the Pleistocene might be a very unusual feature of this last act of the Phanerozoic.

19.9 SAVED FROM INSTABILITY: EARTH'S VERSUS MARS' ORBITAL CYCLE

Dramatic as they are, Earth's orbital swings and associated climate changes are mild compared to those of Mars. Recent computer calculations by French and American scientists of the effect of other planets' gravitational pulls on Mars' tilt and orbit led to a surprising conclusion: The tilt of Mars behaves chaotically.

Chaos is a term co-opted by physicists to describe, not a state of complete randomness, but a common physical property of unpredictability in complex physical systems. Complex physical systems subjected to certain perturbations will not respond in a cyclical or deterministic fashion, but instead will break into a mode of unpredictable behavior or *deterministic chaos*. Certain characteristics of the behavior can be predicted by computer models, but the details of "what happens when" are lost.

For Mars, the chaos lies in the magnitude of the tilt of its axis. Rather than undergoing modest periodic swings in the amplitude of the tilt, the axis behaves chaotically and will swing from very small tilts of 10 degrees to as much as 50 degrees, on a wide range of different timescales. When and for how long a particular tilt excursion will occur cannot be predicted, but the amplitude range and general character of the swings can. Evidence for wide swings in the Martian rotational axis is present in a series of layered deposits thought to be dust and ice, residing near the poles (figure 19.10). Alternating epochs of deposition of these two materials can be achieved through large swings in the tilt of the axis.

Disturbingly, similar computations for Earth show equally dramatic results, but evidence for wild swings is lacking in the geologic record. The stabilizing influence seems to be Earth's Moon. The tidal pull of the Moon on Earth is strong compared to the pull of the other planets (because of their greater distances), and the Moon's pull strongly damps out large excursions in Earth's tilt that might otherwise occur. If indeed the presence of a large natural satellite is responsible for stabilizing Earth's tilt, and hence preventing frequent and drastic excursions of climate, the implications are profound for life elsewhere.

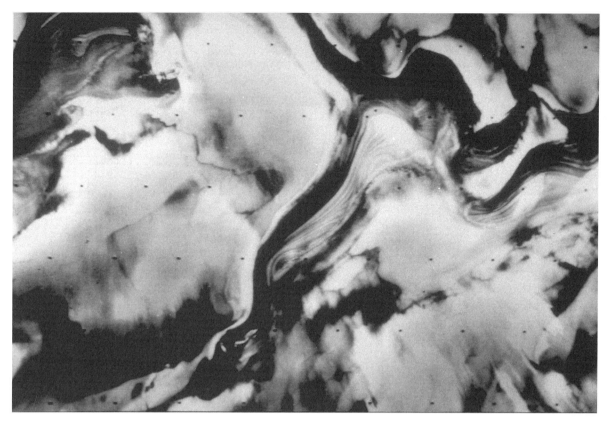

Figure 19.10. Layered deposits of dust and ice at the south pole of Mars. Viking Orbiter photograph courtesy of NASA.

Finding an Earth-like planet around another star would not be enough to buoy the hopes of finding advanced life: There must be a moon or other means to prevent disastrous swings in that planet's tilt.

19.10 EFFECTS OF THE PLEISTOCENE ICE AGE: A PREVIEW

With the onset of the oscillatory ice ages, the less stable climate contributed to species extinctions, extensive migrations, and the development of new species and even genera of animals. Of much interest to us is the coming of human-like creatures and then humans as a part of this 2-million-year time of change. In the next chapter, we explore one of the most startling results of the long evolution of this habitable planet: The coming of the age of humankind.

19.11 QUESTIONS

a. Is a habitable Earth with no large Moon ruled out absolutely? How else might stability be achieved?
b. Could a planet with a more highly elliptical orbit than Earth's recover from glacial swings in which the entire surface area becomes ice-covered?

19.12 READINGS

Barron, E. J. 1983. A warm, equable Cretaceous: The nature of the problem. *Earth-Science Reviews* 19, 305–338.

Barron, E.J. 1992. Paleoclimatology. In *Understanding the Earth: A New Synthesis* (G.C. Brown, C.J. Hawkesworth, and R.C.L. Wilson, eds.). Cambridge University Press, Cambridge, UK, pp. 485–505.

Barron, E., Fawcett, P.J., Peterson, W.H., Pollard, D., and Thompson, S.L. 1995. A "simulation" of mid-Cretaceous climate. *Paleoceanography* 10, 953–962.

Broecker, W. 1985. *How to Build a Habitable Planet*. Eldigio Press, New York.

Cloud, P. 1988. *Oasis in Space: Earth History from the Beginning*. W.W. Norton, New York.

Crowly, T.J., Yip, K-J.J., and Baum, S.K. 1993. Milankovitch cycles and carboniferous climate. *Geophysical Research Letters* 20, 1175–1178.

Marshall, H.G., Walker, J.C.G., and Kuhn, W.R. 1988. Long-term climate change and the geochemical cycle of carbon. *Journal of Geophysical Research* 93, 791–801.

McGoweran, B. 1990. Fifty million years ago. *American Scientist* 78(1), 30–39.

Meert, J.G., and van der Voo, R. 1994. The Neoproterozoic (1000–540 Ma) glacial intervals: No more snowball Earth? *Earth and Planetary Science Letters* 123, 1–13.

Milne, D., Raup, D., Billingham, J., Niklaus, K., and Padian, K. (eds.) 1985. *The Evolution of Complex and Higher*

Organisms. NASA SP-478. U.S. Government Printing Office, Washington, DC.

Murphy, J.B., and Nance, R.D. 1992. Mountain belts and the supercontinent cycle. *Scientific American* **266**(4)(April), 84–91.

Peixoto, J.P., and Oort, A.H. 1992. *Physics of Climate*. AIP Press, New York.

Raymo, M.E., Ruddiman, W.F., and Froelich, P.N. 1988. Influence of late Cenozoic mountain building on ocean geochemical cycles. *Geology* **16**, 649–653.

Rinaldo, A., Dietrich, W.E., Rigon, R., Vogel, G.K., and Rodriquez-Iturbo, I. 1995. Geomorphological signatures of varying climate. *Nature* **374**, 632–635.

Shackleton, N.J., and Opdyke, N.D. 1973. Oxygen isotope and paleomagnetic stratigraphy of equatorial Pacific core V28-238. *Quaternary Research* **3**, 39–55.

Stringer, C., and Gamble, C. 1993. *In Search of the Neanderthals: Solving the Puzzle of Human Origins*. Thames and Hudson, London.

Tattersall, I., Delson, E., and Van Couvering, J. 1988. *Encyclopedia of Human Evolution and Prehistory*. Garland Publishing, New York.

20 TOWARD THE AGE OF HUMANKIND

20.1 INTRODUCTION

Earth's evolutionary divergence from the neighboring planets of the solar system, beginning with the stabilization of liquid water, culminates in the appearance of sentient organisms sometime within the past 1 million to 2 million years. The fossil record is abundant in its yield of creatures intermediate in form and function between the great apes and modern humans; new discoveries seem to be made with increasing pace. But hidden between and among the fossil finds are the details of how and why we came to be. Even as we acknowledge our common origins with the life around us, the singular results of sentience – art, writing, technology, civilization – are surprising and enigmatic.

The story of human origins is not simple, and this chapter attempts only a sketch of the evidence and the lines of thought current in today's anthropological research. It begins with a broad view of the climatological stage on which these events took place. It ends with a focus on the closing act of human evolution, the coexistence of modern humans with a similar but probably separate sentient species in Europe and the Middle East – the Neanderthals.

20.2 PLEISTOCENE SETTING

The earliest fossils along the lineage toward humanity exist in the Pliocene epoch, prior to the Pleistocene, during a time of relative climate stability. The pace of human evolution picks up in the Pleistocene, and species close enough in form to us to warrant assignment to the genus *Homo* (Latin, man in the sense of humans) appear close to, but perhaps slightly before, the time when climate shifted into an ice-age pattern of glacial and interglacial episodes.

The effect of glaciers was profound. During the depths of the glacial episodes, ice sheets stretched across significant parts of North America, Asia, and Europe. These sheets exceeded 3,000 meters in thickness in places, and hence acted like huge mountain ranges in diverting air flow and weather patterns by thousands of kilometers. Ocean currents were affected by changes in the amount of sea ice year round, by alterations in salt content, and by the patterns of rainfall and snowfall. The rise and fall of sea level by more than 100 meters opened and closed overland routes between continents. The amount of plate movement of continents was relatively small, no more than tens of kilometers over a million years (chapter 9), but this was more than made up for by the oscillations associated with the advance and retreat of glaciers. Such oscillatory effects acted to move ecological niches significantly on timescales ranging from 100,000 to 10,000 years, and probably even less. Forests waxed and waned over large areas; food supplies changed dramatically between cold-dry and warm-wet episodes. Animal species encountering such changes either perished or migrated vast distances, and many opportunities for speciation (formation of new species) must have been available as small groups became isolated (chapter 18).

The foment caused by the instability of climate is reflected in the extinction of a number of mammalian species during this time. It also may have served as the stimulus for a dramatic change in the kinds of primate species present in Africa and possibly Asia. The alternate waxing and waning of savanna versus forestland, so different in the kinds of species and survival styles they support, may have been at the nexus of the production of new primate lineages.

20.3 THE VAGARIES OF UNDERSTANDING HUMAN ORIGINS

The fossil record of human origins is woefully incomplete, as is the case with fossils in general. As discussed in chapter 8, the vast majority of living organisms are broken down after death without their body forms being preserved. The very few that die in environments resulting in fossil production must serve as the faint signposts of an evolutionary process involving vastly larger numbers of organisms.

With human evolution, this problem of incompleteness is compounded by another challenge, what might be called the "goldfish bowl" effect. Human origins means *our* origins and, as such, any discoveries are subjected to intense scrutiny by the public. There is a natural tendency, with any announced new fossil find, to hope that it solves "the" puzzle, so that often unjustified conclusions are drawn by the press, as well as by anthropologists themselves. Adding to the emotional foment are the personal and collective religious beliefs that we hold; for many religions the notion of an animal origin for human beings, without supernatural intervention, is heretical and offensive.

For these reasons the history of the search for physical evidence of human origins has been replete with dramas played out in social and religious arenas, beginning with the publication of Darwin's ideas on human origins in his 1871 book *The Descent of Man*. The notorious Piltdown hoax of 1913, a fabricated skull constructed essentially of an ape jaw and human cranium, may have been an interesting scientific Rorschach test but also created a credibility gap with long-term repercussions. The "Scopes Monkey Trial" of 1925 was a famous legal challenge to a Tennessee law restricting the teaching of evolution; it centered on the conflict between Biblical scripture and biological understanding of species origins. More recently, the somewhat unfortunate popularization of the investigations of human origins as the search for a "missing link," a concept as much the fault of anthropologists as journalists, made it difficult for the casual reader to come to an understanding of the increasingly detailed gallery of human ancestors and related animals.

20.4 HUMANITY'S TAXONOMY

To appreciate the search for human origins requires returning briefly to the discussion of taxonomy of chapter 18. All human beings alive on Earth are members of the same species, *Homo sapiens* (Latin, wise man), in turn the sole representative of the genus *Homo*, which in the past has contained a number of other species. We are members of the family *Hominidae*, comprising several now-extinct genera, along with *Homo*, chimpanzees, and gorillas. The inclusion of the African great apes and humans in the same family is the recent resolution of a long-standing taxonomic argument; previous classifications putting apes in a separate family were flawed because physiologically (and genetically) humans are more closely related to chimps and gorillas than any of the three are to the orangutan.

The apparently large gap between ourselves and nearest animal relatives arises in part because many other creatures classifiable in the genus *Homo* are extinct. Whether by climate change or competition from our most successful immediate ancestors, we sit out on a rather isolated limb of the primate family tree.

In what follows, we briefly sketch a picture of human evolution based on key fossil species identified to date, one that is summarized in figure 20.1. As in any such narrative, the simplicity of the results belies the decades of controversy, discovery, and revision that have preceded and will follow this particular moment in anthropology. Consider that you have been given the task of assembling a jigsaw puzzle. You do not know what the final picture will look like, nor do you know how many pieces there are. The pieces are not in a box; they've been scattered around town and you must find them. Some are in such poor condition that their edges are frayed, torn, or missing; nonetheless you must find the pieces and, through trial and error, assemble the final image. Such is the essence of the anthropological search for how humankind came to be.

20.5 THE FIRST STEPS: AUSTRALOPITHECINES

Africa seems to be the source of the most ancient fossils in our ancestral family tree. This continent is rich today in primate species and, particularly in the equatorial regions, would have exhibited relatively gentle environmental fluctuations in response to the overall climate instability of the Pleistocene and preceding Pliocene. Much confusion and uncertainty about whether Africa or Asia was the origin point for the outward radiation of new hominid species seems, for the moment, to be resolved in favor of Africa.

Studies of genes in apes and humans, coupled with estimates of the rate of mutation of such genes, leads to the conclusion that the African apes (chimps and gorillas) and humans had a common and now-extinct ancestor as recently as 5 million years ago, but no earlier than 9 million years ago. Although fossil evidence is scant as to who exactly the ancestor was, paleontologists now generally agree that this genetically estimated timescale is probably right. A possible candidate for the origin of the Hominidae family was discovered in Ethiopia in 1994. *Aridipithecus ramidus* (where the genus name is Latin for "chimp-like"), present in the form of jaw and cranial fragments, bears the signature of the great apes but differs in detail from gorillas, chimpanzees, and ourselves. Ramidus

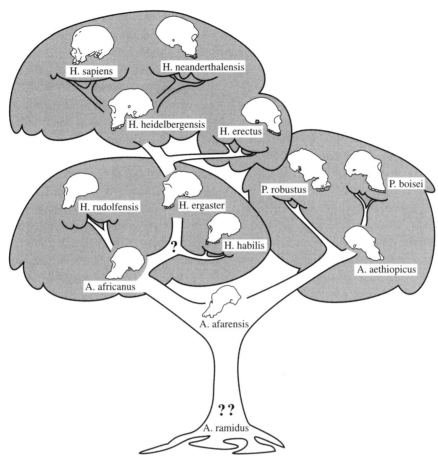

Figure 20.1. Gallery of skulls illustrating the major species on the road to humankind. The tree is meant to suggest branchings off of common ancestors, but the details of such branchings remain extremely uncertain and will continue to change as more fossils are found. H.= *homo*, A.= *australopithecus*, P.= *paranthropus*.

lived some 4.4 million years ago, based on geological dating of the sediments in which the fossils were found (a difficult task unless surrounding or incorporated igneous material can be radioisotopically dated). It is a candidate for the origin of the Hominidae family by virtue of its recognizably ape-like features.

Beyond this point, a variety of hominids (members of the Hominidae family) begin to appear in the African fossil record. Whether driven by fluctuating climate conditions to move onto the African plains from the forest, or for other reasons, a number of such species are seen in the period from 4 million down to 2 million years ago. Over 300 specimens define *Australopithecus* ("southern ape") *afarensis*, a species that lived in Africa over a span of time from 2.5 million to 3.4 million years ago, and perhaps longer.

Still very much apelike, with little to indicate a direction toward human ancestry, *A. afarensis* is distinguished by the large number of specimens, its broad span of time, and its representation in a fairly complete skeleton known popularly as Lucy. Its younger age than the genetically determined split of human from the great apes, a demonstrably upright posture, and a larger brain size than the chimpanzee suggest that it is an early species on the road to humankind. Nonetheless, were we to see a living afarensis today, it would seem no more than a fascinating ape that happened to walk upright and was perhaps a bit smarter than a chimpanzee. Other species of the genus *Australopithecus* existed down to about 1.5 million to 2 million years ago, and an offshoot genus, *Paranthropus* (near man), also is represented in this time by several species. For a period of some 2 million years, the African continent was a far richer tapestry of hominid species than is all of today's world combined.

20.6 THE GENUS *HOMO*: OUT OF AFRICA I

Between 3 million and 2.4 million years ago, the African climate shifted to a dryer, cooler regime than had dominated previously. Whether the changing climate stimulated contemporaneous dramatic changes in the Hominidae line is unclear. A traditional picture is that the human lineage resulted from creatures who moved out from the forests into the plains, leaving behind the lineage that became great apes. This view is now held in lower regard, based on evidence that both *Australopithecus* and *Paranthropus* were adapted to partially open, woodland conditions. But certainly fluctuating environmental conditions caused shifts in the distribution and nature of woodlands, shifts that provided a greater opportunity for isolation of groups, followed by speciation encouraged by environmental stresses.

Between 2.5 million and 2 million years ago, several different species appear in Africa that, according to anthropologists, were too human in appearance to merit inclusion in the genus *Australopithecus*; instead, they are the earliest members of the genus *Homo*. They possessed crania larger and differently shaped than *Australopithecus*. They appeared to fashion crude stone tools to assist their hunting and food preparation. The most successful member of the genus *Homo* in terms of species longevity, *Homo erectus* (upright man), appears around 2 million years ago or a bit later. Erectus had a larger cranial capacity and more human features than the *Homo* species before it. There is evidence for more extensive stone modification and use as tools. Erectus as a species is recognizable for a million years, the longest-lived member of the genus *Homo* to date.

Only shortly after the appearance of *Homo* in Africa, members of this genus began migrations eastward into Asia. Recent finds of *Homo erectus* in eastern Asia that have ages approaching 2 million years suggest a prompt dispersal in that direction. Migrations of *Homo erectus* populations would continue for over a million years, eventually leading to the establishment of groups in Europe as well (with a continuous lineage that extends almost, but not quite, to the present). Hypotheses as to the origin of this propensity for travel include the changing climate, driving many species toward dispersal or extinction, and the tendency, suggested in fossil remains, of African *Homo* to range widely in its scavenging and hunting forays. Whatever the cause, the wandering nature of *Homo* distinguished it from its predecessors.

It is with the Out of Africa I migration that the story of human evolution takes a complex turn. Because *erectus* and similar *Homo* species had spread onto three continents (Africa, Asia, and Europe), the geographical area covered was too large to permit gene transfer by interbreeding among groups. Instead, the fate of the various *Homo* groups became decoupled from one another, and a complex and poorly understood pattern of emergence of various post-*erectus* species is played out over many hundreds of thousands of years. The situation, by 200,000 years ago, was the apparent existence of post-*erectus* species on three continents, with brain sizes approaching or equaling present-day values (figure 20.2), and whom, for want of a better term, we shall call "archaics." The pace of change had accelerated, perhaps because of increased climate fluctuations, the propensity for migration that would naturally produce isolated populations ripe for further speciation, or other causes. That situation could well have persisted to the present-day, when an extraterrestrial visitor might have found at least Africa, Asia, and Europe populated by people quite different from today's humanity, comprising several species, but nonetheless using tools and in the aggregate having many of the characteristics we call human. But such was not to be the case.

20.7 OUT OF AFRICA II

As in all sciences, controversy rages in anthropology over crucial parts of the story of human origins. Two views exist as to what happened to effect a transition from the post-*erectus* populations scattered across Europe, Asia, and Africa to the present situation of a single, modern species, *Homo sapiens*, occupying all the Earth (figure 20.3).

The *multiregional* origin posits that the post-*erectus* populations encountered each other enough to allow interbreeding to maintain a single, archaic-human species, but not enough to erase regional differences. This species evolved separately and semi-independently on the three continents into modern *Homo sapiens*, with the genetic transfer rate sufficient to ensure maintenance of a single species. The origin of races, in this view, is very ancient. The alternative view, *replacement*, posits that a final, late speciation event occurred somewhere in the world, and the new species, modern *Homo sapiens*, spread outward from that point and supplanted the existing archaic peoples on all three continents. Under this hypothesis, the establishment of racial groups is a very recent phenomenon associated with modern *Homo sapiens*' adaptation to regional climate variations.

The multiregional hypothesis relies largely on regional differences in appearance among *erectus* and later archaic specimens that are vaguely similar to those among the various races of modern humans today. However, the hypothesis suffers from very serious drawbacks. It presumes a model of speciation that is at odds with the best such model available today, punctuated equilibria. It also relies

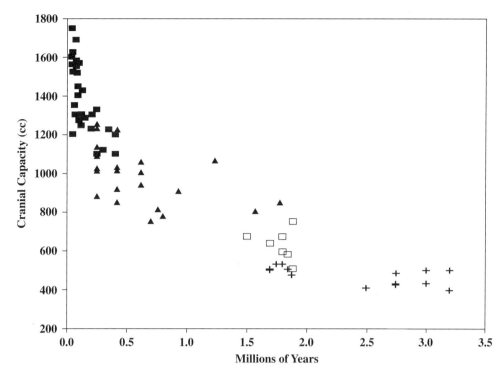

Figure 20.2. Cranial size in hominid species as a function of time, adapted from Mellars (1996). The units of volume of the cranium are cubic centimeters.

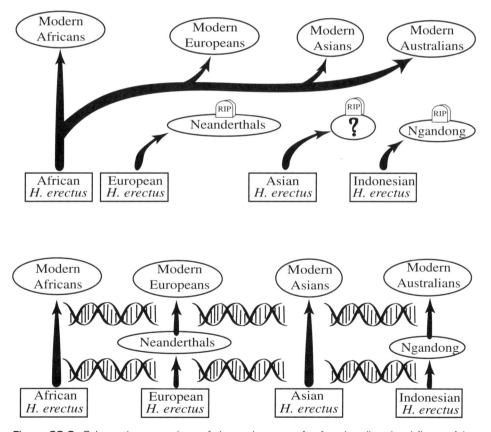

Figure 20.3. Schematic comparison of the replacement (top) and mulitregional (bottom) hypotheses for the origin of modern humans. The double helixes in the multiregional model symbolize gene transfer by occasional interbreeding. Time flows upward in each model. Modified from original figure by Christopher Stringer from Stringer and Gamble (1993) by permission of Thames and Hudson.

on an excessively delicate balance of interbreeding: enough to ensure that humankind did not diverge into separate species (while evolving substantially from *erectus* into *sapiens*), yet not so much that regional differences were erased.

The replacement model, on the other hand, has support from the fossil record in the form of evidence for modern *Homo sapiens* appearing in Africa some 130,000 years ago, then appearing in the Middle East about 100,000 years before present, into Asia and Europe after that. Once spread into these many lands, with their different climates and physical demands, *Homo sapiens* flowered into the many races of today.

Where, then, did the first members of modern *Homo sapiens* arise? The evidence is clear that Africa was again the homeland of this latest innovation, but specifically where cannot be pinned down. Evidence for modern humans, in the form of bones and extensive advanced tool kits, appears in southern Africa and East Africa some 120,000 to 135,000 years ago, and later in Northern Africa. The oldest Asian samples of modern *Homo sapiens* seem somewhat younger – consistent with a migration from elsewhere.

Because the fossil record is too crude to pin down the origin of our modern species, genetic techniques have been applied. The basic idea is to use genetic variations among present-day human populations to trace backward the region and time in which a putative single speciation event occurred. The challenge with this approach, of course, is that every child is the product of the shuffling of genes from a father and a mother. However, genes contained in the cellular mitochondria (chapter 12) are inherited almost always from the mother alone. Provided, then, that there has been a continuous lineage from an initial speciating group to the present, examination of mitochondrial DNA differences might yield the origin location and time.

This technique, pioneered by University of California researchers A. Wilson and R. Cann, required extracting mitochondrial DNA samples from people all over the world. The results suggested two robust conclusions. First, all human mitochondrial DNA is extremely similar, indicating that very little time has elapsed in which racial differences have built up. Second, African peoples show relatively more variation in their mitochondrial DNA than do people from other continents, suggesting the Africans have had somewhat more time to build up such variations – that is, modern humans have existed in Africa the longest.

When taken in total, the fossil and genetic evidence favor a model in which modern *Homo sapiens* first speciated somewhere in Africa, between 100,000 and 200,000 years ago, and quickly spread via migrations throughout the world. What happened in its encounters with the archaic humans already present in Africa, Asia, and Europe was a long story, spanning many tens of thousands of years. There is little in the way of physical clues to reveal the nature of the replacement process. The best documented – and perhaps longest – overlap of modern and archaic humans was in Europe and the Middle East, home of the archaic Neanderthals. It is worth focusing on this last act of human evolution, not merely for its own sake, but as a cautionary tale as we contemplate how our own species might interact or coexist with intelligent life elsewhere in the cosmos.

20.8 FINAL ACT: NEANDERTHALS AND AN ENCOUNTER WITH OUR HUMANITY

All history contains lessons about our own humanity, but these are often couched in ambiguous terms. The same is true for the saga of the Neanderthals in Europe and the Middle East (figure 20.4), over a time (perhaps between 30,000 and 100,000 years) when this interesting species crossed paths with anatomically modern humans.

20.8.1 Climate Setting

The first hints of characteristic Neanderthal features and tools extends back 300,000 years ago or so, in Europe and western Asia. The last physical evidence for Neanderthals is found in southern Spain as recently as 27,000 years ago. This long interval spans two major glacial episodes of the Pleistocene, as seen in the oxygen-isotopic record from sea sediment cores, displayed in figure 20.5. Separating the two is Earth's penultimate interglacial (we are living in the most recent one), extending from 118,000 to 126,000 years ago. During the interglacial time, forests would have covered large areas of continents, including Europe and Asia. A variety of large mammals seen today in restricted regions were present throughout the world; hippopotami and elephants were found in the British Isles, for example.

More recently than 118,000 years ago, temperatures initiate a somewhat bouncy descent toward the most recent glacial, though they do not begin to approach the cold of the previous glacial until 70,000 to 80,000 years ago. In the range of the Neanderthals, broken woodlands existed, probably supporting herds of grazing animals; many of the large mammals disappeared as the glaciation intensified.

From 70,000 to 30,000 years ago, ice sheets advanced and retreated in rapid oscillations. Major cold episodes

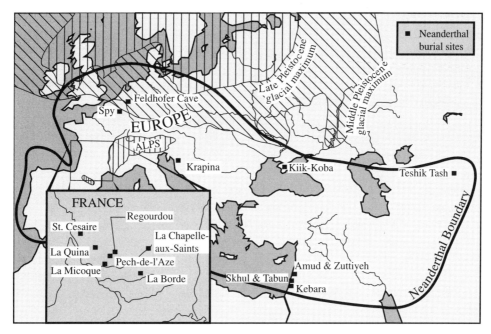

Figure 20.4. General geographic areas occupied by Neanderthals. Hatch marks indicate the extent of the glaciers during the middle and late Pleistocene. Modified from original figure by Annick Peterson from Stringer and Gamble (1993) by permission of Thames and Hudson.

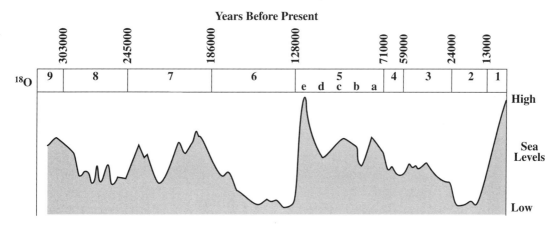

Figure 20.5. Sea level, and hence temperature, over the past 300,000 years from oxygen-18 data in seafloor sediments (see chapter 6 for discussion of technique). Times of high sea level, hence less ice, are warm; low sea level indicates colder, glacial epochs. The numbers from 1 to 9 are standard labels for glacial and interglacial episodes. From Stringer and Gamble (1993) by permission of Thames and Hudson.

(such as 60,000 to 70,000 years ago) must have accelerated extinctions of a variety of animals; forests retreated southward and previous woodlands likely became open tundra in Europe. In the Middle East, periods of clement climate may have existed at these colder times, encouraging migration of Neanderthals down to that geographic crossroad of the world. The final mild climate episode before the peak of the last glaciation occurred 40,000 years ago. The climactic freeze was reached about 19,000 years ago, after the extinction of the Neanderthals.

Early Neanderthals, with somewhat different features than their "late" Neanderthal descendants, existed from perhaps 300,000 to 130,000 years ago. The time after that, up to perhaps 40,000 years ago, was really the heyday of the Neanderthals, with characteristic stone cultures and stable physical features. During this time, Neanderthals made incursions into the Middle East, interleaving with peoples of more modern appearance and stone cultures. From 40,000 years to their extinction, the Neanderthal populations declined in geographic distribution, while innovating through imitation of stone tool types brought by the modern peoples emigrating to Europe.

20.8.2 Physical Features of Neanderthals

Neanderthals were not the stooped over, ape-like, brutish cousins of the depictions of the popular literature. They were short but very robust people, with broader and deeper muscle attachments in their bones, and hence more massive musculature, than the average for any modern populations. Although their hip attachments differ from ours in encouraging more stress on the sides of their thighs than on front and back (making easier the squatting and sideways movements typical of foraging activity), the fossil remains of their skeletons are consistent with fully upright postures.

It is the head of Neanderthal people that most dramatically outlines the difference from all modern humans (figure 20.6). The Neanderthal skull has very heavily enlarged brow ridges; a cranial vault that is low and somewhat flattened relative to that of modern humans; more massive jaws and teeth relative to the rest of the skull than in modern humans; a huge, broad nose, and virtually no chin. However, the cranial capacity of Neanderthals equaled or exceeded that of modern humans. To accommodate the brain in the more flattened skull, Neanderthal heads had a more prominent rear "bun," than do modern humans. Human skulls are constructed such that they grow outward as the brain grows during infancy and childhood. Presumably, Neanderthal's did the same, hence the shape of the skull, which would strike any human being today as being very odd, reflected a differently shaped (and presumably differently functioning) brain, but one on average somewhat larger than ours.

Many of the features of human and Neanderthal heads likely related to the need to support chewing and grinding forces. Our skulls have high front domes, providing adequate support against muscular forces; Neanderthal brow ridges did the same in the absence of the high forehead. Our chins likewise provide structural support during chewing, and are a somewhat unusual innovation in the hominid line; Neanderthal jaw stresses were supported by more traditional heavy bones.

The striking stockiness of Neanderthal bodies (both male and female) and evidence for large muscles could readily be argued as a result of a more strenuous physical lifestyle. However, such features are present in preadolescent Neanderthal children, whose ages at death are easily dated from the state of their dentition. More likely, the stocky build enabled a physically demanding lifestyle, but finds its origin in adaptation to the very cold climate that characterized Europe during much of the Neanderthal heyday. This is the case among modern people who live in very cold climates – but not nearly as extreme as that of the Neanderthals.

Figure 20.6. The author as (left) *Homo neanderthalensis*; (right) *Homo sapiens*.

A further clue to this adaptation lay in the heroic nose. Anthropologists have argued that it could serve two possible (and likely simultaneous) functions: warm the frigidly glacial air as it is inhaled, and allow for a greater volume of inhalation with a consequently higher tolerance for physical exertion. Enthusiasts for backcountry winter sports know the hazards of overexertion and consequent sweating: hypothermia (a loss of body temperature control) and death can result. A bigger nose is an adaptation allowing a high-exertion lifestyle in the cold.

Having emphasized the differences from modern humans, it is now necessary to remark upon how close the Neanderthals are to us in their appearance. Meet one in modern dress in an office and you would likely have to regain your composure: This seems to be a human being, but what a strange head and face! More different than any of the remarkable variety we share as modern humans, one of the great enigmas of the Neanderthals is the juxtaposition of the oddness with the closeness to modern humans. Most anthropologists today hold the view that Neanderthals are *Homo neanderthalensis*, a different species sharing the same genus as modern humans. The physiological differences between Neanderthals and modern humans are larger than between other primates that are, without controversy, classified as different species.

The origin of Neanderthals seems to lie in pre-existing populations of *Homo erectus* or a successor species *Homo heidelbergensis*, resident in Europe as well as western Asia for many hundreds of thousands of years. Many of the traits of Neanderthal features can be seen in fossils from prior to 300,000 years ago in England, Germany, Greece, and France – remains that seem transitional between *erectus/heidelbergensis* and Neanderthal. Far removed from the changes occurring in Africa that led to modern *Homo sapiens*, the Neanderthals were an evolutionary event in and of themselves – a distinct population of *Homo* evolved from ancestors who migrated out of Africa or Asia long before the speciation event that produced modern *Homo sapiens*.

Neanderthal fossil remains show differences from individual to individual. However, these differences are smaller than are the differences between individual members of today's modern humans. Our species has spanned the globe, adapting to a range of climates far greater than those the Neanderthals contended with. It is not surprising, then, that we should be a more varied species than Neanderthal. Equally important is the lack of transitions between Neanderthals and modern humans. With only a few controversial exceptions from the Middle East, the fossil record seems to be telling us that there is no transitional form, no people that reflect a strong heritage of interbreeding between coexisting Neanderthals and modern (or near-modern) humans who lived at the same time.

In the summer of 1997, extraordinary evidence confirmed the lack of substantial interbreeding between Neanderthals and modern-type humans. A team from the University of Munich, Germany, successfully extracted mitochondrial DNA from the remains of a Neanderthal skull, a relatively recent one dated at 30,000 years. The mitochondrial DNA from the Neanderthal differed equally from that of living humans from all continents, a strong indicator that Europeans do *not* possess Neanderthal genes. More profoundly, the differences between human and Neanderthal mitochondrial DNA suggest a split between the lineages leading to Neanderthals versus modern humans as early as 600,000 years ago. This is consistent with the indications from the fossil record of the break being at least 300,000 years ago, since older Neanderthal-like fossils may still lie undiscovered, and the mitochondrial mutation rate, or "clock" is likely uncertain by a factor of at least two. The story of *Homo neanderthalensis* is a separate one from our species, played out over the same time period and sometimes in the same locations, but a separate one nonetheless – at least until the end.

20.8.3 Neanderthal Lifestyle

Neanderthal cultures have been exaggerated in the popular literature in both directions – emphasizing the primitive and exaggerating the achievements. Neanderthals buried their dead, but the extent to which the burials were ceremonial remains in dispute. (The arrangement of artifacts and animal bones is not much removed from accidental, in most cases). They left no cave paintings, unlike the prolific European artists, Cro-Magnons, who replaced them, but the Neanderthals did leave evidence that they used pigments for some purposes. They had distinctive tool styles, yet variety and innovation are extremely limited: Neanderthal tool types remain similar for blindingly long expanses of time (tens of thousands of years). The sophistication of the tools, compared to those of Cro-Magnon, is low and would have provided relatively limited assistance in a physically demanding environment.

In some cases, a handful of different tool styles will be seen in a limited area (about 100 km in extent) for thousands of years. This, combined with other evidence that Neanderthal population densities were always very low compared with that of modern humans, suggests that Neanderthal populations didn't interact with each other. Groups would come and go across a landscape, rarely or never encountering each other. This is very different from all modern human cultures; modern humans are a traveling species characterized by the continual interaction of different tribes, cultures, and nations.

Part of the reason for such noninteraction may be that Neanderthal groups ranged over very limited areas. Analysis of tools and animal remains suggests that Neanderthals were not, first and foremost, hunters, but instead largely foragers and scavengers. Hunting occurred, but not on the scale practiced by even early tribes of modern humans.

Details of Neanderthal social life are at best sketchy; at worst, fictional. The anatomy of the skull and neck area suggest that Neanderthals could not be as articulate as modern humans; whether that meant that speech was not heavily employed is unclear. The arrangement of family groups is also speculative. Some anthropologists argue that the characteristics of Neanderthal hearths and other structures in caves imply a very different arrangement from most or all modern humans; in particular, one in which males lived separately from females in day-to-day existence. Other anthropologists argue that such inferences constitute overinterpretation.

At the heart of such musings lies the question of the Neanderthal mind. Although the size of the Neanderthal brain equaled or exceeded ours, on average, its different shape makes uncertain conclusions regarding intelligence. Some anthropologists argue that the large brain size was a function of the bigger body mass, or another adaptation to cold climates but that, in fact, the complexity of the brain, based on its shape, was less. Given that we do not understand well the nature of our own brain, such speculations are dangerous ones. Undoubtedly there were differences in the behavior, capabilities, and skills of Neanderthals relative to moderns; unfortunately, the nature of those differences is so faintly hinted at by the physical evidence that they remain wholly mysterious.

20.8.4 Interaction of Neanderthals with Moderns

Neanderthals and Moderns overlapped in geographic range for almost a third the duration of Neanderthal's existence. Modern forms of *Homo sapiens* moved into the Middle East from Africa by about 90,000 years ago. Neanderthal, under pressure during especially cold periods to move south, is found as early as 120,000 years ago and as recently as about 50,000 years ago in the Middle East. Remains of modern humans and Neanderthals interleave but do not overlap in time; hence there is no evidence for interaction between the species.

As modern humans pushed outward from Africa, they began to appear in Europe and western Asia by about 40,000 years ago, spurred on perhaps by episodes of unusual warmth around that time. Unlike the Middle East, a geographic crossroads from which both species came and went, Europe is a continental cul-de-sac. As Moderns spread across Europe, bringing sophisticated tools and weaponry, efficient hunting techniques, and a lifestyle that included much contact and interchange between tribes, the Neanderthals began to be pressured. It would take over 14 millenia for the Neanderthals to succumb; at any given time it might well have looked like the two species were coexisting peacefully.

A sign of the pressure on Neanderthals is a change in their monotonous stone tool culture. Later tool sets associated with Neanderthals show much more variety than do their earlier classic tool types, and a resemblance to the kind of tool kits the Moderns were using. Whether Neanderthal tried to imitate the Moderns to gain the latter's hunting advantage, or for other reasons, the change in tool types occurs only after modern-type humans arrived in Europe.

From 40,000 to 27,000 years ago, the geographic range of Neanderthals shrinks progressively, ending in southern Spain. This area is geographically distant from natural migration routes, and represents a logical "last refuge" for a people who are succumbing to whatever pressures the Moderns were bringing to bear. Extinction need not have been caused by war or other direct suppression. Only a very small reduction in breeding success is required to eventually drive a species to extinction. For a typical human generational interval (20–30 years), a roughly 2% difference in successful child-rearing between Neanderthals and Moderns could have led to Neanderthal extinction in a millennium.

The Moderns who first migrated to Europe and, by their advanced hunting techniques and gregarious lifestyles, drove Neanderthal to extinction, were not the Europeans of today. They were Cro-Magnon, a tall and slender race that does not resemble any of the modern peoples of Europe. They were, however, anatomically modern in essentially all respects, and the differences in features from today's Europeans is racial in nature. Successive migrations to Europe over the millennia brought other peoples to Europe; it is possible to trace many such waves just as one can for other parts of the world. The most ancient Europeans living today are thought to be the Basque people, both on linguistic grounds and through analysis of mitochondrial DNA. Well before them, however, came Cro-Magnon and others who have left their legacy in cave paintings, animal sculptures, musical instruments, elaborate burials, advanced tool kits, evidence of highly organized settlements, and perhaps a genetic contribution to later peoples of Europe.

20.8.5 Who Were the Neanderthals?

The bulk of the anthropological evidence indicates that Neanderthals were a separate species of humans that evolved more or less in place from earlier *erectus*, or closely related, species. This evolution occurred during a time when various other Archaic populations, less well understood from the fossil record, arose from *erectus*-type populations in Africa, Asia, and possibly Australia. The Neanderthal speciation resulted in a people who had a cranial capacity similar to or larger than modern humans, but with significantly different physical and cultural attributes, reflecting perhaps substantial behavioral and intellectual differences as well. Displaced by modern humans who originated much later than they did, the Neanderthals are considered to be a separate and older natural experiment in the speciation of human beings, one that lived a long time and nearly made it to the present day.

The focus here on the Neanderthal story is not meant to imply that it is the most important episode in human evolution. It is, instead, the best documented of the interactions between Archaic human populations – those derived from the ancient *Homo erectus* migrations out of Africa – and Moderns, those peoples resulting from the much later speciation event in Africa that produced modern *Homo sapiens*. The replacement of Archaics by Moderns occurred elsewhere around the world (excepting the Americas and Antarctica, where Archaics were absent), but nowhere else is the physical evidence so extensive and clear.

It must be emphasized that not all anthropologists accept the views presented here. Some argue that Neanderthals represent most or all of the gene pool from which Cro-Magnons arose; others assert that substantial interbreeding between Neanderthals and migrating Moderns produced European stock. The former seems very implausible, given the evidence for substantial human migrations, the profound physiological differences between Cro-Magnon and Neanderthal, and the recent analysis of mitochondrial DNA from Neanderthal bones indicating a very ancient split from the lineage of modern humans. The latter, less extreme view, that modern Europeans possess some dose of Neanderthal genes, is problematic given the virtual absence of remains of transitional people and the circumstantial evidence that Neanderthals kept to themselves. We do not even know whether Neanderthals and Moderns could have produced fertile offspring; in fact we will never know.

We yearn to meet ancestors who will tell us where we came from and why. We people our myths with giants and elves, ogres and trolls, beings who are not quite human, and whose imagined existence allows us to hold a mirror up to ourselves, to evaluate what it truly means to be human. The occasional encounter of modern humans with Neanderthals between 40,000 and 27,000 years ago

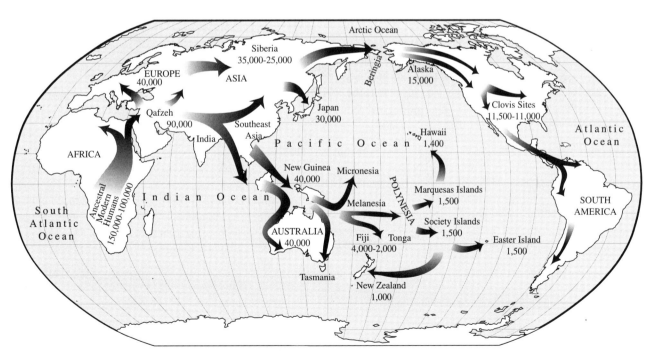

Figure 20.7. Map of the world showing approximate times, in years before present, for the first migrations of modern humans to various continents. Modified from original figure by Simon S.S. Driver from Stringer and Gamble (1993) by permission of Thames and Hudson.

might have carried with it some of that mythic quality, a reckoning with another intelligent species whose common origin in the distant past could have been guessed at but not understood. Those of us alive today missed the chance for such an encounter by no more than a quarter of the span of time of modern humans, and less than 2% of the Pleistocene epoch.

20.9 THIS MODERN WORLD

With the demise of the last Archaics – the Neanderthals – modern humans became the singular branch of the hominid family to inherit Earth. Once begun some 150,000 to 100,000 years ago, migrations did not stop, and never have; over and over again from one continent to another waves of human migrants have traveled and made new homes (figure 20.7). Eastern Asia was reached perhaps 70,000 years ago, from which the first modern humans touched Australian soil 40,000 to 50,000 years before present. Northern Asia did not see modern humans until 25,000 years ago, and the Americas were entered from there no later than 12,000 years before present. The mid-Pacific islands were reached by humans only a few thousand years ago, Antarctica a century ago. The most recent human landfall on a hitherto untouched place was on the Moon in 1969. Much of Earth's ocean floor and all planets of the solar system have yet to see human habitation, though our mechanical and electronic proxies prowl these alien terrains.

The story of Earth takes a new turn with the spread of modern *Homo sapiens*, one in which the progressive growth of agricultural and industrial societies creates novel impacts on land, oceans, and atmosphere. To understand this present time, we must begin in the last glacial episode, with the details of the climate and vegetation record that provide the baseline from which anthropogenic influences can be evaluated.

20.10 QUESTIONS

a. If climate instability stimulated the development of humans, why were not similarly sophisticated creatures the product of earlier epochs of climate instability?

b. Given the fate of the Neanderthals at the hands of humans, what might be humanity's fate should we ever encounter a similar, but technologically more advanced, intelligent species beyond Earth?

20.11 READINGS

20.11.1 General Reading

Shreeve, J. 1995. *The Neanderthal Enigma*. William Morrow and Company, New York.
Tudge, C. 1996. *The Time Before History*. Touchstone Books, New York.

20.11.2 References

Kimball, W.H., Johanson, D.C., and Rak, Y. 1994. The first skull and other new discoveries of *Australopithecus afarensis* at Hadar, Ethiopia. *Nature* **368**, 449–451.
Krings, M., Stone, A., Schmitz, R.W., Krainitzki, H., Stoneking, M., Paabo, S. 1997. Neanderthal DNA sequences and the origin of modern humans. *Cell* **90** 19–30.
Larrick, R., and Ciochon, R.L. 1996. The African emergence and early Asian dispersals of the genus *Homo*. *American Scientist* **84** (Nov.-Dec.), 538–551.
Mellars, P. 1996. *The Neanderthal Legacy*. Princeton University Press, Princeton, NJ.
Stevenson, V. (ed.) 1983. *Words: The Evolution of Western Languages*. Van Nostrand Reinhold, New York.
Stringer, C., and Gamble, C. 1993. *In Search of the Neanderthals: Solving the Puzzle of Human Origins*. Thames and Hudson, London.
Tattersall, I. 1995. *The Last Neanderthal: The Rise, Success, and Mysterious Extinction of Our Closest Human Relatives*. Macmillan, New York.
Tattersall, I., Delson, E., and Van Couvering, J. 1988. *Encyclopedia of Human Evolution and Prehistory*. Garland Publishing, New York.
Thorne, A.G., and Wolpoff, M.H. 1992. The multiregional evolution of humans. *Scientific American* **266**(4), 76–83.
Waddle, D.M. 1994. Matrix correlation tests support a single origin for modern humans. *Nature* **368**, 452–454.
White, T.D., Suwa, G., and Asfaw, B. 1994. *Australopithecus ramidus*, a new species of early hominid from Aramis, Ethiopia. *Nature* **371**, 306–312.
Wilson, A.C., and Cann, R.L. 1992. The recent African genesis of humans. *Scientific American* **266**(4), 68–73.

IV THE ONCE AND FUTURE PLANET

21 | CLIMATE CHANGE OVER THE PAST 100,000 YEARS

21.1 INTRODUCTION

Humankind's present-day dilemma with respect to global warming often is viewed with virtually no temporal perspective at all. The decade of the 1980s was among the hottest in the short century that reliable weather records have been kept. But how does this century compare to other centuries, or this millennium to others? In the third part of this book, we explored extremes of Earth climate far more profound than those experienced in modern times, or even through the short span of human history.

To really put global warming in perspective, however, we need to understand how the climate has varied during the most recent geologic epoch, the Pleistocene, a time when all of Earth's geologic processes, and the chemistry of the atmosphere, are fully modern in all respects. The time since the last interglacial, through the last ice age to the present interglacial, is recent enough that evidence is available by which very detailed records of climate can be constructed. The most thorough records can be assembled for the past 10,000 years of Earth history, the Holocene. In this chapter, techniques for assembling detailed climate information are summarized, and we compare the climate in this interglacial with that in the last, a kind of "Jekyll and Hyde" story.

21.2 THE RECORD IN ICE CORES

As discussed in chapter 6, the stable heavy isotopes of both hydrogen and oxygen exist in ocean water, and the resulting heavy water tends preferentially to exist in liquid form as opposed to vapor. Thus, water evaporated from equatorial oceans and moved poleward in storm systems is progressively depleted in heavy water, and this effect is more pronounced in colder climates. The ice sheets deposited in polar latitudes over the past few hundred thousand years therefore contain a record of warmer and colder times through the amount of depletion of deuterium and oxygen-18 in the ice. Together, the hydrogen and oxygen isotopes in ice cores allow a record of temperatures to be assembled with quite high fidelity, showing century or even decadal variation, back through the last age, and even farther. The ice cores also record carbon dioxide content in the atmosphere at the time each layer is laid down, because carbon dioxide is trapped in air bubbles in the ice during deposition each winter. Other gases that may contribute to the trapping of atmospheric heat are found in the bubbles. The cores also contain a record of the amount and kinds of dust that blew across and were deposited annually on the ice sheets. Glacial epochs seem to be not only colder but also dryer, on a worldwide basis, so that broader deserts and hence more airborne dust are a signature of those times.

Ice core records dating back through the last interglacial must be collected from sites that have remained glaciated even through that warm time, and have survived intact through the Holocene. A combination of high latitude and high altitude is required. However, many such sites are very dry, and hence the ice layers deposited are thin. Pressure from the continuing addition of annual layers eventually squeezes the layers to the point where they cannot be sampled. Periods of warmth cause a different problem: The diffusion of oxygen and other isotopes through the softening ice eliminates the annual layers and may even smooth out the decadal or century-scale variations. Furthermore, correlating core depth with dates is not easy. For cores in which annual ice accumulation is large, the annual cycles may be counted directly. Nearer the bottom of cores or in dryer regions, the annual variations are smeared out and a model of ice accumulation that is tied to the inferred temperatures must be

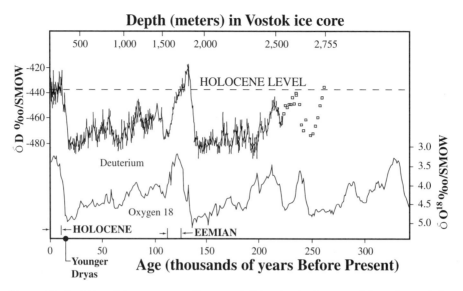

Figure 21.1. A Vostok ice core extending over 2.7 km depth. Amount of deuterium relative to hydrogen in the ice is a measure of ocean temperatures; higher deuterium (lower negative number) means higher temperature at the time of deposition. For comparison, an oxygen-isotope record from Pacific ocean sediments also is shown. (Remember that, for the seafloor sediments, a higher oxygen-18 value corresponds to cooler temperatures; the record has been inverted for easy comparison with the ice record.) The dashed line corresponds to the average Holocene deuterium value in the ice. After Vostok Project Members (1995).

applied, or sea core sediments can be used to correlate ages.

Two regions of Earth that have produced excellent records are Antarctica and Greenland; their positions in opposite hemispheres of Earth have allowed a determination of how widespread various climate changes might be. Figure 21.1 is an ice core from the Vostok station in Antarctica showing 2.7 km of the total 3.4 km depth that has been drilled as of 1997. Two glacial cycles are represented in the data in the figure; the total core extends over 4 cycles or 400,000 years. For comparison, an ^{18}O record from seafloor sediments is shown, and the two track each other very well. The ice core, however, clearly is more detailed, showing shorter-duration variations. The ice core temperature record also tracks the carbon dioxide record, as shown in the Vostok core in figure 21.2. Lower carbon dioxide values seem to correspond to lower temperatures. Whether the carbon dioxide is responding to, or forcing, the temperatures is a key puzzle in the study of Pleistocene climates that we return to in chapter 22. The carbon dioxide record is much less accurate than the isotopic record because of the problems of diffusion of the carbon dioxide through the ice. It is very important, however, not just for correlating temperature changes with carbon dioxide variations, but also because of the possible direct effects on plant communities of changes in the carbon dioxide content of the atmosphere.

The basic pattern over the past 100,000 years or so begins as the last interglacial – the Eemian interglacial – gives way at 115,000 years before present to the last of the Pleistocene glacials. An initial period of extreme cold,

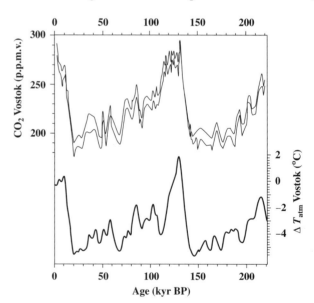

Figure 21.2. A record from the Vostok ice core showing atmospheric carbon dioxide abundance (upper curves) and temperature (bottom curve) over the past 220,000 years. Temperature is derived from oxygen-18 and deuterium isotopic abundance in the ice. The doubled carbon dioxide curves reflect uncertainty in the actual value at a given time, which lies within the envelope bounded by the two curves. Reprinted from Jouzel et al. (1993) by permission of Macmillan Magazines, Limited.

blurred in the sea sediment record, rapidly retreats and a mild glacial time oscillates in warmth, until a second deep glacial some 60,000 years ago is reached. Climate then moderates, but cools again progressively – with oscillations seemingly on all timescales resolvable by the core – until 19,000 years ago when the glacial climax is reached. The glacial snow line – where ice exists year round – dropped some 1,000 meters (3,300 feet) from today's value, and glaciers pushed down through much of northern Europe, Asia, and North America. Glaciers were even present in some mountainous parts of North America equatorward of 35° latitude. Some 5,000 years later, temperatures began to rise quickly, and the present interglacial began.

Careful examination of the ice core record in figure 21.1 reveals that the Eemian and Holocene interglacials are very different in their character, a fact not evident in the coarser Pacific Ocean sediment core. The onset of the Eemian 135,000 years ago is characterized by a time of extreme warmth, exceeding anything in the Holocene, and an apparent progressive decline through average Holocene levels until the precipitous dropoff into the glacial. The Holocene is characterized by an equally sudden rise, but to a value only somewhat above the average temperature for the past 10,000 years. Following this rise, the temperature seems to settle to a plateau that is broken only occasionally by modest excursions. This interglacial, the one in which human civilization began and has flourished, seems to be more stable than the previous one. The Eemian-Holocene difference becomes more startling in the higher-resolution ice core record collected in Greenland. In that moister climate, the annual deposition of ice is greater, hence the resolution of the temperature record is higher. However, to reach the Eemian requires drilling deeper. The start of the Eemian is almost 2.9 km below the surface in the Greenland core, by which depth the Vostok core has reached twice that age. However, the former's resolution is remarkable, as shown in oxygen isotope data from the Greenland core (figure 21.3).

The nature of the climate in the Eemian interglacial is strikingly different from that in the Holocene. Even though Holocene climate wanders toward warmer and colder times (the most recent of which were the Medieval Warm Period followed by the Little Ice Age, ending in the nineteenth century), these are modest fluctuations around a rather stable, typical climate. The ice core reveals that the Eemian had three climate states. One is similar to our Holocene climate. A second is quite a bit colder, and the third much warmer. The warmer state appears to be roughly 2°C warmer than the average global temperature

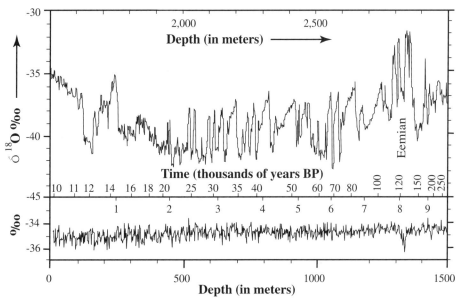

Figure 21.3. Greenland ice core showing ^{18}O concentrations versus time. Colder periods are characterized by less ^{18}O (more negative numbers) because relatively more ^{16}O is being sequestered from the oceans into the ice. Time in the past, corresponding to depth in the core, is labeled on the horizontal axis within the panel. For compactness, the core is broken up into 1.5-km segments: The lower figure covers the past 10,000 years; the upper figure represents core material much more compressed by the weight of the ice overburden, hence extending back 250,000 years. Adapted from Dansgaard et al. (1993) by permission of Macmillan Magazines, Limited.

today – comparable to the predictions for warming due to doubling of atmospheric carbon dioxide discussed in chapter 22.

Even more striking is that the Eemian climate seemed to shift from one state to another on a scale of just one or two decades, whereas the states themselves might last thousands of years or just a few decades. In such a chaotic and fickle environment, human civilization would not have been capable of sustained food production by agriculture. During the Eemian time, of course, there was no human agriculture: modern *Homo sapiens* was about to make, or had just made, its appearance in Africa, while archaic forms dominated in Europe (the Neanderthals) and Asia. One is left to speculate what might have happened to modern humans had the Eemian possessed the character of the Holocene. Might humans have invented organized agriculture 130,000 years ago, rather than 8,000 years ago? Might they have spread more rapidly through the Asian and European continents, rather than lingering for some tens of thousands of years close to the African homeland?

The striking character of Eemian interglacial climate has been challenged on the basis of evidence that the Greenland ice cores showed distortion near their bases; some of the extreme variability might therefore be an artifact. This issue is not fully resolved: Deep-sea cores containing isotopic information on climate do not correlate well with the ice core data, but European pollen core data seem to support the higher variability of the Eemian climate. Should the last interglacial prove more variable than the present one, the question is raised as to how climate states might become unstable. We address this question in chapter 22.

21.3 CLIMATE FROM PLANT POLLEN AND PACKRAT MIDDEN STUDIES

Plants are sensitive indicators of climate. Different plants thrive in different locations. Dry, cold conditions might produce grassy steppes whereas a change to warm and wet circumstances will bring trees to the same area. In Europe and similar climates, cold conditions favor evergreen trees while in warmer periods deciduous trees are more common. In the American Southwest and other arid but mountainous parts of the world, vegetation is a sensitive function of altitude: Desert plants give way to scrub woodlands and then to forests as one climbs upward from plains to mountain heights. While separating the effects of rainfall from temperature is not easy, plants have proved to be a unique means of directly sampling local and regional climate changes.

Plant remains are surprisingly durable if they are stored under the right conditions. Sediments on lake bottoms or in dried lake beds preserve plant pollen in conditions that are often anoxic, or otherwise favorable for the long-term preservation of material. Dating of the material through the Holocene is possible using the radiocarbon technique, i.e., determining the $^{14}C/^{12}C$ ratio to infer a date, as described in chapter 5. For earlier times the dates become increasingly uncertain, and for the Eemian (well beyond the 70,000-year useful limit for such techniques) less precise estimates must suffice.

Figure 21.4 is an example of the occurrence of pine and alder pollens in lake bottom sediments in France and Germany, with ages dating to the Eemian. Fluctuations in occurrence of pollen, if they translate with fair fidelity to climate, suggest that the Eemian was indeed characterized by strongly variable climates. During warmest Eemian times, temperatures may have been 5°C higher in the winter than on average today in Europe, and the coldest of Eemian episodes were comparable to those in glacial conditions. Further, the extent and timing of variations may have differed in different parts of Europe, and hence almost certainly differed in geographically more distant locations.

Pollen studies also have been used to track climate changes in continental locations as the last ice age came to a close. In arid regions of the world, such as the southwestern United States and northern Mexico, these studies are supplemented by information from packrat middens.

Packrat middens are well-preserved fragments of plants, and less commonly insects, accumulated locally by packrats (also called woodrats). Their preservation is the result of being encased, amber-like, in crystallized urine, a reflection of the habit of packrats to urinate on their

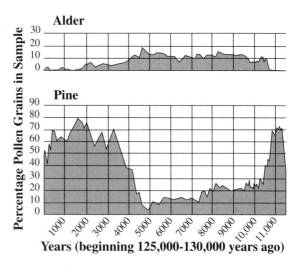

Figure 21.4. Occurrence of pine and alder pollens in lake-bed sediments from the Eemian interglacial in France and Germany. More pines suggest colder conditions; more alders, warmer conditions. Adapted from Field et al. (1994) by permission of Macmillan Magazines, Limited.

caches of collected material. The dry climate of the Southwest can preserve these *amberates* for tens of thousands of years. Dating is done by radiocarbon techniques on the organic material. Packrat middens are common enough in caves and crevices throughout the southwestern United States that a useful climate record over 40,000 years has been assembled through studies beginning around 1960. The same kind of studies have more recently been performed in South America, Africa, the Middle East, Asia, and Australia.

The source of the climate information lies in the fact that packrats collect material from a relatively limited area around a midden site, and enough is collected over the life of the animal that the midden becomes representative of the local plant environment at the time the animal lived. Corrections must be made for the variation in types of plants collected by different packrat species (a function of their diet).

The result of dated amberat samples is a map in space and time of the plant communities, and hence climate shifts, of a given area, dating back halfway through the last ice age. Figure 21.5 (color plate VIII) shows an example of this kind of work in the form of vegetation maps of the southwestern United States. This region, stretching from California to western Texas, encompasses large parts of three great Mexican and American deserts: Mohave (California), Sonoran (Arizona), and Chihuahuan (New Mexico and Texas). They differ in their elevation and amount and seasonal timing of rainfall. The deserts give way in the uplands and mountains to other plant communities as a function of altitude and rainfall, which can differ greatly depending on the orientation of the mountain ranges. Roughly, though, today's grasslands appear above 1,000 meters, woodlands above 1,500 meters, pine parklands above 2,000 meters, conifer forests above 2,500 meters, spruce-fir forests above 3,000 meters (subalpine conditions), and tundra above treeline (alpine) above 3,500 meters.

In the depths of the ice ages, conditions were different. The same region 18,000 years ago looked as if one had raised the general altitude of the land 500 meters or so. Tucson, Arizona, at an elevation today of 700 meters, is in the Sonoran desert; 18,000 years ago, Juniper and Pinyon pine woodlands interspersed with grasslands occupied the Tucson basin. The Santa Catalina Mountains north of Tucson, where today big conifers grow typically above 2,000 meters, contain ice age packrat middens with conifer seeds at 1,500-meter altitude. The upper mountain reaches of the Catalinas, just shy of 3,000 meters, were "too high" then for trees; today that summit is covered with Douglas fir and White fir trees.

An important complexity in this story, revealed by the amberat data, is that the change from glacial to interglacial was not simply a matter of moving the same vegetative communities up in altitude; latitudinal shifts occurred as well. Species that were not adapted to glacial climates became more abundant in the Holocene, and some species widespread in the last glacial were squeezed out by new plant species. Many plant types of subtropical origin only colonized the Sonoran desert of southern Arizona and northern Mexico during the Holocene. In the nearby mountains, Douglas fir did not have to compete during the last glacial with Ponderosa Pine, a tree that today occupies much the same elevation regime. Ponderosas apparently are particularly adapted to the moisture pattern of the Holocene, and moved southward to colonize large areas of the Arizona uplands only after the last glacial ended.

The importance of the packrat midden and pollen records goes beyond the estimation of climate variation based on vegetation occurrence. They show, by example, the tremendous ecological displacements associated with climate changes. Not just a theoretical inference, the massive displacement of vegetative types and associated animals is preserved as physical evidence in lake sediments and packrat middens. Such records are a reminder that, if indeed human activities today are precipitating or accelerating global warming, we should expect significant changes in occurrence of forests, woodlands, and agricultural belts as a result of such activities.

21.4 TREE RINGS

Many species of trees grow in such a way that in cross-section their trunks display rings, reflecting an annual or seasonal cycle to their growth. In very wet climates, or in soils close to the water table in arid regions, the growth of the trees is little affected by year-to-year variations in precipitation. But in soils on dry hillsides in arid regions, the amount of precipitation in a given year may have a substantial effect on the amount of increase of trunk thickness. In those regions, the thickness of the annual or semi-annual growth ring may reflect the precipitation conditions in that locale, that is whether the year was wet or dry (figure 21.6). Temperature also plays a role in ring thickness, primarily through its effect on soil moisture. Different parts of each tree ring can be examined to provide information on seasonal variations of rainfall within a given year.

Tree rings provide a remarkable record of climate because they show details on timescales as short as seasons. The great age of trees allows an individual tree to record

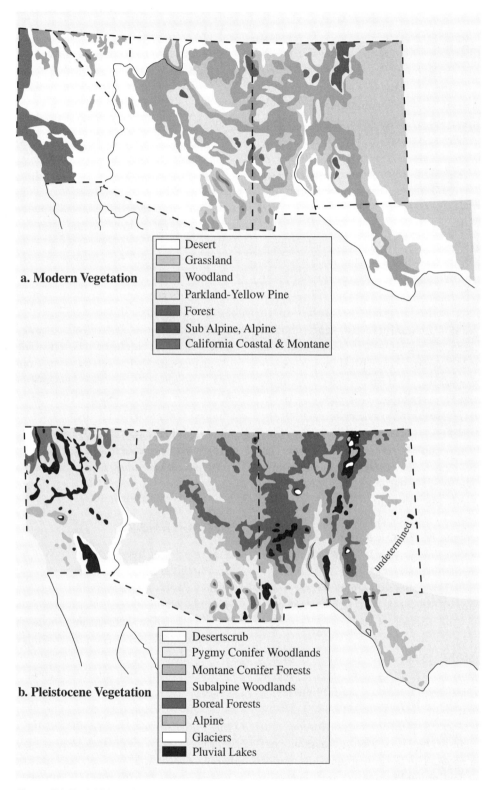

Figure 21.5. (a) Map of southwestern United States vegetation from today and (b) a similar map from 18,000 years ago, based on packrat midden and pollen studies. Note that keys are slightly different, in part because vegetative associations and species types changed in addition to the altitude of occurrence. Roughly, "Parkland" and "Forest" in (a) correspond to "Montane Conifer Forests" in (b). Panels (a) and (b) are based on Betancourt et al. (1990).

Figure 21.5. (Continued) (c) Typical Sonoran desert vegetation seen today in Tucson, Arizona. Panel (c) is photograph by the author.

Figure 21.5. (Continued) (d) Juniper trees some 80 km southeast of, and 600 meters above, Tucson. Junipers similar to these (though of a different species) were present in the Tucson basin during the last ice age. Panel (d) is photograph by the author.

the local climate changes, from one growing season to another, over centuries, even millennia in the case of the Bristlecone pine. It is possible to correlate rings from living to dead trees by matching patterns of narrow and wide rings, or *cross-dating*. By accumulating tree-ring records from a large number of trees in a given region that have experienced a common climate history, one can use overlapping tree-ring patterns from living to increasingly older specimens of dead trees to push the chronology far back in time. By using techniques of statistical analysis on many samples, sources of error such as trees with missing rings or doubled rings (an occasional occurrence) can be minimized. The American astronomer Andrew Ellicott Douglass, who began his tree-ring work at the turn of the century and went on to found the pioneering Laboratory of Tree-Ring Research at the University of Arizona in Tucson, first put this technique on a firm quantitative footing.

Because of the great aridity of the mountains in the southwestern United States, samples of bristlecone pines of enormous antiquity are available, and the tree-ring record of annual climate has been extended back using crossdating techniques some 9,000 years, 80% of the duration of the Holocene. Some fragments of trees dated by carbon-14 techniques to 11,000 years may eventually allow the entire Holocene to be characterized by this technique. Similar efforts are being made at other sites around the world where conifers are available in mountainous climates, for example Tasmania, Chile, Argentina, China, and Tibet. Techniques for coring into living trees without killing them make it possible to sample trees in areas where logging is not permitted.

The tree-ring thickness often is not a simple function of the total annual precipitation in the immediate area, but is a complicated function of several factors. High temperatures in the growing season may increase evaporation and hence soil dryness, even if the year exhibits normal amounts of precipitation. The distribution of precipitation seasonally may be important in affecting the growth of a tree, because the growing season does not extend through the entire year. These lead to ambiguities as to whether a thick ring means a wet year, a cold year but

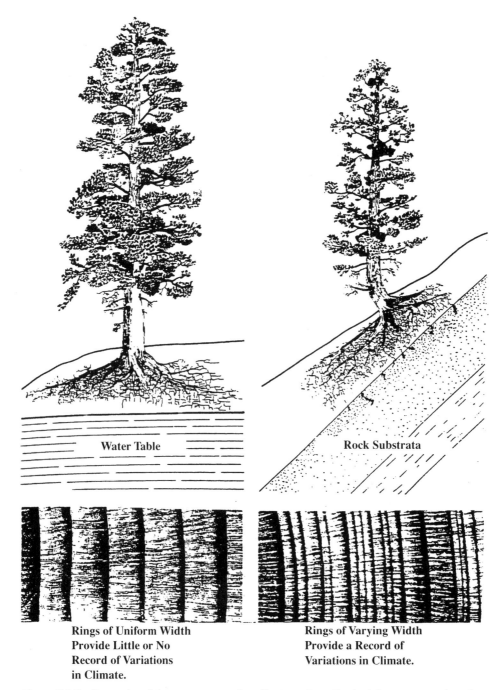

Figure 21.6. Illustration of the appearance of conifer tree rings. On the left, a tree growing where the water table is near-surface is insensitive to annual precipitation variations and will leave regular growth rings. On the right, a tree growing on a dry hillside, dependent on precipitation, will show variations in the thickness of its growth rings, reflecting year-to-year variations in moisture. From the University of Arizona Laboratory of Tree-Ring Research.

normal in precipitation, etc. A rough, experimental rule of thumb is that high-latitude, high-altitude trees tend to be most responsive to temperature, whereas low-altitude, low-latitude trees (but outside of the tropics) are most sensitive to precipitation.

An example of tree-ring data is shown in figure 21.7. This data set and others from Giant Sequoia trees in the Sierra Nevada of California have been used to create a detailed chronology of droughts in California extending from the present back to 100 B.C., a span of over 2,000 years. Time periods when droughts were rare (such as around A.D. 1,000) are preceded and followed by episodes of repeated droughts, which undoubtedly had a severe effect on the people living in the region at the time. Similar

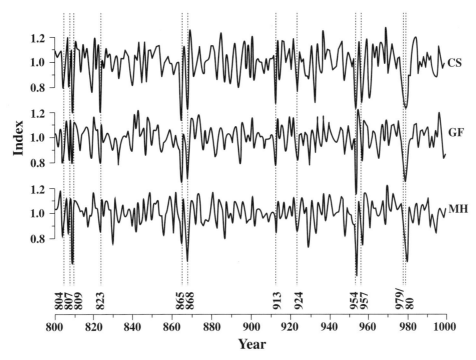

Figure 21.7. Plot of years of wetness and drought in the Sierra Nevada, from A.D. 800 to 1000. Cores from living and dead sequoias were collected at three sites [Camp Six (CS), Giant Forest (GF), and Mountain Home (MH)] within 100 km of each other. The ring index is a measure of thickness; the higher the index, the thicker the ring and hence the wetter the year. Years in which all three stations show thin rings are marked and are likely to have been ones of severe drought. Reproduced from Hughes and Brown (1992) by permission of Springer-Verlag.

tree-ring data from around the world combine to provide a detailed view of regional climate changes unobtainable with other techniques.

Further information contained in tree rings includes stream-flow data and a history of forest fires, because conifers often survive such fires but retain evidence of the charring that can be dated to a single year. Stream-flow data are obtained by examining annual ring widths in trees growing in the major watershed areas of a given river: Wide rings mean wet years with high stream flow, thin ones dry years with low flow. Furthermore, the ability to derive absolute dates from the cross-dating of living and dead trees makes tree rings an important tool for calibrating radiocarbon dates: The age of the wood itself, from the ^{14}C technique, can be compared with the absolute chronology from cross-dating rings.

21.5 CLIMATE VARIABILITY IN THE LATE HOLOCENE

The techniques discussed above are only some of many tools that paleoclimatologists have used to assemble a record of the Holocene climate worldwide. The detail both in time and space of such a record is unique to this most recent time in Earth's history.

The last major climate perturbation of the Holocene, the Little Ice Age, best documented in Europe, extended from the sixteenth to the early nineteenth century (figure 21.8). At its peak, annual temperatures dropped 1.5°C from present norms in Europe; in at least some parts of the world it is one of the coldest episodes of the Holocene. During the coldest of those times in Europe there were years with essentially no growing seasons; crops failed, starvation occurred, and rivers known today to remain largely ice-free year round (for example, the Thames through London) often froze over.

A portion of this cold period, from 1645–1715, coincides with a well-documented and unusual absence of sunspots on the Sun, the so-called "Maunder Minimum." The significance of the sunspot absence for the total brightness of the Sun is unknown, but another indicator of decreased solar activity at the time is a peak in the amount of radiogenic carbon-14 seen in contemporary tree rings. The production of atmospheric carbon-14 depends on the supply of high-energy galactic cosmic rays (chapter 5). High solar activity pumps up the Sun's magnetic field which tends to deflect such particles from entering the solar system; low solar activity does the opposite. The increased carbon-14 seen in tree rings from the Little Ice Age is consistent with lower solar activity. This, in turn, suggests that

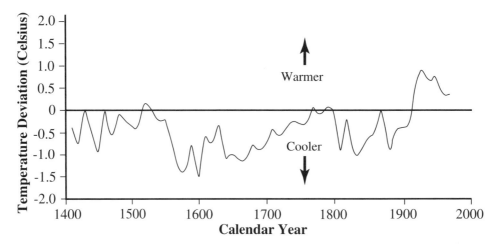

Figure 21.8. Changes in summer temperatures in the northern hemisphere, averaged decade by decade, from medieval times to the present. Deviations from a reference temperature are plotted versus calendar year. The cold period known as the Little Ice Age is evident from the 1600s to the end of the nineteenth century. The figure is drawn from data of Bradley and Jones (1993), who used a variety of information to reconstruct temperatures.

the Sun might have shined less brightly during that time, for reasons that are unknown. A decreased solar output might have been just one contributor to this very recent and unusual cold spell, whose origin remains poorly understood.

Prior to the Little Ice Age, during a time in Europe known as the Medieval Warm Period, major upheavals in agricultural communities were occurring in what is today the southwestern United States. Prolonged periods over a century or two in which the crucial summer rains failed, punctuated by floods that may have damaged or destroyed irrigation canals, seems to be roughly coincident with the decline of organized civilizations that later came to be known as the Anasazi, Mogollon, Hohokam, and others. Whether climate was the direct cause of the dispersal of the peoples who created these cultures, or was only an additional stress on top of others, may never be understood.

We do not know, at the Holocene level of detail, what the Eemian climate variability was like; we have only the tantalizing hints from the oldest ice core and pollen studies which were subject to degradation over some hundred millennia of time. But it should be sobering to us that the historical variability of climate in the Holocene, as significant as it has been for human history, seemingly pales in comparison to the wild swings of the Eemian. Further, we do not understand the origin of the comparatively "minor" climate shifts in the Holocene. Too short-lived to be caused by orbital variations, they may have their origins in small changes in the Sun's heating, altered patterns of ocean currents and sea surface temperatures, volcanic eruptions putting sunlight-absorbing aerosols into the stratosphere, or other effects. In considering the possible impacts of human influences on climate, it is important to remember that we are a long way from understanding natural climate variability.

21.6 THE YOUNGER DRYAS: A SIGNPOST FOR THE OCEANIC ROLE IN CLIMATE

Perhaps the most striking episode of climate variability occurred right at the beginning of the Holocene. Named after an arctic flower, the *Younger Dryas* can be seen in pollen core and ice-core records as a sudden drop in temperature back toward ice-age values, and the resurgence of glaciers, perhaps in less than a century. This cold snap seems to have been restricted to continental areas around the North Atlantic, such as Europe and Canada, and was less significant elsewhere. It lasted a thousand years, and then, in the space of perhaps a decade or two, temperatures rose sharply toward typical Holocene values and the glaciers resumed their retreat.

What is striking about the Younger Dryas is that it occurred during a time when the glaciers were still in the process of retreating from lands that in the full flush of the Holocene are entirely unglaciated. Although the shrinking of the glaciers themselves, in reducing continental albedos, should have reinforced the warming, there are other effects of the retreat that could have played more unpredictable roles.

Geoscientists W.S. Broecker of Columbia University and G.H. Denton of the University of Maine have identified the Younger Dryas as a kind of "smoking gun" for the importance of oceanic-atmospheric interactions in determining climate. During the retreat of the glaciers that began 14,000 years ago, glacial meltwater in North America was accumulating in southern Manitoba and flowing down the Mississippi River in an impressive torrent comparable to today's massive Amazon river. To the east, glaciers blocked freshwater flow into the North Atlantic. About 11,000 years ago, the ice to the east had retreated sufficiently that much of the meltwater began flowing eastward across what would become the Great Lakes, along the St. Lawrence River and into the Altantic Ocean.

This massive influx of freshwater diluted the normally salty water in the upper layers of the North Atlantic. Water with dissolved salt is denser than fresh water; this is why we float so much better in the ocean than in a swimming pool. In the Atlantic ocean today, a current of water flows northward at depths of about 800 meters below the ocean surface, until it reaches roughly the latitude of Iceland. There surface winds blow colder surface water aside, the warm water at 800 meters rises, releasing heat. Importantly, it then *sinks* because of its relatively high salinity, creating a heat pump whereby heat is delivered to the North Atlantic ocean and the water that delivers it then sinks to great depths.

The freshwater influx of the Younger Dryas, in diluting the salty waters of the North Atlantic, would have prevented their sinking and hence interfered with the transport of heat by the subsurface northward current. Because the moderate climate of Europe is dependent on the release of heat from the North Atlantic waters, this Younger Dryas cap of diluted, cold water should have had a profound chilling effect on the climate for as long as the substantial flow of freshwater continued to pour into the North Atlantic.

The oceanic-atmospheric connection in climate should not come as a surprise. The mass of Earth's oceans is some 500 times that of the atmosphere; the oceans are therefore capable of holding vastly larger quantities of heat than the atmosphere. Similarly, the oceans can hold 60 times as much carbon dioxide as is in the atmosphere today. What has prevented climatologists from understanding the role of the oceans in climate is the lack of knowledge of ocean circulation, and the inherent difference in circulation timescales between ocean and atmosphere. Much of the deep ocean may not mix with the shallower waters on timescales of interest to human global warming issues (decades): just how much does and how it does so are critical to understanding the interaction between the atmosphere and the ocean.

Much of the deepest insight into how the ocean exchanges heat, carbon dioxide, and other important climate quantities with the atmosphere has come from trying to understand the puzzle of glacial cycles: Why does Earth become glaciated, why do interglacials occur, and what is the role of carbon dioxide?

21.7 INTO THE PRESENT

The end of the last ice age came as summer sunshine in the northern hemisphere began to approach a maximum (8% greater at 11,000 years ago than at present), the result of the orbital swings of the Milankovitch cycle described in chapter 20. This may have been the stimulus for a series of changes in ocean circulation patterns that led to worldwide, virtually simultaneous retreat of the glaciers. The precise mechanism by which the increased northern hemisphere summer heating triggered oceanic changes remains unknown, but the release of additional carbon dioxide stored in the seas represents part of the answer. Worldwide warming of the ocean and melting of glaciers during the early Holocene caused sea level to rise by roughly 130 meters relative to its value at the peak of the glaciation. Higher sea levels reduced the amount of continental shelf exposed above the sea, isolating continents previously connected by land bridges, and contributed to changes in regional climate patterns and migration routes.

As climate changed around the world, the vegetation and animal life changed with it. Large animals began to disappear from widespread regions of the continents, existing extensively now only in Africa and parts of Asia. By 11,000 to 12,000 years ago, the *Clovis* hunters of North America, with their well-honed spearpoints, may well have played a primary role in driving big game to extinction – mammoths, native horses, ground sloths, saber-toothed cats, and camels that all once lived there. Certainly the changing climate played a role too – but the association of the trademark Clovis spearpoint with remains of large American mammals, and their very rapid extinction close to the time of appearance of these people, seems to be compelling evidence for the human hand at work, as scientist Paul Martin of the University of Arizona's Desert Research Laboratory has argued for several decades.

The relatively stable Holocene climate allowed elaborate forms of agriculture to be invented by humans on all continents. Cities grew up as agricultural and trade centers, perhaps first in the Middle East around 8,000 years ago, then in Europe and the Americas some 2,000 to 3,000 years later. Human population numbers increased steadily as new agricultural techniques and improved transportation technologies were invented. In the past few centuries, humans have harnessed reserves of hydrocarbons trapped in the sedimentary layers of Earth as sources of energy. By the early twentieth century, the use of such fossil fuels was prodigious and had a measurable effect on the total carbon dioxide in the atmosphere. By the late twentieth century, the possible effects of atmospheric carbon dioxide increase on the climate became a matter of worldwide debate. An examination of the scientific and human issues behind this debate is the focus of the next chapter.

21.8 QUESTIONS

a. How might one use tree rings in a forest of different species of conifers to infer the outbreak of a large insect infestation sometime in the past?
b. What flora and fauna existed in your home area during the coldest part of the Pleistocene?

21.9 READINGS

21.9.1 General Reading

Ward, P.D. 1996. *The Call of the Distant Mammoths*. Copernicus (Springer-Verlag), New York.

21.9.2 References

Barnola, J.M., Raynaud, D., Korotkevich, Y.S., and Lorius, C. 1987. Vostok ice core provides 160,000-year record of atmospheric CO_2. *Nature* **329**, 408–414.

Berger, A., and Loutre, M.F. 1991. Insolation values for the climate of the last 10 million years. *Quaternary Science Reviews* **10**, 297–317.

Betancourt, J., Van Devender, T.R., and Martin, P.S. 1990. *Packrat Middens: The Last 30,000 Years of Biotic Change*. University of Arizona Press, Tucson.

Bradley, R.S., and Jones, P.D. 1993. Little Ice Age summer temperature variations: Their nature and relevance to recent global warming trends. *The Holocene* **3**, 367–376.

Broecker, W.S., and Denton, G.H. 1990. What drives glacial cycles? *Scientific American* **262**(1) (Jan.)49–56.

Brown, P.M., Hughes, M.K., Baisan, C.H., Swetnam, T.W., and Caprio, A.C. 1992. Giant sequoia ring-width chronologies from the central Sierra Nevada, California. *Tree-Ring Bulletin* **52**, 1–14.

Crown, P.L. 1990. The Hohokam of the American Southwest. *Journal of World Prehistory* **4**, 157–256.

Dansgaard, W., Johnsen, S.J., Clausen, H.B., Dahl-Jensen, D., Gundestrup, N.S., Hammer, C.U., Hvidberg, C.S., Steffensen, J.P., Sveinbjornsdottr, A.E., Jouzel, J., and Bond, G. 1993. Evidence for general instability of past climate from a 250-kyr ice-core record. *Nature* **364**, 218–220.

Field, M.H., Huntley, B., and Muller, H. 1994. Eemian climate fluctuations observed in a European pollen record. *Nature* **371**, 779–783.

Greenland Ice Core Project Members. 1993. Climate instability during the last interglacial period recorded in the GRIP ice core. *Nature* **364**, 203–207.

Hughes, M.K., and Brown, P.M. 1992. Drought frequency in central California since 101 B.C. recorded in giant sequoia tree rings. *Climate Dynamics* **6**, 161–167.

Hughes, M.K., Touchan, R., and Brown, P.M. 1996. A multi-millenial network of giant sequoia chronologies for dendrochronology. In *Tree Rings, Environment and Humanity* (J.S. Dean, D.M. Meko, and T.W. Swetnam, eds.), Radiocarbon, University of Arizona, Tucson. pp. 225–234.

Jouzel, J., Barkov, N.I., Barnola, J.M., Bender, M., Chappellaz, J., Genthon, C., Kotlyakov, V.M., Lipenkov, V., Lorius, C., Petit, J.R., Raynaud, D., Raisbeck, G., Ritz, C., Sowers, T., Stievenard, M., Yiou, F., and Yiou, P. 1993. Extending the Vostok ice core record of paleoclimate to the penultimate glacial period. *Nature* **364**, 407–412.

Loaiciga, H.A., Haston, L., and Michaelsen, J. 1993. Dendrohydrology and long-term hydrological phenomena. *Reviews of Geophysics* **31**, 151–171.

Martin, P.S. 1963. *The Last 10,000 Years*. University of Arizona Press, Tucson.

Maslin, M. 1996. Sultry interglacial gets sudden chill. *EOS* **77**, 353–354.

McCulloch, M., Mortimer, G., Esat, E., Xianhua, L., Pillans, B., and Chappell, J. 1996. High resolution windows into early Holocene climate: Sr/Ca coral records from the Huon Peninsula. *Earth and Planetary Science Letters* **138**, 169–178.

Petit, J.R., Basile, I., Leruyuet, A., Raynaud, D., Lorius, C., Jouzel, J., Stievenard, M., Lipenkov, V.Y., Barkov, N.I., Kudryashov, B.B., Davis M., Saltzman, E., Kotlayov, V. 1997. Four climate cycles in Vostok ice core. *Nature* **387** 359–360.

Rotberg, R.I., and Rabb, T.K. (eds.). 1981. *Climate and History: Studies in Interdisciplinary History*. Princeton University Press, Princeton, NJ.

Steadman, D.W., and Mead, J.I. (eds.). 1995. *Late Quaternary Environments and Deep History: A Tribute to Paul S. Martin*. The Mammoth Site of Hot Springs, South Dakota: Scientific Paper No. 3, Hot Springs, SD.

Swetnam, T.W. 1993. Fire history and climate change in giant sequoia groves. *Science* **262**, 885–889.

Tudge, C. 1996. *The Time Before History*. Touchstone Books, New York.

Vostok Project Members. 1995. International effort helps decipher mysteries of paleoclimate from Antarctic ice cores. *EOS* **76**, 169.

22 HUMAN-INDUCED GLOBAL WARMING

22.1 THE RECORDS OF CO_2 ABUNDANCE AND GLOBAL TEMPERATURES IN MODERN TIMES

Ice cores contain trapped bubbles of air which, provided they can be properly dated, represent a record of the composition of air over time. Because of the weight of overlying layers of ice, compressing the pores in the ice, it is very difficult to extend the record back as far as that for temperature derived from the isotopic composition of the water itself. In fact, the manner in which the air bubbles were originally trapped in ice results in their movement upward or downward relative to the ice itself, making age determination a challenge.

Figure 22.1 displays CO_2 values from an ice core collected in Greenland. The dating of the air was achieved by taking advantage of a byproduct of nuclear weapons testing: The isotope ^{14}C reached a peak in Earth's atmosphere, from the detonation of nuclear bombs, in 1963. Using this peak in heavy carbon, geochemists M. Wahlen of Scripps Institution of Oceanography and colleagues determined that the trapped air was displaced by the equivalent of 200 years relative to the ice surrounding it.

With this important correction, the figure shows that, during the Little Ice Age, CO_2 values were fairly constant. Beginning in the mid-1800s, carbon dioxide began to increase. Direct measurements from a station in Hawaii, selected to be high above any local industries and hence sampling worldwide CO_2 borne by the trade winds, show that the increase accelerates after World War II.

Today, the carbon dioxide abundance is 30% higher than it was during the Little Ice Age. Some of the increase, particularly that in the mid-nineteenth century, may be ascribed to the general warming that occurred as climate moved out of the Little Ice Age; other ice cores suggest that CO_2 20,000 years ago (near the last major ice age peak) was half that at present. However, some of the nineteenth century CO_2 increase also was likely caused by changes in land-use patterns, including deforestation: During their lifetime, trees are an important *sink*, or removal agent, of atmospheric carbon dioxide.

In the twentieth century, there is little disagreement that industrial activities, that involve the burning of carbon-rich *fossil fuels* (see chapter 23), are primarily responsible for the increased atmospheric carbon dioxide. Here, industrial is defined broadly to include use of automobiles, home heating systems, as well as agriculture involving burning of forests for clearing. Adding all of these activities together, one expects an even larger atmospheric carbon dioxide increase than is seen in the top panel of figure 22.1; some of the excess likely is taken up in the oceans and perhaps forested regions of the continents.

Other atmospheric gases have increased during the twentieth century relative to preindustrial values. Methane (CH_4) is 100% more abundant today than it was in the early nineteenth century. Again, this increase seems most readily accounted for by increased industrial activity and development of intensive agricultural techniques. Nitrous oxide (N_2O) is 10% higher than in preindustrial times; chlorofluorocarbons (CFCs) used in air conditioning and other applications have no natural sources and are appearing in the atmosphere for perhaps the first time in Earth's history. These and other compounds represent a significant perturbation to the background composition of Earth's atmosphere, where background is defined as the preindustrial Holocene atmosphere.

Projections for the future increase of CO_2 are also shown in figure 22.1. Four cases are considered. The baseline "business as usual" assumes no change in world dependance on fossil fuels while economies and population continue to grow, at least through the first half of the twenty-first century; the current rate of worldwide

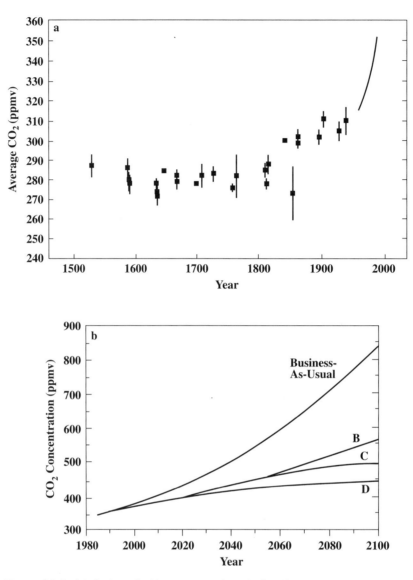

Figure 22.1. (a) Carbon dioxide concentrations in Earth's atmosphere over time from the European Middle Ages to the present. Concentration is expressed in parts per million; hence, 1 ppmv represents 0.0001%, or 10^{-6}, of the total air. (b) Projected increase in carbon dioxide levels, neglecting uptake by the ocean or continental biomass, for four possible cases described in the text. From Mortensen (1996).

deforestation also is assumed. Case B is obtained by a shift toward fuels with higher energy output per unit carbon dioxide produced, i.e., natural gas (see chapter 23), along with cessation of deforestation and imposition of tight emission controls. Case C assumes that renewable energy sources (solar) and nuclear power take over from much of the fossil fuel use during the second half of the twenty-first century. Case D is the result of such a shift in the *first* half of that century, so that industrialized countries experience no growth in their emission of carbon dioxide. Although some uptake by oceans and continental biomass is expected, such buffers do not depress completely the atmospheric rise in CO_2, and are only temporary in any event. Sometime over the next century, atmospheric carbon dioxide will be double the nineteenth century values, and possibly even quadruple that baseline.

In the rest of the chapter, we examine the physical links between such atmospheric chemical changes and alterations to the overall thermal balance of Earth's surface and atmosphere. Unfortunately, a record of temperature detailed enough in space and time to document possible changes extends no further back than about 1850; proxy records of temperature must be relied upon in spite of their less-definitive nature. Even direct temperature measurements have their limitations; oceanic and continental stations have moved over time, and the expansion of cities often creates localized warmings around weather stations associated with increased concrete and less vegetation. In Tucson, Arizona, the U.S. National Weather Service Station had to be moved after it was observed that construction of new airport terminal facilities in the area created a jump in recorded daily high temperatures; the presence of new concrete buildings altered the local environment around the thermometer.

Figure 22.2 is a record of Earth's global mean surface temperature over the past 140 years, for the northern and southern hemispheres combined, from a number of continental and oceanic stations. The temperature is obtained by averaging records from these stations over the entire year. Although there are dips and plateaus in the curve, overall the climate has warmed during this time period. After 1970, temperatures climbed dramatically in both hemispheres and remained high through the first half of the 1990s. Global temperatures in the past two decades exceed those in the nineteenth century by about 0.3° to 0.6°C. Because the rise encompasses both hemispheres in a record obtained over a large number of stations, it cannot be primarily a result of urbanization in the large cities.

A check on the direction and the magnitude of the rise comes from studies of valley glaciers around the world, essentially all of which have retreated up the valley hundreds of meters or kilometers since the middle-1800s. Careful measurement of the retreats, combined with models of how much temperature increase is required to produce a given amount of retreat, allows an estimate of the past century's worldwide temperature increase independent of weather stations. Such glacial studies by Dutch climatologist J. Oerlemans indicate a worldwide temperature rise of roughly 0.7°C, with an estimated error of plus or minus 0.2°C. This number is close to, and consistent with, the globally averaged temperature rise derived from weather stations.

The 0.3–0.6°C temperature rise cannot be ascribed immediately to human-induced atmospheric changes, because comparable or higher global temperatures are indicated for the early Holocene in proxy records discussed in chapter 21. Although the twentieth century global mean temperature is at least as warm as any in the past 500 years, we do not know whether the global average temperature will remain constant, fall, or rise further in the coming decades. Our understanding of the physical processes by which Earth's climate is maintained above the water freezing point provides a guide by which we can evaluate the potential effects of human-caused changes in atmospheric composition.

22.2 MODELING THE RESPONSE OF EARTH TO INCREASING AMOUNTS OF GREENHOUSE GASES

22.2.1 Review of Basic Greenhouse Physics

The basic physics of the greenhouse effect was described in chapter 14. As the amount of infrared-absorbing gases is increased, Earth's atmosphere becomes more opaque to infrared photons. The altitude above the surface at which such photons are finally free to escape (the *mean radiating level*) therefore moves upward, toward lower air density, as greenhouse gas concentration increases. However, as figure 22.3 shows, because the temperature falls with altitude in the troposphere, the new mean radiating level is colder than the old one, and hence less efficient at removing energy. Its temperature must increase, raising the entire temperature profile of the troposphere. In this way, increasing greenhouse gas concentrations raise the mean surface temperature of Earth.

The basic physics of the greenhouse process described above is straightforward enough that there is little argument about its validity. We know, for example, that Earth's neighboring planet, Venus, receives less sunlight at and near its surface than does Earth because of a layer of bright, reflecting clouds. However, the surface temperature of Venus is over twice that of Earth's, and the atmosphere is possessed of a carbon dioxide pressure of 90 bars,

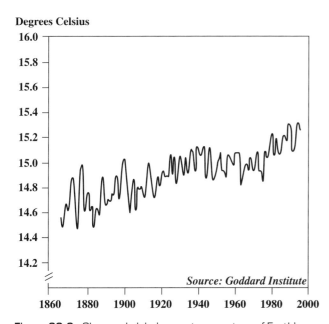

Figure 22.2. Observed global mean temperature of Earth's surface from 1860 to the present. From Bright (1997) by permission of W.W. Norton.

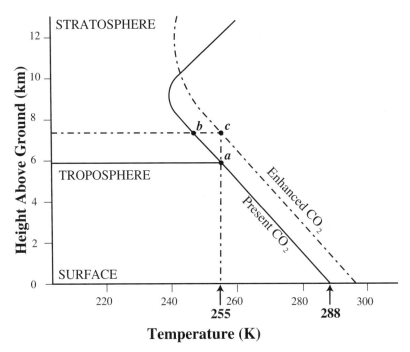

Figure 22.3. Schematic illustration of how increasing the amount of greenhouse gas in Earth's atmosphere can increase the surface temperature. The solid line represents the present temperature profile in Earth's atmosphere, and the mean radiating level (point "a") is shown. Increasing greenhouse gas concentration makes the atmosphere more opaque to infrared photons, forcing the mean radiating level upward to an altitude (point "b") where the temperature is lower. To rid the atmosphere of the same amount of heat, the temperature at the new mean radiating level must increase to point "c," forcing the whole temperature profile to increase (curve labeled "Enhanced CO_2"). Modified from Mitchell (1989).

300,000 times the amount of CO_2 in our atmosphere. It is not too great a leap to infer that the Venusian atmosphere is in a state in which the enormous amounts of carbon dioxide create a greenhouse effect much larger than Earth's, and models show that the surface temperature and CO_2 abundance are indeed consistent with each other.

22.2.2 Some Complications

As simple as the basic physical concept is, it does not fully describe the actual situation. The most fundamental complication is that radiation (transport of photons) is not the sole means of the movement of heat energy outward. Particularly in the lower part of the atmosphere, the temperature profile becomes so steep (decreases so sharply with altitude) that bulk air movement (that is, convection) plays an important role. *Dry convection* involves bulk movement of air without condensation of water to form clouds; it occurs in the lowermost part of the atmosphere and particularly in dry regions. *Moist convection* includes the effects of cloud condensation and evaporation, which add and delete heat from the surrounding air. Cloud formation most often is a result of air containing water vapor rising, expanding as the surrounding pressure drops with altitude, and then cooling until the air can no longer hold the water as a gas. The *dew point* is thus reached, and water condenses out to form small liquid drops or solid ice particles.

Moist convection is a sufficiently energetic process that it alters the environment around it and the consequent transport of energy. Large amounts of water in an atmosphere initially unstable (tending toward bulk air motions to remove heat) can create large thunderstorm complexes, in which updrafts and downdrafts may reach all the way up to and beyond the tropopause (defined in chapter 15). This is particularly the case in the tropics, but large storm complexes also dominate weather in mid-latitude continental regions. The convective transport of heat, particularly involving moist convection, alters the relationship between greenhouse gas increase and the temperature response of the atmosphere; by how much (and even in what direction) remains a matter of dispute.

Formation of clouds also alters the radiative balance of the atmosphere, aside from the effects of moist convection.

Clouds can reflect, scatter, and even absorb incoming solar visible radiation; they also may absorb infrared radiation moving upward from the deeper atmosphere. The overall effect of clouds on global climate is complicated. The difficulty arises from the wide range in shapes of clouds, size of the cloud droplets or ice particles, the breadth of altitudes over which clouds form and extend, and conditions under which precipitation (rain, hail, sleet, or snow) forms. Some cloud types may lead to a net warming of the atmosphere, whereas others will cool it. Hence, if global temperatures increase because of enhanced greenhouse gases, and the resulting increased moisture (from more vigorous evaporation of ocean water) creates more cloudiness, the net effect of that cloudiness depends largely on the types of clouds and their mean altitude. Recent satellite and aircraft measurements of the amount of visible and infrared radiation coming out of, and moved around within, clouds are beginning to untangle these very complicated effects.

Snow, continental ice sheets, and sea ice provide very highly reflective surfaces that prevent much sunlight from being absorbed at high latitudes on Earth's surface. As global temperatures increase, the amounts of land and sea ice and snow will decrease, causing more sunlight to be absorbed and amplifying the greenhouse warming. How much of an amplification will occur depends on the details of the response of the ice and snow to warming. Increased precipitation at high latitude, another likely result of warming, could actually increase snow and ice cover in winter at high latitudes and/or altitudes, providing a moderating effect to the amplification.

Variability in the output of the Sun affects the amount of energy the atmosphere must transport back out, and has the potential to obscure the signature of human-induced global warming. Measurements of the Sun's luminosity taken over the past couple of decades show that it has varied only by plus or minus 0.02%. Compared to the effects of increased CO_2 over the same period, this number is quite small and, although some climate amplifications of the solar variability are possible, they are unlikely to reverse or dominate global warming associated with increasing carbon dioxide. On longer timescales, the Sun's luminosity varies more significantly (chapter 14), but projections of human-induced global warming are concerned primarily with the next half-century, a time not much longer than that over which detailed solar measurements have been made.

Perhaps the most important uncertainty lies in the role of the oceans. A thorough discussion of this is deferred to section 22.5, because of its complexity. Figure 22.4 illustrates graphically how the processes discussed above fit together and emphasizes that climate is not simply a matter of the vertical structure of the atmosphere, but also of what is happening from place to place on Earth's surface and in its oceans. We know from our experience with weather patterns that the three-dimensional nature of the planet is important. To capture this aspect of the problem requires rather involved computer models, to which we now turn.

22.2.3 General Circulation Models

One-dimensional climate models simulate the transfer of energy and matter only in one direction, namely, up and down. However, on a planet, energy and matter also move sideways in the atmosphere and on the surface. It is useful to define the sideways direction parallel to a line of latitude as *zonal*, and parallel to a line of longitude as *meridional*. Because Earth is roughly spherical, different latitudes receive varying amounts of sunlight; even though Earth's axis is tilted, the equator receives the largest amount of heat averaged over the year. As a consequence, heat tends to be redistributed by the oceans and the atmosphere in a meridional direction, that is, from the equator to pole. Warm tropical air rises, moves away from the equator, and sinks; this cycle is repeated at higher latitudes.

The Earth also spins on its axis, and this spinning motion modulates the transport of heat from equator to pole. Sinking air in the northern hemisphere is forced to spin clockwise, and in the southern hemisphere counterclockwise. Regions in which air is drawn inward by low pressure, forced to rise and form clouds and precipitation, will rotate counterclockwise in the north and clockwise in the south.

These systems of high and low pressure produce much of the weather with which we are familiar at middle and low latitudes. Their sense of rotation, induced by Earth's spin, interacts with the distribution and shape of continents to produce complex patterns. Low pressure spiraling counterclockwise as it moves eastward across the central United States draws moisture off the Gulf of Mexico to produce the well-known severe thunderstorms that often plague Texas, Oklahoma, Arkansas, and other midland states. The positions of high and low pressure systems in the Pacific and Indian oceans determine each summer the strength of the south Asian monsoon rains, critical to food production cycles for billions of human beings.

To simulate such complex weather patterns, computer models must do more than calculate how photons are absorbed and re-emitted on their way out of the Earth's atmosphere. They must also keep track of how energy (heat), moisture, and bulk air flow from one region to

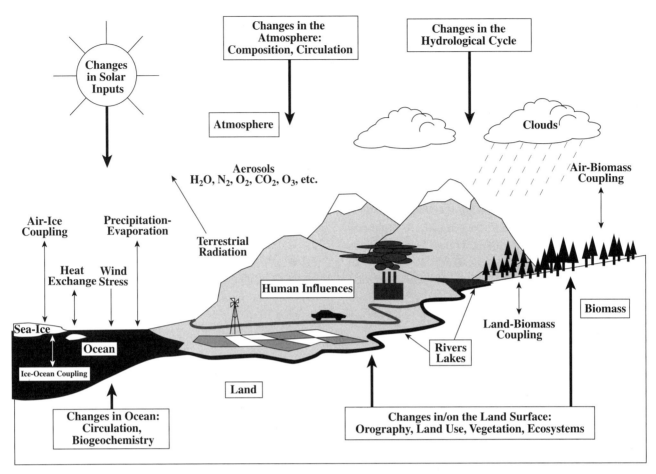

Figure 22.4. Processes affecting the nature of climate today, with an emphasis on the changes that might result from human influences. *Wind stress* is the movement of ocean water caused by the action of wind; *Biomass* refers to biological organisms both living and dead, that interact chemically with the atmosphere, land, and oceans. From Trenberth et al. (1996).

another. Models that do this are called *General Circulation Models*, or GCMs. The strategy is to divide Earth into a checkerboard in which each square, or grid point, is as small as possible; smaller gridding requires faster computers to handle the more numerous grid points. Newton's laws of motion, along with the laws of thermodynamics, are applied to the air, water, and heat in each grid, and both matter and energy are allowed to flow from one grid point to another. Sunlight shines according to latitude, season, time of day, and amount of cloudiness. In this way the flow of moisture, winds, and heat around Earth can be simulated on large, fast computers. Such GCMs, relying on detailed temperature, wind, pressure, and moisture information from thousands of weather stations worldwide, are used to predict weather several days or more in advance. GCMs have also been adapted to predict atmospheric circulation patterns on other planets, as well as the nature of the climate at earlier times in Earth's history. They are the basic computational tool for evaluating climate change caused by increasing abundance of greenhouse gases.

As carefully constructed as they are, GCMs have limitations. The first of these is intrinsic to the nature of climate itself. The ocean, atmosphere, and land form a coupled, *nonlinear* physical system. In recent decades the properties of such systems have been investigated and found to exhibit chaos. Chaos does not imply complete randomness (chapter 19). However, such systems can evolve into many different states, unlike simpler systems. A simple system, started out in two slightly different configurations, will diverge rather slowly in appearance. A chaotic system, started out in two different states, will *exponentially* diverge in its characteristics – an almost explosive parting of the ways between the two slightly different starting states. Insight into the difference between a simple and a chaotic system is not easy to gain, but figure 3.3 may be of help: it shows that an exponential function of a parameter always grows more rapidly than a power-law function of the same parameter. The general nature of a chaotic system can be described from a probabilistic point of view, but not its details. Climate has this nature. Thus, although GCMs are very good at using data to predict trends in climate over various periods of time (months, years, decades), they cannot completely capture the details of climate fluctuations (which we perceive as "weather"), and may fail to identify when Earth's climate could shift into a drastically different state.

The second limitation has to do with grid size. Most GCMs today are limited to grid sizes of a few hundred kilometers north-south and east-west. However, weather is affected by processes on much smaller scales; mountains, shapes of coastlines, and changing surface characteristics may occur on scales of tens of kilometers or less. Moist convection cloud and rain formation must be characterized on kilometer and smaller scales. These *subgrid* processes play key roles in determining the movement of air, moisture, and heat around Earth, yet they cannot be explicitly computed in GCM models. The strategy is to try to grossly characterize such processes computationally so that, on the scale of a grid point, they produce the same effects that the real processes do. Studies of how well GCMs account for the effects of moist convection, for example, suggest that, as yet, this strategy is only partly successful. With rapidly increasing computer power, grid sizes are decreasing, allowing for more accurate simulation of localized processes.

The third limitation of general circulation models lies in the coupling of atmospheric and oceanic processes. Because the nature and causes of ocean circulation patterns are only incompletely understood, no model exists today that fully characterizes how the atmosphere and the ocean interact. GCMs may be particularly sensitive to this limitation because of their large demand for computing power and the difficulty of handling simultaneously the short timescales of the atmosphere (days) and the long timescales associated with ocean mixing (centuries). However, much effort over the past few years has been put into improvements in the accuracy of the air-ocean interaction in such models, based on better understanding of oceanic circulation, the detailed physics of the exchange of material between ocean and air at the sea surface, and increased computing power. The most recent GCMs, to emphasize their more sophisticated incorporation of coupled ocean and atmosphere processes, are called atmosphere-ocean global circulation models (AOGCMs).

Despite their shortcomings, AOGCMs represent the most detailed and accurate simulations of Earth's climate that is available with present-day computing power. As computers continue to improve in speed and memory, the challenge will be to incorporate physical processes with greater fidelity. It is part of the nature of scientific research to test models of physical systems against their real behavior, based on observational data. With expanded means of collecting data on the current Earth environment, as well as on those of other planets and the Earth in its past, the reasonable expectation is that AOGCM models will continue to improve in their capability to elucidate the behavior of Earth's climate and make forecasts of future climate changes. By way of example, figures 22.5 and 22.6 illustrate climate-change predictions of some state-of-the-art GCMs.

22.3 PREDICTED EFFECTS OF GLOBAL WARMING

The uncertainties in predictions of GCM models have made difficult the debate over potential impacts of human-induced global warming and appropriate corrective or preventive efforts. It also has provided ammunition for those who would argue that no action should be taken at this time, given such uncertainties. However, as with the modeling of any complex physical system, some results are more robust than others, and it is crucial to identify those predicted impacts that are both most likely and potentially most harmful to human societies.

Committees of scientists are convened at the behest of governmental agencies to assess the work of various groups in global climate change, and to make a consensus determination as to what predictions of climate modeling are highly certain as opposed to speculative. The international Intergovernmental Panel on Climate Change (IPCC) has served in this capacity for over a decade. Very recently, a U.S.-convened Subcommittee on Global Change Research conducted a similar exercise. The results of these groups serve as a useful summary of potential effects of the human-induced increase in greenhouse gases, by providing an ordering of effects from most to least probable. Although some experts caution against the idea of "science by committee," because of the intrinsic compromises, the IPCC committees have studied carefully the results of multiple climate simulations. Some of the potential impacts are summarized below, in rough order of decreasing certainty in their occurrence.

22.3.1 Large Stratospheric Cooling

The stratosphere, the region above 10- to 15-km altitude, is an important contributor to the heat balance of the atmosphere. Because the air is so thin at those altitudes,

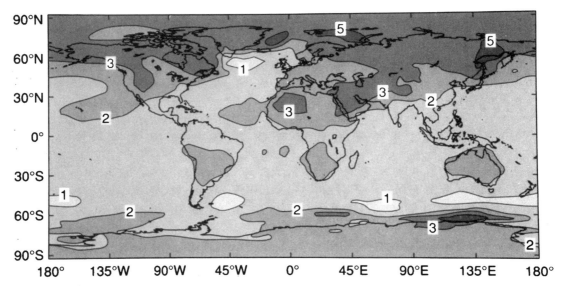

Figure 22.5. Predicted rise in winter surface temperature for CO_2 doubled over a 100-year time interval, based on a GCM. Selected contour lines are labeled with the corresponding temperature increase in degrees Celsius. As a guide to the eye, areas with the darkest shading suffer the greatest temperature increase. From Kattenberg et al. (1996).

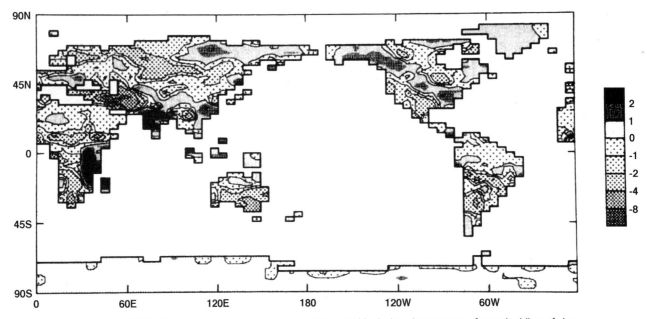

Figure 22.6. Changes in soil moisture around the world, during the summer, for a doubling of the CO_2 abundance over 100 years. Scale is change in water abundance expressed as equivalent depth in centimeters; negative signs mean decreased soil moisture. Hence a drop in soil moisture of 1 cm means that the amount of water in the soil is equivalent to a centimeter less rain each summer after doubling atmospheric carbon dioxide. From Kattenberg et al. (1996).

infrared photons at many wavelengths are free to move upward and out to space. Temperatures in this part of the atmosphere increase steeply with altitude, however, because ozone (O_3) and other gases can absorb ultraviolet photons from the Sun. As shown in figure 22.3, the increase in infrared opacity of the lower atmosphere shifts the minimum temperature point (tropopause) upward; in effect, the level at which infrared photons become capable of escaping moves upward. The stratospheric increase of temperature with altitude thus is delayed to higher levels, and the net result is a cooling of much of the stratosphere. The importance of this cooling is that it provides a test of a fundamental aspect of the greenhouse atmospheres. If stratospheric cooling were not occurring, basic aspects of the theory might be wrong. Complicating the signature of the cooling are the depletion of ultraviolet-absorbing ozone from introduction of industrial CFCs into the atmosphere, and the warming effect of sulfate aerosols that are injected into the stratosphere by volcanic eruptions. The cooling effect of decreased ozone may, in fact, dominate over the effect of increased carbon dioxide. Satellite and balloon data show that stratospheric temperatures have been decreasing since 1979, but sudden warmings due to volcanic eruptions are apparent as well.

22.3.2 Global Mean Surface Temperature Warming

Almost certainly, the introduction of carbon dioxide and other greenhouse gases will increase the global mean temperature of the planet. The "almost" derives from the fact that the coupled system of Earth's atmosphere, oceans, and land area represents a very complex physical system. Such systems have the capability to shift suddenly into new modes or ways of behavior. For Earth's climate, shifts in atmospheric processes that affect ocean circulation could have surprising feedbacks on the climate, and we discuss one such possibility in section 22.5. Nonetheless, in the near term, the system's response to increased greenhouse gases must be an overall warming of Earth's surface temperature if the basic physics of atmospheric energy balance is correct. How much warming is another matter; for a doubling of carbon dioxide, climate models predict warmings in the 1° to 4°C range. A doubled CO_2 level, and concomitant increase in other emissions tied to fossil fuel and other common industrial processes, almost certainly will be achieved by the middle of the coming century; the resulting temperature rise will exceed that experienced in the twentieth century and might even exceed that from the last ice age to the middle Holocene.

22.3.3 Global Mean Increase in Precipitation

Increased sea surface temperature means increased evaporation rate over the oceans, leading to an increase in precipitation averaged over the globe. However, the distribution of this increased precipitation over the globe will be highly variable and may be difficult to predict accurately with current climate models, primarily because of the large area covered by each grid point. Furthermore, although precipitation may increase, many continental areas will have dryer soils because the higher temperatures also will increase local continental evaporation rates. Models suggest that, in most locations, the increased precipitation will not compensate for this increase of evaporation, and desertification (conversion to a more arid regime) might result over food-producing areas.

22.3.4 Northern Polar Winter Surface Warming

Evidence from studies of the Cretaceous and other paleoclimates described in chapter 19 suggest that the polar regions of Earth experience amplified climate change relative to lower latitudes. All current model simulations show a maximum surface warming in high northern latitudes in winter, but much less arctic warming in summer. It is thus very probable that sea ice in the northern hemisphere (where no polar continent exists) will be reduced substantially in extent in the coming decades. Predictions are less certain for the southern hemisphere because of the complicated pattern of sea ice and ice shelves associated with Antarctica.

22.3.5 Rise in Global Mean Sea Level

The sea will rise because of the expansion of the water as it is warmed. This is the most certain but by no means the most important component of sea-level rise. Shrinking ice caps, continental glaciers, and the major Greenland and Antarctic ice sheets may play important but less certain roles. If the trend over the past century were to continue, an additional 15-cm of sea-level rise would occur between now and 2100. Folding in the possible accelerated melting of ice sheets due to global warming effects leads to a less certain, but larger, value up to 95 cm by the year 2100. An even larger value could obtain if the West Antarctic ice sheet collapses, but this is a highly speculative possibility. The effects of sea-level increases are somewhat controversial and complex, because much of the damage occurs during storms that locally raise sea level due to the low pressure and effects of onshore winds. However, given the large number of human habitations in coastal areas, essentially at sea level, increased disruptions and property damage are all but inevitable.

22.3.6 Summer Continental Warming and Increased Dryness

This category and the ones below are all much less certain that those above, because of the dependence on details of physical processes that are complicated and not yet fully simulated by computer models. As noted above, increased precipitation worldwide does not mean wetter continents. Most computer models show a decrease of soil moisture in the interior mid-latitudes of southern Europe and North America during the summer. This is mostly due to earlier springtime melting of snow in the warmer mid- and high-latitude springtime, higher summertime temperatures leading to increased evaporation from soils, and (in some regions) decreased summer rainfall.

22.3.7 Regional Vegetation Changes

Changes in temperature, rainfall, and soil moisture patterns will drive species poleward or to higher elevations. The extent of the process is uncertain, but the rapidity of the global warming compared to natural climate change will cause major ecological disruptions and possible species extinctions. Preserves set up today for endangered plant and animal species will be abandoned by those species as they move poleward, possibly being interrupted in their retreat by major urban or developed areas. Additionally, in mountainous country, species will be forced to move upward in response to increased warming and drying; uppermost mountain ecosystems may become locally extinct. The process of such extinctions may not be a gradual one: Unusually hot and dry summers in the American Southwest contributed in the mid-1990s to devastating mountain forest fires and consequent shrinkage in the distribution of rare species.

22.3.8 More Severe Precipitation Events

The ability to predict the nature and distribution of precipitation events using GCM's is limited. Warmer temperatures lead to a more vigorous cycle of water evaporation and rainfall, leading to more severe droughts or floods in some regions. Several models show an increase in intensity of rainfall or snowfall on a warming globe, suggesting the possibility of an increased frequency of extreme precipitation events. A warmer, wetter atmosphere sitting atop a warmer ocean might trigger more frequent tropical storms and hurricanes, as well as increase their intensity. Because such storms are affected by many factors, however, the extent of this effect is uncertain; what is known is that such storms are formed today over warmer ocean waters and lose their strength over cooler surfaces.

22.3.9 Changes in Climate Variability

As global temperatures warm, the response of the oceanic-atmospheric system may change in unpredictable ways. Because temperature increases into the next century potentially make that time the warmest yet in the Holocene, the character of the climate could change from fairly low variability to high variability. The impetus for such a suggestion is the as-yet uncertain record from the last interglacial, the Eemian, discussed in chapter 21. Evidence seems to lead toward that interglacial possessing a slightly warmer and much more variable climate. Is there a trigger for variability that is reached at worldwide climates slightly warmer than today's? The speculation is intriguing but as yet unsupported.

22.3.10 Regional-Scale Changes Will Look Very Different from the Global Average, but Their Nature Is Uncertain

Prediction of how climates will change within distinct regions of hundreds of kilometers extent is less reliable than predictions of globally averaged changes, but will improve as more powerful computers allow finer grid sizes in AOGCMs. How worldwide climate feeds into changes in localized deserts, mountain regions, grain belts, etc. is crucial to forecasting the economic impacts of human-accelerated climate change.

22.3.11 Biosphere-Climate Feedbacks

In contrast to the interaction of ocean and atmosphere, interaction between the biosphere and the climate system remains poorly characterized. In particular, how both oceanic and continental photosynthesizers will react to enhanced levels of CO_2, as well as warming, remains a matter of debate. Experiments involving subjection of plants to enhanced levels of carbon dioxide show both positive and negative effects, dependent on the species of plant, the amount of increase in carbon dioxide, and the protocol of the experiment. Even if ongoing experiments provide a complete understanding of the physiology, the large-scale interaction among oceans, atmosphere, and organisms will remain a challenge to quantify.

22.3.12 Details of Life in the Next Quarter Century

Because of the oceans' capability for temporarily storing heat and carbon dioxide, we have already bought into a significant amount of global warming over the next 25 years, associated with the increase to date in carbon dioxide – even if we stopped producing greenhouse gases today. Possible effects outlined above have human implications, of which some are:

(i) lower food production due to loss of agricultural land;
(ii) increased damage or loss of coastal areas due to high sea level and storms;

(iii) loss of treasured ecological preserves, recreational areas, and species due to rapid climate change;
(iv) increased discomfort and energy consumption due to more frequent occurrence of extremely hot days;
(v) decreased health and increasing disease vulnerability due to heat stress and poorer nutrition;
(vi) slowing of population growth, or even population decline.

22.4 THE DIFFICULTY OF PROOF: WEATHER VERSUS CLIMATE

How can we detect the onset of human-induced global warming? There are natural and human-induced effects that can obscure the temperature signature of increasing CO_2 and other greenhouse gases. First, the oceans tend to slow and smooth out the effects of increased greenhouse gases over many decades. Second, weather obscures climate trends. The day-to-day, season-to-season, and year-to-year variations that make up what we call weather are only part of the natural variations in climate extending from hours up to millions of years (figure 22.7). Possible human-induced increases in global mean temperature over the next decade will be partially obscured by the inherent natural variability of Earth's climate.

Yale climatologist M.E. Mann and colleagues have assembled climate data over the past five centuries from a number of sources, including ice cores, isotopic ratios in sea corals, tree-ring records, and historical accounts. They find that climate over the past half-millennium shows particular variability on timescales of several decades, as well as on timescales of several hundred years. The shorter of these timescales is troublesome for watchers of potential global warming because reliable instrumental temperature and precipitation records are available only for the past hundred years. This is really too short a span to allow

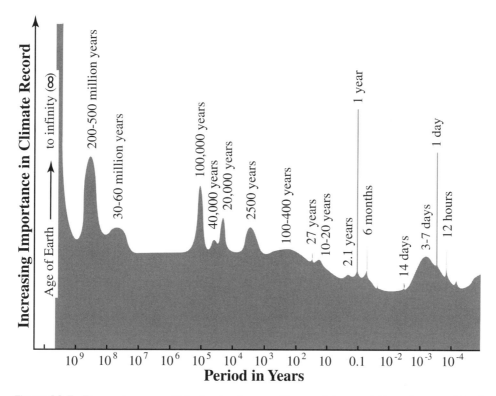

Figure 22.7. Timescales over which climate changes. The graph is a model based on a variety of data revealing climate change. Shown is the "importance" of the temperature variation, averaged over Earth, as a function of timescale. Where Earth's climate shows a tendency to fluctuate or change on a particular timescale, the value of the temperature importance in the climate record will be high. Because of Earth's rotation, daily and semidaily variations in temperature are a very strong component of climate, as are seasonal effects. There are variations also on decadal timescales, having to do with the pattern and distribution of droughts. The Little Ice Age shows up as a bump on the timescale of several centuries. Longer-term climate changes that show up on the graph include those due to the Milankovitch variations in the orbit of Earth, and those over millions of years caused by the movement of the continents. Modified from Peixoto and Oort (1992) by permission of Springer-Verlag.

removal of a natural climate variation that might cycle once every several decades. The longer cycle of several centuries also produces ambiguities in interpreting the temperature records over the coming decades; we appear to have recently come out of a particularly cold period (the Little Ice Age), and it is unclear how much additional warming to expect apart from that caused by human input of CO_2 into the atmosphere.

Figure 22.8 illustrates this effect. The thin solid line represents what computer models predict: the slow, gradual warming of Earth due to the increased greenhouse gases. The horizontal dashed line is the average temperature that we might have in the absence of human greenhouse gas increase. Now add the weather: The short-dashed line is the temperature variation one might get year to year in a climate that does not have human-induced greenhouse warming. The thick solid line is the year-to-year global temperature for an Earth with a human-caused greenhouse warming. Given plausible year-to-year and decadal variations it may take many years to notice possible human-induced global warming, as fluctuations in temperature obscure the rise in global mean temperature.

Finally, particulates in the atmosphere (aerosols) can act to obscure the signature of global warming. It has been proposed that the cool epoch from the 1950s to the 1970s (figure 22.2) might have been the result of an accumulation of industrial sulfate aerosols in the atmosphere, blocking some of the sunlight from reaching the surface. Tighter pollution restrictions cleaned up many sulfate-producing factories, leading to a reduction of such aerosols and, perhaps, an increase in sunlight reaching the surface. Major eruptions of volcanoes spew sulfur compounds into the stratosphere, where they condense as sulfate aerosols that block sunlight for several years until they sediment out. Such aerosols produce spectacular, red sunsets worldwide because the stratospheric winds blow the material quickly around the globe after injection into high altitudes by the eruption. Recent eruptions such as that of Mt. Pinatubo appear to have cooled the globe by a few tenths of a degree for several years. More such events, although they cannot be predicted, are to be expected. The AOGCMs incorporate the effect of aerosols, and predict less warming and less precipitation change (for a given carbon dioxide increase) than did the aerosol-free models prior to them. Unfortunately, deliberately introduced aerosols are not a good answer to slowing global warming; when dissolved in raindrops they are the primary source of acid rain which itself has been shown to cause serious damage to forests and surface waters.

22.5 ROLE OF THE OCEANS IN EARTH'S CLIMATE

Hundreds of times more massive than the atmosphere, the world's oceans exchange heat, moisture, and carbon dioxide with the atmosphere. Oceans play a key role in the

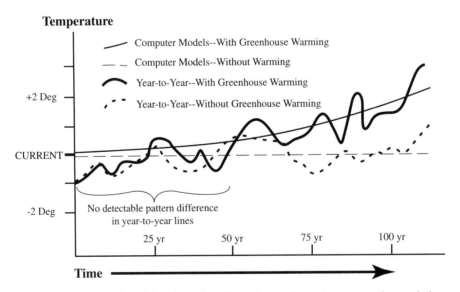

Figure 22.8. Illustration of the obscuring effect of natural year-to-year weather variations on global climate change. The lines labeled "computer models" are intended to represent global mean temperatures, in which year-to-year variations are smoothed out, for the cases of global warming and no-global warming, respectively. The "year-to-year" lines include annual and decadal variations. Such variations could prevent small amounts of global warming (be it human-induced or otherwise) from being detectable.

nature of and changes in climate, but a full understanding of that role is not yet at hand. Much of the problem lies in the difficulty of observing ocean circulation patterns; unlike the atmosphere, which can be well stocked with balloons, aircraft, and other measuring devices, the deep ocean remains relatively inaccessible. Additionally, coupling ocean circulation, moisture, and heat exchange models to those of the atmosphere is not easy. Because the ocean is much more massive than the atmosphere, the timescales over which changes occur differ greatly between the two. AOGCMs that couple the two require significantly more computing power than atmospheric GCMs alone.

As we saw with the proposed mechanism for the Younger Dryas episode, the continents play a role in ocean circulation, particularly with regard to movement of freshwater but also through modification of rain-producing weather patterns. In turn, the oceans affect continental weather, and we will explore one well-known example, the El Niño phenomenon. The ability to model these interactions, and ultimately a full coupling between ocean, atmosphere, and land, remains as yet an unachieved goal of climate studies.

22.5.1 Basics of Ocean Circulation

Surface oceanic circulation is well mapped (figure 22.9), but the ocean's vertical movements are incompletely understood. The ocean circulation pattern is driven by two basically different forces, but the relative importance of the two remains controversial. *Wind stress* is the motion of air over the ocean, coupling to the surface waters by friction, which in turn transfers kinetic energy to ocean waters beneath the surface. Such effects, acting over large distances, can drive circulation patterns within the bulk of the ocean, not just at the surface. The *buoyancy force* is associated with variations in density of ocean water, which leads to rising or sinking motions that drive circulation patterns. Two sources of density variations exist in the oceans. Temperature differences lead to warm water of relatively low density, and cold water of relatively high density. Ocean water contains large amounts of dissolved salts, and these also create density differences: Saltwater is denser than freshwater. Wind stress and buoyancy do not operate in isolation from each other. Wind stresses can act to remove the very uppermost layers of surface water, revealing ocean water of a different temperature beneath. In the North Atlantic, such exposed water is warm, and releases heat to the atmosphere. As the water cools it becomes denser, and sinks.

Ocean circulation is complicated by the great depth of the oceans. Deep ocean water may be isolated from surface circulation patterns on very long timescales, for hundreds or thousands of years. Such deep water may store large amounts of carbon dioxide accumulated from the atmosphere and stabilized by the large pressures near the ocean bottom. Drastic changes in ocean circulation patterns, perhaps driven by climate change, could bring deep water rapidly to the surface, where it would release this CO_2 and further perturb climate.

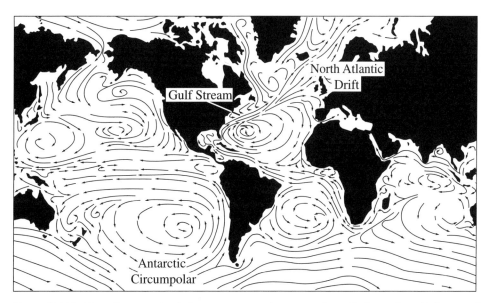

Figure 22.9. Map of ocean circulation patterns at the sea surface. Redrawn from Peixoto and Oort (1992) by permission of Springer-Verlag.

Ocean circulation can be tracked by sampling amounts of tritium in ocean water at various depths and locations; because tritium occurs only as a result of nuclear bomb tests beginning in the 1950s, it provides a measure of the mixing of shallow and deep ocean waters over the past few decades. However, such sampling does not address the movement of the deepest and most sluggish ocean water. Instead, the content of radioactive carbon-14, which varies somewhat with time, can be sampled in deep-ocean water to determine how long it has been isolated. It appears at present that water between 1,000- and 3,500-meter depth in the Indian and Pacific ocean represents the "oldest" deep-ocean water, i.e., that which hasn't been mixed with surface waters for the longest period of time.

The specific mechanisms (wind stress or buoyancy) by which different parts of the ocean are mixed remains in dispute. Recent models suggest that the ocean current around the broad-latitude circle surrounding the Antarctic is driven by wind stresses. Upwelling occurs in the so-called Drake passage region, and this upwelling is linked with very distant return circulations in the North Atlantic where new deep water is formed by sinking motions. Recalling that this sinking is accompanied by release of heat from shallow Atlantic waters, which plays a major moderating effect on climate there, it becomes clear that worldwide connections in ocean currents drive or modify important features of the present-day climate.

The issue of what drives ocean currents is crucial to an understanding of the role they play in climate change. Shifts in weather patterns alter both winds and precipitation. Wind shifts will change the ocean wind stress pattern, and precipitation changes can drive changes in ocean salinity and hence buoyancy. The details of these changes, the consequent response of ocean circulation, and the resulting alteration of climate remain challenging and unsolved (but important) problems.

One aspect of oceanic-atmospheric interactions involving changes in sea surface temperatures has received much attention because of its effects on worldwide weather patterns on several-year cycles. Examination of this, El Niño, phenomenon is worthwhile because it illustrates the complexity of oceanic-atmospheric interactions.

22.5.2 El Niño Phenomenon

On intervals of roughly 2 to 10 years, an anomalously strong ocean current pushes unusually warm and fresh surface waters against the western coast of South America, where it moves southward carrying tropical fauna. This phenomenon is referred to as *El Niño*, a co-option of a traditional term used by Peruvian fisherman to describe the annual warming of offshore waters right after Christmas (hence, the reference to "the Child," i.e., the Christ Child of the Christian religion). Meteorologists now use the term to refer only to the unusually intense warmings.

Accompanying the warming of eastern Pacific waters is an associated warming of air temperatures. A strong El Niño may persist for more than a year. Furthermore, strong episodes involve perturbations to Pacific ocean currents over many thousands of miles to the west of the South American coast. The atmospheric disturbance is, if anything, more widespread. During El Niño years, severe droughts occur in Australia and northern Brazil, and unusually heavy rains may occur in Ecuador and northern Peru, among other parts of the world. The Spanish conquistadors were aware of these *años de abundancia* (years of abundance) in the western parts of then-New Spain but did not connect them with changes in ocean currents. In fact, the alteration in weather patterns had been understood as a separate phenomenon involving shifts in positions of seasonal high and low pressure patterns, known as the *Southern Oscillation*. It was not until the 1960s that the link between it and El Niño was recognized. This link is now accepted and the combined phenomenon is referred to somewhat clinically as *ENSO* (El Niño/Southern Oscillation).

The mechanics of the El Niño phenomenon involve waves generated in the Pacific ocean that push warm water toward the east, raising sea level there slightly and depleting the amount of warm water in the western Pacific. The onset and dissipation of this event involve a complex, not wholly understood, series of feedbacks between the changed sea-surface temperature distribution and the atmospheric temperature pattern. What is not understood at all is the origin of the ENSO phenomenon in its entirety. Proposals range from the idea that the ocean circulation pattern itself drives the oscillation, all the way to the somewhat exotic notion that undersea eruptions contribute heat to the deep ocean and stimulate the effect. Also poorly understood is whether oceanic-atmospheric oscillations comparable to those in the equatorial Pacific might exist elsewhere. Finally, the speculation has been made that ENSO represents the signpost of a relatively warm climate on the edge of instability, and that not all times during the Holocene experienced such oscillations.

Much research is devoted to understanding all of the oceanic and atmospheric connections associated with ENSO, as well as what its true recurrence pattern (cyclical or chaotic) is. These connections may be subtle. For example, a very recently proposed effect of ENSO is a temporary slowing in the growth of atmospheric CO_2 abundance, through an alteration in ocean circulation that inhibits the usual movement of CO_2-rich deep waters to the surface. Larger shifts in ocean circulation caused by processes other than ENSO have the potential for more drastic changes in carbon dioxide abundance and hence climate, but on long timescales that are harder for humans to observe. Thus ENSO and its diverse accompanying phenomena provide us with short-term, but dramatic, illustrations of the ocean's effect on climate.

22.5.3 Prolonged Global Warming and Ocean Circulation Shutdown

It is natural to speculate what the long-term impacts of global warming might be on the oceanic-atmospheric system. As the Younger Dryas episode suggests, the complex and chaotic system that we call climate could flip-flop between states if perturbed strongly enough. Broecker and others have proposed that prolonged global warming, in inducing extensive melting of polar icesheets, might flood the North Atlantic with buoyant freshwater, preventing the release of heat and the sinking of North Atlantic warm waters, and shut down the circulation pattern that extends all the way down to Antarctica.

In such a circumstance, the climate in Europe could become substantially colder, and other climate perturbations might be expected worldwide. It has even been suggested that the increased rainfall in the North Atlantic predicted by some AOGCMs could inhibit or shut down this circulation. Once shut down, it might takes tens of thousands of years for a gradual rearrangement of the ocean salinity and heat to reestablish a circulation, and there is no guarantee it would be the same as today.

What would be the result of such a shutdown? Colder temperatures along the North Atlantic would encourage the buildup of ice sheets over a larger area for a longer fraction of a year. It is tempting to suggest that such a process could push the climate system in the other direction, that is, toward an ice age. However, the interrupted circulation of the ocean might inhibit the absorption of the large, human-induced carbon dioxide abundance in the atmosphere; it might even force release of additional CO_2 from the ocean. This would push the climate toward warmer states. Such a tug of war between the tendency toward glaciation versus even warmer overall climate likely would be accompanied by a climate variability not experienced today.

Our limited understanding of oceanic-atmospheric interactions makes such ideas at best speculative, but the principal notion that ocean circulation could drive sudden shifts in the climate regime of Earth must be taken seriously enough to deserve further study. Too much evidence exists of shifts from the ice ages to interglacials, with sudden and brief additional swings here and there, to ignore the possibilities.

22.6 GLOBAL WARMING: A LONG-TERM VIEW

Although society is naturally most concerned about possible effects over the next 50 years (one to two generations), the longer-term legacy of fossil-fuel burning is also of interest. Once released, CO_2 produced from fossil fuels will remain in the oceanic-atmospheric system for many centuries or longer, until it is sequestered on the ocean floor by biological processes and then subducted. The fossil fuel "bank" from which we make our withdrawals of carbon does not readily accept deposits. Hence, even if we significantly slow fossil-fuel burning worldwide, elevated CO_2 levels will persist over many human lifetimes, and a return to preindustrial values will not occur, by some estimates, for many thousands of years.

By the middle of the next millennium, that is, around A.D. 2500, atmospheric carbon dioxide levels could be as high as 1,000 to 2,000 parts per million, compared to the 350 parts per million of today. Simplified models that ignore the effects of oceans yield a climate in that time frame as warm as the Cretaceous climate described in chapter 19. Of course, the oceans are likely to strongly modify the warming, as described in the preceding section, through absorption of carbon dioxide and transport of heat. Regardless, however, of the effect of oceans, humans will still have made a lasting impact on the climate system by introducing into the oceanic-atmospheric system, on geologically short timescales, up to 10 times the preindustrial amount of carbon dioxide. Other than episodes of high volcanic activity, and large impacts, this change to our atmosphere has no precedent in the Phanerozoic. Like the photosynthesizing oxygen-producers of the Proterozoic, we are creating something of an atmospheric revolution – and like those earlier revolutionaries, we will have to adapt to our changes.

22.7 POSTSCRIPT: HUMAN EFFECTS ON THE UPPER ATMOSPHERE – OZONE DEPLETION

Another human-induced perturbation of the atmosphere involves the thinning of the layer of ozone high in Earth's atmosphere (10 to 35 km above the surface). This layer absorbs ultraviolet photons from the Sun which are lethal to plant and animal life near the surface. The natural process of ozone destruction and formation involves the breakup of ozone (O_3) molecules by sunlight to form atomic and molecular oxygen (O and O_2, respectively), and reformation of O_3 in a chemical chain involving oxygen, water, and sunlight. Certain reactive elements such as chlorine can accelerate the breakup of ozone by acting as a catalyst: A single atom of chlorine (Cl) reacting with ozone produces ClO and stable O_2. The ClO is also reactive and quickly combines with another oxygen atom to produce atomic chlorine and an O_2 molecule. The chlorine atom is now free to attack another ozone molecule.

Eventually the chlorine may react with other molecules in the atmosphere to produce more stable compounds such as HCl or $ClNO_3$ that are removed from the stratosphere – but only after a large number of cycles in which the chlorine has destroyed ozone. Human-produced refrigerants called chlorofluorocarbons, or CFCs, contain chlorine. These compounds drift up into the stratosphere, where the chlorine is released by solar ultraviolet radiation. Since the mid-twentieth century this source of chlorine has exceeded natural sources such as sea salts, which is why CFCs are accelerating the loss of ozone from the stratosphere.

Monitoring by satellites and aircraft now provides a worldwide picture of stratospheric ozone variations. Since the mid-1970s an ozone hole has appeared over Antarctica, indicating enhanced destruction of ozone there. Apparently the cold winter temperatures and sluggish atmospheric circulations above the southern polar region encourage the formation of water-ice clouds in the stratosphere. The ice particles act as chemical sites where the HCl and $ClNO_3$ react with each other to form Cl_2. As spring approaches and sunlight hits the Antarctic atmosphere again, the Cl_2 is broken apart into atomic chlorine, Cl. Thus, instead of being removed from the atmosphere, the chlorine is recycled into its active catalytic form again, enhancing the destruction of ozone and producing the deep springtime ozone hole around the south pole.

The particular climate conditions over the Antarctic conspire to make that region more sensitive than other parts of the globe to the introduction of chlorine into the stratosphere. However, this does not mean the rest of the world's ozone is immune to the effects of enhanced chlorine. Ozone depletion is seen at all latitudes in the northern and southern hemispheres. Because the residence time of active chlorine in the stratosphere is decades, enhanced destruction of ozone worldwide will continue even as international agreements force the phasing-out of CFCs. The effects of ozone depletion on humans include increased skin cancer rate, additional stresses on immune systems, and cataracts. More serious is the potential for damage to the food chain, through particular classes of organisms showing high sensitivity to increased ozone exposure. Additionally, ozone depletion slightly aggravates global warming through changes in the temperature of the upper atmosphere, and through increased lifetime of methane in the atmosphere, the destruction of which is aided by ozone.

22.8 QUESTIONS

a. Consider how you, as a policymaker, would weigh the economic consequences of various responses to global warming predictions. Would you take aggressive action now or a wait-and-see attitude?
b. What might the response of climate be to the oceans if, hypothetically, they extended no more than 100 meters deep (as opposed to the actual, deep-ocean situation on Earth)?

22.9 READINGS

Bright, C. 1997. Tracking the ecology of climate change. In *State of the World 1997* (L.R. Brown, C. Flavin, H.F. French, and L. Starke, eds.). W.W. Norton, New York, p. 22.

Broecker, W.S. 1995. Chaotic climate. *Scientific American*, **273**(5) (Nov.), 62–68.

Broecker, W.S., and Denton, G.H. 1990. What drives glacial cycles? *Scientific American* **262**(1) (Jan.), 49–56.

Crowly, T.J., and Kim, K-Y. 1995. Comparison of longterm greenhouse projections with the geologic record. *Geophysical Research Letters* **22**, 933–936.

Enfield, D.B. 1989. El Nino, past and present. *Reviews of Geophysics* **27**, 159–187.

Houghton, J.T., Callander, B.A., and Varney, S.K. (eds.). 1992. *Climate Change 1992: The Supplementary Report to the IPCC Scientific Assessment*. Cambridge University Press, Cambridge, UK.

Houghton, J.T., Meira Filho, L.G., Callander, B.A., Harris, N., Kattenberg, A., and Maskell, K. (eds.). 1996. *Climate Change 1995: The Science of Climate Change*. Cambridge University Press, Cambridge, UK.

Kattenberg, A., Giorgi, F., Grassl, H., Meehl, G.A., Mitchell, J.F.B., Stouffer, R.J., Tokioka, T., Weaver, A.J., and Wigley, T.M.L. 1996. Climate models – projections of future climate. In *Climate Change 1995: The Science of Climate Change* (J.T. Houghton, L.G. Meira Filho, B.A. Callander, N. Harris, A. Kattenberg, and K. Maskell, eds.). Cambridge University Press, Cambridge, UK, pp. 285–357.

Kiehl, J.T. 1994. Clouds and their effect on the climate system. *Physics Today* **47**(11) (Nov.), 36–42.

Manabe, S., and Stouffer, R.J. 1993. Century-scale effects of increased atmospheric CO_2 on the ocean-atmosphere system. *Nature* **364**, 215–218.

Mann, M.E., Park, J., and Bradley, R.S. 1995. Global interdecadal and century-scale oscillations during the past five centuries. *Nature* **378**, 266–270.

Meyers, S.D., and O'Brien, J.J. 1995. Pacific ocean influences atmospheric carbon dioxide. *EOS* **76**, 533.

Mitchell, J.F.B. 1989. The "greenhouse" effect and climate change. *Reviews of Geophysics* **27**, 115–139.

Mortensen, L.L. (ed.). 1996. *NOAA Global Change Education Resource Guide*. U.S. Dept. of Agriculture, Washington, DC.

Oerlemans, J. 1994. Quantifying global warming from the retreat of glaciers. *Science* **264**, 243–245.

Peixoto, J.P., and Oort, A.H. 1992. *Physics of Climate*. AIP Press, New York.

Stone, P.H., and Risby, J.S. 1990. On the Limitations of General Circulation Models. Center for Global Change Science, MIT, Report 2, unpublished.

Subcommittee on Global Change Research. 1995. Forum on global change modeling. *U.S. Global Change Research Program*, USGCRP Report 95–02.

Toggweiler, J.R. 1994. The ocean's overturning circulation. *Physics Today* **47(11)** (Nov.), 45–50.

Trenberth, K.E., Houghton, J.T., and Meira Filho, L.G. 1996. The climate system: An overview. In *Climate Change 1995: The Science of Climate Change* (J.T. Houghton, L.G. Meira Filho, B.A. Callander, N. Harris, A. Kattenberg, and K. Maskell, eds.). Cambridge University Press, Cambridge, UK, pp. 51–65.

Wahlen, M., Allen, D., Deck, B., and Herchenroder, A. 1991. Initial measurements of CO_2 concentrations (1530 to 1940 AD) in air occluded in the GISP ice core from central Greenland. *Geophysical Research Letters* **18**, 1457–1460.

23 LIMITED RESOURCES: THE HUMAN DILEMMA

Security is mostly a superstition. It does not exist in nature, nor do the children of men as a whole experience it.

Helen Keller

23.1 THE EXPANDING HUMAN POPULATION

Overpopulation is the root cause of human-induced global warming and depletion of resources for future generations. From the beginning of humankind to just over 100 years ago, the world's human population was less than one billion. Our planet now holds between 5 and 6 billion persons with a growth rate that will take us over 10 billion by the middle of the next century (figure 23.1). The present net increase in population amounts to about 90 million people a year. Growing population is a two-edged sword. Increasing numbers of people, supported in adequate living standards by advancing technology, represent an expanding reservoir of personalities, innovative ideas, and the creative seedcorn for future developments in both technological and humanistic spheres of existence. On the other hand, unbridled population growth that outpaces technological developments designed to stem its negative impacts could push humanity into a downward spiral of resource depletion, decreased overall living standards, and ever more profound alteration of natural systems by human activities.

Approximately 30 countries – most of Europe along with Japan – have achieved a roughly zero population growth rate (actually, an annual growth rate of less than 0.3%, as defined by the Washington, DC-based Worldwatch Institute). Many of these countries did so without a deliberate effort: The fall in birth rate reflects rising living standards and increasing career opportunities for women. These countries represent 14% of the total human population. In other cases, populations are declining in the context of failing health and living standards. Russia is the principal example, with a negative annual growth rate of 0.6%. With the fall of the Soviet Union, Russia is facing severe economic depression and a plethora of environmental problems stemming from the practices of the former Soviet empire. Although Russia is experiencing the most severe population shrinkage of the former Soviet republics, it is not alone in its woes.

The benefits to a given country of a stable population can include the ability to increase exports of consumables which otherwise would always be "catching up" with the increasing demands within the country itself; potential export of European grain is one example. Not all prosperous nations have achieved near-zero population growth – that of the United States continues to rise at a rate of 1% per year, and there is as yet no economic incentive to trim the growth rate. Other nations have taken various steps to encourage or impose reduced or zero population growth on their citizens, and in some cases harsh measures have been used that have justifiably been criticized on human rights grounds. In most developing countries, reduction in birth rate is tied to the economic status of women. Many traditional societal structures leave women bereft of decision-making or economic power and force them into the role of bearing and raising large numbers of children. So-called "microenterprise" programs provide seed loans for women of developing countries, enabling them to begin their own businesses. In at least some cases the result has been to break the vicious cycle of poverty and large families.

In the remainder of the chapter we focus on issues connected to population growth that illustrate the dilemmas humankind faces as its numbers continue to expand: food, energy, and material production. These are intimately coupled to the question of human-induced global warming, because consumption of fossil fuels may drive global climate change, and food production is sensitive to the net

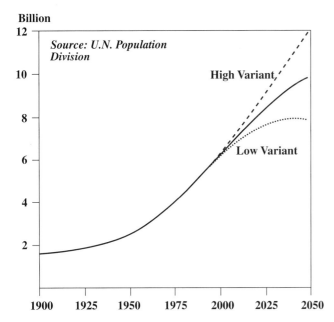

Figure 23.1. Human population increase in the twentieth century and several projections of population growth into the middle of the twenty-first century. From Flavin (1997) by permission of W.W. Norton and Company.

results. At the start of the twenty-first century, for the first time in history, our problems of adequate food and energy supply are global in nature, with no prospect that a long-term solution lies in finding "other places on Earth" where new land and new resources are available.

23.2 PROSPECTS FOR AGRICULTURE

The invention of agriculture occurred sometime in the late Pleistocene or early Holocene, and undoubtedly in several different places at various times. Major gains in agricultural productivity have occurred at various times throughout human history. Cro-Magnon in the Europe of 30,000 years ago may have required 1,000 hectares (or 10 square kilometers) of land to support a single family, largely by hunting. By the early Holocene, agricultural techniques using domestic animals brought this down a factor of 100, to 10 hectares per family. Medieval European farming practices supported one family on a hectare of land. Nineteenth century Asian rice-growing techniques reduced this by another factor of 5. Among today's most intensive agricultural systems, Japanese rice farmers support a family on 0.1 hectare of land – 10,000 times less area than the Pleistocene humans required.

The Industrial Revolution fundamentally changed the techniques of agriculture beginning in the middle-eighteenth century in England. The development of engines to harness coal and eventually oil created a transportation infrastructure that accelerated movement of goods, including agricultural commodities and tools, between urban and rural areas. It also enabled the mechanization of agriculture itself, improving yields. The most recent agricultural innovation is the "Green Revolution," beginning in the mid-1960s and characterized by development of novel high-yield varieties of dwarf rice and wheat, along with expanded use of irrigation, chemical pesticides, and fertilizers. The resulting increase in yield up through the late 1970s was dramatic (figure 23.2), and the fraction of humanity facing chronic hunger today is less than 20%, compared with over 30% in 1969.

However, the increasing yield of grains began to plateau in the late 1970s and measured per capita has been declining since the late 1980s (figure 23.2). The increasing yield was aided by a continuing expansion in the area of land on the planet brought under agricultural cultivation. This increase has now ceased in Eurasia and the Americas, and in Africa further dramatic increases in farmland are unlikely. In much of the world, intensive agricultural practices have led to severe soil erosion and loss of topsoil, so that the ability to retain today's cultivated farmland is in serious doubt. In developing countries, loss of agricultural land to industrialization is significant. This is particularly evident in China, where 100 million rural workers migrated during the 1990s into cities hoping for better jobs and a higher living standard. A conservative estimate of the amount of farmland paved over to accommodate these new urban dwellers is 435,000 hectares: sufficient to feed millions of people. In wealthy countries such as the United States, a different social phenomenon is destroying arable land: Significant amounts of the best farmland are being lost to development of new suburban communities populated by a middle-class trying to escape the problems of urban environments.

Current food production is delicately balanced with increasing population pressure and desire for higher living standards worldwide. Large increases in food production come with a price: increased energy consumption and increased release of both carbon dioxide (from energy consumption associated with mechanized farming and transport of agricultural hardware and products) and methane (from agricultural activity), which may intensify global warming. Progressive soil erosion due to intensive farming practices reduces arable land. It appears that worldwide agricultural productivity is not capable of further increase with the technologies available today. In fact, after 30 years of grain output outpacing population growth, we now face the situation that our food production is losing ground against worldwide population growth.

Perhaps most ominous, worldwide agricultural production potentially is among the most sensitive of human

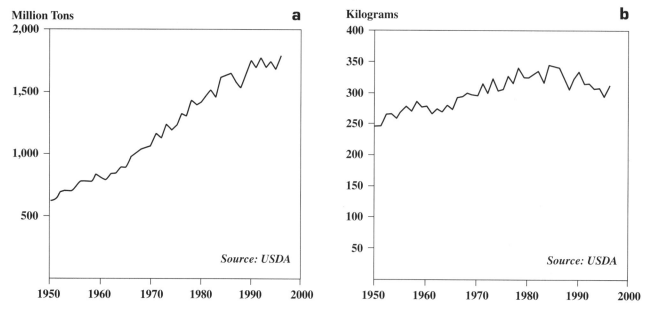

Figure 23.2. (a) Estimates of total worldwide grain production over the last half century. (b) Gr person per year over the same time period. From Brown (1997), by permission of W.W. Norton and Company.

endeavors to accelerated global warming. As discussed in chapter 22, a general prediction of global circulation models is increased soil dryness, particularly in equatorial regions, due to rising temperatures and an inability of increased rainfall to compensate for the greater amount of evaporation. Consequent loss of productivity of agricultural lands could accelerate the trend of decreasing harvestable area per person, and blunt the positive impacts of new agricultural technologies, such as development of new crops by genetic engineering. Although some have argued that higher latitude regions could be developed into farmland in a warmer global climate, much of this land contains very thin, poor soils, the yield of which is likely to be limited. It is difficult to predict the potential for net loss of agricultural land over the next half century because of the sensitivity to details of the climate models, but the trend is clear. The next agricultural revolution of necessity will be characterized by techniques designed to protect the current world inventory of agricultural land against loss both by destructive farming techniques and by human-driven or human-accelerated global climate change.

Earth's surface area is finite, and humanity's ability to feed itself in the face of its ever-increasing numbers depends on driving up the food yield per acre. In some respects, our entire post-Pleistocene history has been characterized by innovations that have made increasing yields possible. Although we are nowhere near any fundamental limit on producing food from the overall inventory of surface organic material, neither are we guaranteed a future of increasing abundance. Human ingenuity may or may not find new ways to squeeze even more food from the Earth. In a general sense, our ability to do so may depend on utilizing energy sources that have less impact on the global climate than those we now employ. It is to the issue of energy sources and their limitations to which we now turn.

23.3 ENERGY RESOURCES

23.3.1 Fossil Fuels

Coal and oil, along with natural gas, have been primary sources of energy since the nineteenth century. These *fossil fuels* are derived from long-dead organisms whose content of carbon and hydrogen is buried in the earth. Coal is largely derived from the decay of organic matter from forests of the Carboniferous period. Large quantities of dead plant matter collected in swamps, undergoing slow decomposition. The pressure of accumulated layers of sediments over geologic time increased pressures and temperatures in the organic layers, forcing out moisture and other materials. What was left were beds of compact carbon-rich material, *coal*, interleaved with sandstones, shales, or other sediments. The highest temperatures and pressures produced the highest grades of coal, such as anthracite, which is nearly pure carbon. Lower heating and pressurization produced bituminous coal, an industrial fuel with less carbon per gram than anthracite.

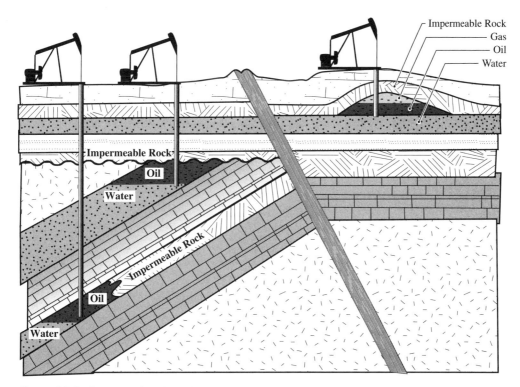

Figure 23.3. Cross section through sedimentary layers showing reservoirs of oil and gas. Here the hydrocarbons have been transported by water from shales into porous sandstones. The oil floats above the water and is trapped. Methane has come out of the oil to form a gas pocket in the shallow reservoir on the right.

The development of coal as an energy source was pioneered in Great Britain. By the end of the nineteenth century, scarcity in wood supply coupled with the design of increasingly efficient steam engines drove the large-scale production of coal. British coal production was two-thirds of the world's production capacity of this fuel through the middle of the nineteenth century, eventually to be succeeded by the United States. Coal is a relatively "dirty" fuel, producing large amounts of emission per unit of energy derived. Problems with air pollution from coal occurred in England as early as the thirteenth century, when it was burned directly in homes as a source of heating and cooking; Queen Elizabeth at the time banned its use in brewhouses within a mile of the court.

Oil production began in the United States in the mid-nineteenth century, its first use being limited to oil lamps and a few other purposes. However, the invention of the automobile spurred the increased production and refinement of oil, to be followed by natural gas as a somewhat cleaner substitute for many applications. The widespread use of oil and gas transformed the worldwide economic landscape, creating rich empires out of formerly impoverished countries that happened to sit atop major oil reserves. The formations of oil and natural gas are somewhat more complicated than for coal, and are illustrative of the natural processes associated with formation of energy reserves (figure 23.3):

(i) Living organic matter is produced at the surface by biological processes, beginning with photosynthesis.
(ii) The organisms die and the organic matter must be buried in sediments before it is destroyed by reaction with oxygen at the surface.
(iii) Parts of the organic material undergo reactions to form petroleum and/or natural gas (methane and other volatile carbon-hydrogen compounds).
(iv) The oil and gas migrate into permeable (porous or cracked) beds of rock.
(v) The beds of rock are folded or faulted in such a way as to trap the oil and gas.
(vi) The above must happen recently enough so that the traps have not been heated by tectonic activity, which would effectively burn the oil and gas into a nonusable form. Structural deformation also must be limited, because this could cause cracks that would allow the oil or gas to escape. Thus most oil and gas geologically is relatively young (post-Cretaceous).
(vii) The beds must remain buried until the present day because exposure dissipates the oil.

(viii) Recoverable oil comes from beds that retain a high permeability. Clays or other mineral cements, if they get into the pores before recovery, prevent the oil from being pumped out.

23.3.2 The Challenges of Fossil Fuels

Fundamentally, fossil fuels represent stored solar energy. Fossil fuels have their origin in plants and animals that lived on Earth in the past. These organisms, of course, were born and thrived as part of a food chain whose base is the photosynthetic production of sugars by plants, using sunlight as energy, and primarily carbon dioxide, water, and nitrogen as raw materials. The vast majority of Earth's crustal carbon is stored as limestone (carbonates) and in shales, so that fossil fuels represent a very tiny fraction of the buried crustal organic matter.

The use of fossil fuels by human beings represents an acceleration of the portion of the carbon cycle that returns buried carbon to the atmosphere as carbon dioxide (chapter 14). Chemically, we are extracting reduced carbon (hydrogen-rich carbon, or hydrocarbons) from the crust, oxidizing it (combining it with oxygen) by burning it to yield energy, and then depositing the oxidized carbon (carbon dioxide) into the atmosphere. We are, in effect, greatly speeding up the portion of the carbon cycle wherein carbon trapped in sediments is subducted to high temperatures, converted to carbon dioxide, and emitted from volcanoes. Our contribution is a dominant one: In a matter of decades we will double the carbon dioxide content of the atmosphere. How long we can do so depends on the supply available, which is highly uncertain. The known reserves of oil as of the middle 1980s are shown in figure 23.4. Estimates of roughly a century or so to deplete the world's proven oil reserves, at current usage rates, might be extended by a factor of several depending on the amount of undiscovered petroleum resources.

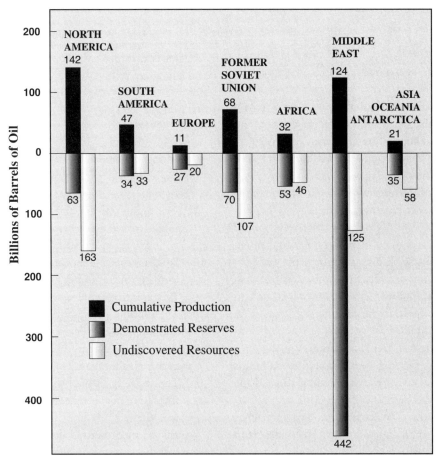

Figure 23.4. Reserves of oil worldwide, as of the mid-1980s, expressed in billions of barrels of oil. Total production to date in the various regions of our planet is shown on the upper part of the graph; demonstrated and estimated reserves are given on the lower part. The estimated undiscovered reserves are very uncertain, ranging from conservative values about half that shown on the graph to speculative numbers several times larger. Figure redrawn from Masters et al. (1986).

This human contribution to the carbon cycle is a novel one that began only in the last few centuries. Prior to the Industrial Revolution, the use of wood as a fuel predominated. Because wood is a product of biochemical processes in living plants that directly extract carbon dioxide from the atmosphere, burning of wood adds no net carbon dioxide to the atmosphere as long as the replenishment of trees balances their consumption. In fact, this balance probably has not been achieved for many centuries, the deforestation of the eastern United States beginning in the eighteenth century being a prime example of net consumption of wood. The post-Industrial Revolution world simply could not rely on wood as a principal energy source, because demand would enormously outstrip supply. Thus, we are committed to the use of fossil fuels, and hence to steady increase in the amount of carbon dioxide introduced to the oceanic-atmospheric system, or we must consider sources of energy that do not rely on the oxidation of reduced carbon extracted from the Earth's crust. Such *alternative energy sources* have their own pitfalls and limitations.

23.3.3 Alternative Energy Sources

Solar energy involves materials that absorb sunlight and store the energy through chemical reactions. Solar energy is nonpolluting, but the amount of energy generated depends on the area of the solar panels. To make solar energy the primary contributor to electrical energy in North America could require thousands of square kilometers of solar farms. One could imagine the deserts of the western United States arrayed in this way, but the initial infrastructure investment is too costly at present. Further, the idea of covering vast natural areas with solar arrays probably would meet with public disapproval. Likewise, space-based arrays would be very expensive, and the beaming of energy to Earth, by microwaves, could cause environmental or health hazards. Individual use of solar arrays by homeowners to reduce their dependence on community electric power grids is popular in dry, sunny locations, and may become more so as technological improvements drive down the cost of such systems and increase their efficiency.

Geothermal energy can, in principle, be relatively nonpolluting, but the known geothermal reserves that can be economically tapped are not sufficient to supply current world energy needs. *Hydroelectric power* also falls short by a wide margin of supplying world energy needs.

Nuclear fission, the splitting of uranium to generate heat, does not contribute to carbon dioxide production and hence is not a contributor to global warming. Also, the supplies of uranium oxide are such as to provide more energy than the world's coal reserves. However, nuclear reactors that are safe from the kinds of accidents and near-accidents such as Chernobyl and Three Mile Island came on line too late to stem a tide of antinuclear sentiment in the United States and Europe beginning in the 1970s. In spite of the use of such new designs, public acceptance of fission as an energy source remains relatively low. There is also the unresolved problem of disposal of waste products from such reactors. Nonetheless, a number of countries successfully and safely rely on nuclear fission as a primary energy source.

Fusion reactors mimic the Sun's energy-generating nuclear fusion mechanisms and leave behind much less radioactive waste per unit of energy produced than does fission. Hydrogen can be fused to form helium, but it is easier to fuse deuterium. Deuterium is available from seawater in huge quantities and would provide a supply of clean energy to satisfy the world for centuries to come. These reactions require enormous temperatures, however, and sustained generation of energy by fusion is beyond the forefront of today's technology. Fusion technology in the United States and Russia is perhaps 100 years away from generating energy cheaply enough to drive industry in that direction. In fact, current experiments still require more energy to run than they make during the fusion reaction, and fundamental physical problems associated with efficiently extracting energy from the fusion reaction (while sustaining it) may never be solved in the view of some experts.

23.3.4 Energy Use in the Future

Potential changes in the use of different types of energy sources is shown in figure 23.5. Coal offers, by far, the largest amount of usable energy resource, but extraction involves environmental damage (strip mining) and health hazards to workers. Also, coal tends to produce at least twice the amount of carbon dioxide per unit of energy, when burned, as does oil. Natural gas (primarily methane) produces the least amount of oxidized carbon per unit of energy generated, but known reserves are limited.

There is an intriguing possibility that large amounts of methane gas might be locked in seafloor sediments and permafrost regions as *gas hydrates*. These are compounds of water ice stabilized by the trapping of other molecules, such as methane, in void spaces in the ice. If sufficient methane (produced, for example, biologically by microorganisms in seafloor sediments) is in contact with water at the pressures found near the ocean floor, the water will freeze even though the temperature is above the usual freezing point of $0°C$. The presence of methane induces the freezing, the methane becomes trapped in the ice, and this gas-suffused ice is stable in the sediments of

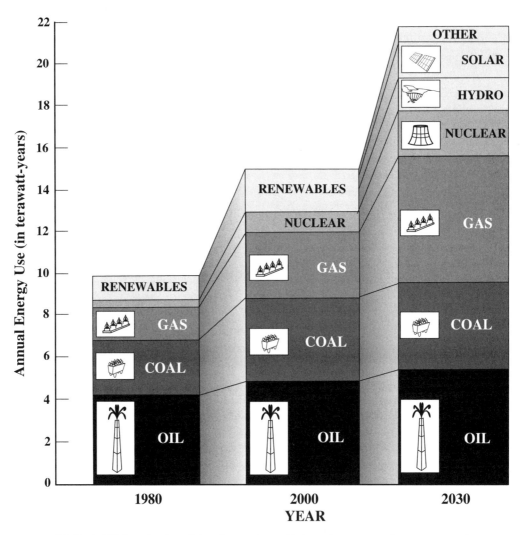

Figure 23.5. A 1986 projection of the change in world dependence on various sources of energy from the year 1980 through 2030. Renewable energy is broken out into solar, hydroelectric, and other sources for the year 2030. Annual energy use is given in units of terawatt years, which is the energy expended in generating a trillion watts of power for a year, or roughly the consumption of a billion tons of coal. From Rogner (1986).

the seafloor until extracted by drilling. A similar effect occurs in cold continental regions where permafrost conditions exist (permanent freezing year-round) and in glaciers. How much methane exists to be tapped as a natural gas energy source is controversial. Also controversial is the issue of safety – some extraction techniques could trigger large landslides in seafloor sediments, releasing large quantities of methane suddenly and exacerbating global warming.

Figure 22.1, showing the increase in atmospheric carbon dioxide for several scenarios of economic growth, illustrates that demand for and burning of fossil fuels will continue for many decades even under the assumption of stringent conservation measures. Present growth in the demand for energy worldwide is roughly 2.5% per year. Based on the fraction of fuels that emit carbon, the growth in energy demand corresponds to a doubling of the present atmospheric carbon dioxide content in less than a century. Increased use of low-carbon fuels or alternative energy sources can decrease the rate, but the prospect for dramatic near-term reduction in carbon emission by alternative energy sources is dim.

Even if the Earth's population were to stay fixed, there would be increasing demand for energy. Energy consumption is roughly proportional to a nation's gross national product. Currently, developing countries are limited in their energy consumption, but dramatic increases will occur as populations in those countries move toward higher living standards. The current political situation is such that hydrocarbon energy supplies are plentiful; the oil crises of the early 1970s and mid-1980s are a dim memory in

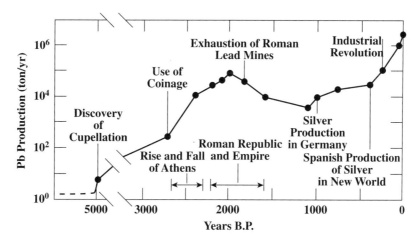

Figure 23.6. Lead production over the past 5,000 years, from analyses of Greenland ice cores. Hong et al. (1994) by permission of American Association for the Advancement of Science.

places like the United States, where people in significant numbers have disposed of small fuel-efficient cars in favor of large sport-utility passenger trucks, with significantly higher fuel consumption. Free-market economies, which have been an engine for the tremendous increases in living standards experienced in industrialized nations, do not respond well to tight controls, such as those proposed by some economists as a means of controlling carbon dioxide emission. Barring a sudden political reconfiguration that limits energy export to developed countries, it seems likely that we will not see significant decreases in per capita energy use in the next couple of decades at least.

23.4 ECONOMICALLY IMPORTANT MINERALS

In addition to energy, our civilization requires a suite of metals and other elements as the raw material for products and processes. Mining and processing of metals extends back to earliest organized civilizations. The record of such activities is preserved in ice sheets and peat bogs. Cores extracted from, for example, the Greenland ice sheet have been analyzed for the presence of copper, lead, zinc, etc., using sensitive laboratory techniques. The origin of these metals in high-latitude ice sheets lies in the efficient transport of pollutants through the atmosphere from low-latitude sites (where the highest concentration of humans have existed) to other latitudes. Atmospheric circulation models suggest that the Greenland deposits of metals in ice provide a good record, as a function of time, of the amount of these metals introduced into the atmosphere.

The records of metal pollution in the atmosphere compare favorably with the timing of civilizations that were heavy users of metals, such as the Roman Empire, the Sung dynasty of China, and the world of the post-Industrial Revolution. Lead, copper, arsenic, antimony, and other metals all begin to exceed their natural background levels beginning about 2500 years ago. Figures 23.6 and 23.7 summarize from a suite of sources the history of lead and copper production, as examples. Interestingly, the ice-core data show a difference between the two metals in terms of modern-day versus ancient atmospheric pollution. In the case of copper, modern smelting techniques are much more efficient than ancient ones, and cumulative large-scale copper pollution of the atmosphere thus does not track the actual use of the metal. The same is not true for lead; a massive increase in atmospheric lead pollution accompanied tremendous growth in use of the automobile after the Second World War. This increase has been reversed in the past couple of decades with the introduction and widespread use of unleaded gaso lines.

Large-scale mining operations, essential to extracting the raw materials that are the foundation for the material goods of our civilization, carry with them environmental and health costs. Smelting copper, for example, although more efficient today than in the past, produces substantial hydrocarbon and other emissions that represent significant sources of pollution in some parts of the world. Open-pit mining, used to access deep copper sulfide deposits that cannot be extracted economically by other techniques, leaves behind a large pit and piles of excavated earth. A medium-size copper mine might involve a pit one to several square miles, surrounded by perhaps 5 square miles occupied by piles of removed and processed dirt.

In the United States, an example of a mineral-rich industrialized nation, significant conflicts arise as mining companies seek to develop ore bodies that happen to be located in areas of high scenic value and heavy recreational

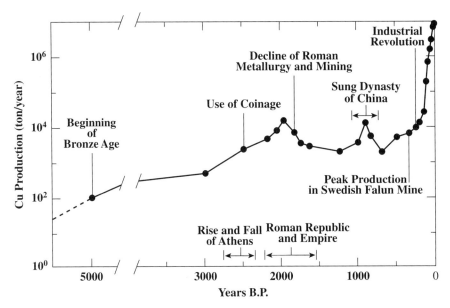

Figure 23.7. Copper production over the past 5,000 years, from analyses of Greenland ice cores. From Hong et al. (1996) by permission of American Association for the Advancement of Science.

use. Such conflicts are becoming more common and more ironic: A wealthy nation that has the resources to set aside large areas as public lands, particularly in the western United States, and the time to enjoy outdoor recreational activities in such places, finds itself faced with the necessity of extracting mineral resources from those same lands, usually with destructive consequences. Current environmental laws do not require mining companies to backfill open pits with the overburden piles, on the basis that such requirements would be highly burdensome economically. Mining in developing nations, where environmental laws are less strict, carries with it even more risks for the health of nearby residents and workers. Recycling of the current inventory of metals may reduce the demand somewhat, but ultimately new technologies must be applied to make hard-rock mining less destructive of the land, and to enable reclamation techniques that are at once more economical and more thorough. Recent progress in reclamation of areas strip-mined for coal demonstrates that technology can be applied to lessen the impact of mining activities on the land.

23.5 POLLUTION

All processes create waste products by virtue of their adherence to the second law of thermodynamics. A steam engine cannot extract all of the heat from the source of hot water. Organisms cannot utilize all of the organic materials they ingest as food; some must be removed as waste. Likewise, all industrial processes emit some form of waste, no matter how efficient the processes may be. These wastes, emitted into groundwater or the air, represent sources of pollution that carry potential health risks.

In the case of air pollution, there are two major kinds of pollution causing health risks. Automotive and industrial emissions create nitrous oxides, (NO_2) which, when combined with molecular oxygen (O_2), produce NO and ozone. Ozone in the stratosphere blocks harmful solar ultraviolet radiation from reaching the lower atmosphere and surface. However, ozone produced in the lower atmosphere, and breathed by humans, can cause health problems. Ozone is a strong oxizider (donator of oxygen), that can produce very reactive *free radicals* which directly alter biological structures. High concentrations of ozone (exceeding 300 parts per billion, or ppb) modify the structure of the mucous membranes in human respiratory tracts, particularly in the smaller, terminal airways that are crucial for transferring oxygen to the blood. In moderate concentrations (100–300 ppb), which are experienced in large cities with pollution problems such as Mexico City, inflammation of respiratory tract linings and some modification of mucous is seen. The second source of health risk is the industrial and automotive emission of fine particulates, a product of the combustion of hydrocarbon fuels, particularly diesel. These fine particulates (figure 23.8), lodged in the lungs, aggravate existing respiratory diseases and may contribute to lung cancer.

Over the past several decades, regulatory processes have lowered emissions per unit of energy, but increasing populations and greater awareness of the link between

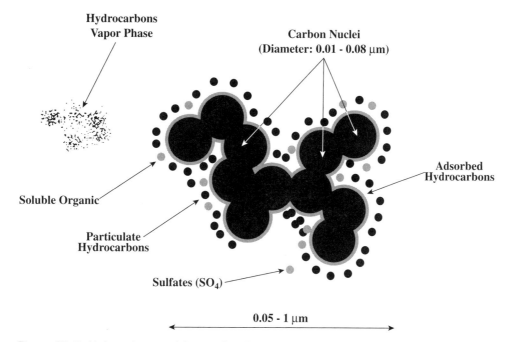

Figure 23.8. Hydrocarbon particles produced by the burning of fossil fuels. The term "Carbon Nuclei" refers to carbon-bearing sites for attachment of other molecules, and not to the atomic nucleus of carbon itself (which is vastly smaller). From Zmirov (1996) by permission of Les Editions des Physique.

air pollution and pulmonary disease have heightened concerns over health risks of such pollution. In Mexico City, two days of ozone levels above 110 ppb lead to an increase in asthma-related hospital visits by 68%. The costs of such increases in health care needs are borne by the public at large in most countries, because health care costs are distributed among the populace by governmental or private health insurance structures. Thus, the issue of air pollution becomes an economic one: Should industry bear the costs of air pollution through requirements to implement schemes to reduce emissions, or should the general public assume the costs through pollution-related increases in health care needs? This difficult issue continues to be debated, but the trend is a gradually increasing imposition of requirements on industry, driven in part by public dislike of the "brown clouds" that hover over most industrial cities of the world. As energy use increases in a growing world population, political pressure will be sustained to find technologies to continue to increase efficiency and reduce pollutive waste output.

23.6 CAN WE GO BACK?

Some have argued that the solution to our problems is to "go back to nature." The current 5-billion-person population is sustained against collapse by technologically based agricultural and transportation systems. To abandon technology is to regress at least back to the first agrarian societies (10,000 years ago), which were capable of sustaining less than 1% of the present human population at a much lower average standard-of-living than today's. The enormous and rapid increase of death rates relative to birth rates that would of necessity accompany such a regression would be regarded as a tragedy whose enormity has never before been experienced. The whole history of humankind is a story of the invention and application to the problems of living of technologies of increasing sophistication.

The very reasonable demand for a good standard of living by all the world's people will require coordination of energy, material, and technological resources among the nations. Although some economists argue that conservation measures will unduly hurt the economic engines of the industrial countries, in the long term such engines will run down unless new sources of energy are discovered and new technologies invented. Efforts now in developing countries to establish energy-efficient, high-technology production and transportation systems are mitigating the global impact of their increasing standard of living; such efforts will need to be made in the industrialized countries as well. Populations in both types of countries will need to educate themselves so as to be able to make reasoned choices with respect to new technologies, and to avoid the excessive risk-aversion that too often replaces intelligent decisionmaking.

The last half of the twentieth century was witness to two great threats to the human family: global nuclear war and overpopulation. With the former perhaps behind us, humankind's greatest challenge will be to face the latter humanely.

23.7 QUESTIONS

a. Why do you suppose that fusion is such a challenging mechanism for generating large-scale energy for human use?
b. Develop a set of arguments in favor of a policy of encouraging continued human population expansion, under the assumption that technology will solve standard-of-living problems.

23.8 READINGS

23.8.1 General Reading

Brown, L., Flavin, C., French, H.F., and Starke, L. (eds.) 1997. *State of the World: 1997.* W.W. Norton, New York.
Williams, G.R. 1996. *The Molecular Biology of Gaia.* Columbia University Press, New York.

23.8.2 References

Brown, L. 1997. Facing the prospect of food scarcity. In *State of the World 1997* (Brown, L.R., Flavin, C., French, H.F., and Starke, L. eds.). W.W. Norton, New York.
Brown, L., Flavin, C., French, H. F., and Starke, L. 1996. *State of the World: 1996.* W.W. Norton, New York.
Brown, L. Flavin, C., Postel, S., and Starke, L. (eds.) 1993. *State of the World: 1993.* W.W. Norton, New York.
Campbell, C.J., and Laherrère, J.H. 1998. The end of cheap oil. *Scientific American* **278**(3), 78–83.
Criqui, P. 1996. Energy and climate change: Socioeconomic aspects. In *ERCA–Volume 2: Physics and Chemistry of the Atmospheres of the Earth and Other Objects of the Solar System* (C. Boutron, ed.). Les Editions des Physique, Les Ulis, France, pp. 277–298.
Flavin, C. 1997. The legacy of Rio. In *State of the World 1997* (Brown, L.R. et al., eds.). W.W. Norton, New York, pp. 4–22.
Hartnett, J.P. 1976. The case for alternative energy sources. In *Alternative Energy Sources* (J.P. Hartnett, ed.). Academic Press, New York, pp. 19–38.
Hong, S., Candelone, J-P., Patterson, C.C., and Boutron, C.F. 1994. Greenland ice evidence of hemispheric lead pollution two milenia ago by Greek and Roman civilizations. *Science* **265**, 1841–1843.
Hong, S., Candelone, J-P., Patterson, C.C., and Boutron, C.F. 1996. History of ancient copper smelting pollution during Roman and Medieval times recorded in Greenland ice. *Science* **272**, 246–249.
Kvenvolden, K. 1993. Gas hydrates – geological perspective and global change. *Reviews of Geophysics* **31**, 173–187.
Margulis, L., and Sagan, D. 1986. *Microcosmos.* Summit Books, New York.
Masters, C.D., Root, D.H., and Dietzman, W.D. 1986. Distribution and quantitative assessment of world crude-oil reserves and resources. In *The Changing Carbon Cycle: A Global Analysis* (J.R. Trabalka and D.E. Reichle, eds.). Springer-Verlag, New York, pp. 491–507.
Mellor, J.W., and Riely, F.Z. 1989. Expanding the green revolution. *Issues in Science and Technology* **VI**(1), 66–74.
National Research Council. 1991. *Policy Implications of Greenhouse Warming.* National Academy Press, Washington, D.C.
Press, F., and Siever, R. 1978. *Earth.* W.H. Freeman and Company, San Francisco.
Rogner, H-H. 1986. Long-term energy projections and novel energy systems. In *The Changing Carbon Cycle: A Global Analysis* (J.R. Trabalka and D.E. Reichle, eds.). Springer-Verlag, New York, pp. 508–533.
Shotyk, W., Cherburkin, A.K., Appleby, P.G., Fankhauser, A., and Kramers, J.D. 1996. Two thousand years of atmospheric arsenic, antimony, and lead deposition recorded in an ombrotrophic peat bog profile, Jura Mouintains, Switzerland. *Earth and Planetary Science Letters* **145**, E1–E7.
Zmirov, D. 1996. Some issues on health impacts of air pollution. In *ERCA Vol. 2: Physics and Chemistry of the Atmospheres of the Earth and Other Objects of the Solar System* (Boutron, J.T., ed.). Les Editions de Physique, Les Ulis, France, pp. 265–276.

24 CODA: THE ONCE AND FUTURE EARTH

The origin and evolution of Earth involved physical processes that operate on all matter and energy in the universe. The formation of stars is a common phenomenon in galaxies, and we are on the threshold of confirming that the formation of planetary systems is a common result of star formation. Planets, most likely, are extremely common throughout the universe, and the technology to detect planets around other stars is just now available. In the year between October 1995 and September 1996, for example, planets were discovered around six other stars similar to the Sun in our galactic neighborhood.

In our solar system, three rocky planets had the potential early on for supporting life. Venus, Earth, and Mars were all endowed with carbon dioxide atmospheres, and at least Earth and Mars received large influxes of organic materials and water. The presence of a watery ocean was a key early step toward regulating and retaining the atmosphere. The absence or early demise of an ocean on Venus is causal to its present state: With no sink for carbon dioxide in the form of carbonates, all of the carbon dioxide remained as a massive atmosphere supporting a super-greenhouse warming: perpetually too hot to ever permit liquid water to exist.

The evidence is abundant that Mars had a mild, wet climate early on; the absence of plate tectonic recycling of carbon dioxide allowed carbonate formation to permanently lock up carbon dioxide in the crust, progressively cooling the surface and atmosphere until liquid water froze completely. Additional loss of atmosphere by impacts completed the thinning of the Mars environment. Mars went the cold way because it is smaller than Earth, which discouraged plate tectonics from occurring and accelerated the loss of atmosphere by impacts. Venus went the hot way because it began with less water than Earth (discouraging ocean formation), or because its closer proximity to the Sun made runaway loss of an early ocean unavoidable.

On Earth, the abundance of water and initiation of plate tectonics set up a relatively stable environment in which the formation of life took place. Early life had little effect on the environment, but once sufficient biomass accumulated and photosynthesis became an important biological process, the buildup of oxygen profoundly changed the evolution of the surface-atmosphere system of Earth. The replacement of carbon dioxide by oxygen as the major active gas in the atmosphere enabled aerobic respiration to take place, a powerful energy source that drove the flowering of species toward increasing diversity and biomass. This increase was modulated and interrupted over the past billion years by climate changes induced by large impacts, orbital variations, plate tectonic movement of continents, and variations in the pace of geologic activity (as well as other possible processes).

The emergence of humankind was favored by an unusual climatic epoch in which ice ages and warmer periods oscillated on 10,000- to 100,000-year cycles. Human beings, in the past 10,000 years of stable interglacial climate, have developed technologies that have enabled an explosive population growth which now strains resources, is changing the chemistry of the atmosphere, and possibly altering the climate. In the end, these effects will be small compared to past changes over geologic timescales, but they are so sudden that they pose severe strains on the delicate balance between our civilization and the natural world upon which we depend.

The history of Earth is unique to our solar system, but it involves physical and chemical processes that are seen to operate on the other planets. Given similar initial conditions, a subset of planets orbiting other stars should have the potential for stable earth-like conditions. The greatest uncertainty lies in the origin of life: Is it the result of a set of common physical and chemical processes that could work on any number of suitable planetary surfaces? There is weak evidence that such is the case, but it is weak indeed: We know far less about how life formed than how Earth has evolved over time.

If indeed life arose in planetary systems elsewhere in the universe, intelligence may have arisen elsewhere as well. Interstellar travel is daunting from the standpoints of time and energy, but communication with other technical civilizations is possible. How many exist now depends on the lifetime of such civilizations: We have had radio telescopes capable of interstellar communication for 30 years out of the 4.5 billion years Earth has existed. If our civilization lasts no more than 100 years, and is typical, the chance of different planetary civilizations overlapping in time is vanishingly small.

In any event, intelligence is uniquely defined by how we think and perceive the world around us. Humankind may become extinct in the next million years, because the average lifetime for vertebrate species is no more than this. Other "intelligent" forms of life may subsequently arise on this Earth or in other planetary systems, but they will not be human beings. Intelligence as defined by humans might be a rather distinct specialization; equally complex mental capability exhibited by other species could be quite different in operation and outlook.

The uniqueness of human beings at this time and on this planet is speculation only, but it should lead to a profound loneliness and an urgency to clean up our act and survive. Should we do so, and should we then turn our attention to the immense cosmos from whence we came, what extraordinary experiences lie in store for our species beyond planet Earth?

INDEX

f designates figure

absolute chronology, 47
   of solar system events, 69–73
accretion
   and Moon formation, 128–9
   and planet formation, 69, 76, 106–7, 123–4, 125f, 127–8
acid rain, 236
adaptive optics, 113
adenosine triphosphate (ATP), 162–3, 215
aerosols, 292. *See also* chlorofluorocarbons
Aescraeus Mons, 61f
Africa, 196, 273
   hominid evolution and, 256–60, 265
   and plate tectonics, 85, 91
African plate, 95, 96
AGB stars, *see* Asymptotic Giant Branch stars
age(s), geologic, 82
age dating, 47, 60, *see also* radioisotope dating
   carbon-14, 49–50
   of craters, 63–9
   geology and, 75–84
   and parent-daughter systems, 51–2
   of solar system events, 69–73
agriculture, 3, 281, 290, 299–300
air movement, 284–6
air pressure, 285
Alabama, sinkhole in, 61f
alanine, 136f, 155
Alba Patera, 185f
alchemy, 17
ALH84001, 149–50
alkali metals, 18
Alpha Centauri, 14
alpha (α) particles, 28, 40
Alps, 95

Altman, S., 159
aluminum, 40, 117, 122, 197, 199
amberates, 273
Americas, humans in, 266, *see also* North America; South America
amino acids, 132, 136, 155, 156, 161
ammonia, 118, 131, 155, 189
amoebas, 220
anaerobes, 219, 220
anaerobic processes, 144, 215
andesite, 95, 200–201
Andes mountains, 201, 203
Angel, Roger P., 113
animals, 221, 224f, 290. *See also* birds; mammals
anions, 197, 198
Antarctica, 56, 96, 149, 189–90, 270, 296
AOGCMs. *See* atmosphere-ocean global circulation models
apes, 256
Apollo missions, 24, 64, 94, 120, 127
Appalachian Mountains, 96
Arabia, 95
Arcadia Planitia, 187f
Archaea, 151–2
Archean eon, 82, 134, 163, 165, 171, 174, 176, 196, 208
   continent formation during, 202f, 203–4
   oxygen levels during, 215, 216
Archosauria, 232
Arecibo radio telescope, 113
Ares Valles, 184
Argentina, 96
Aridipithecus ramidus, 256–7
Aristarchus of Samos, 4, 14
Aristotle, 3, 4
Arizona, 273
Arkansas, 285

artificial elements, 27–8
artificial life, 136, 143
Asia, 266, 271, 273, 299
Asian plate, 95, 240, 248, 249
asteroid belt, 71
asteroids, 6, 73, 109, 223, 232, 233
astrology, 3
astrometry, 112, 113
astronomical unit (AU), 5
astronomy, 3
   Stone Age, 10–12
Asymptotic Giant Branch (AGB) stars, 41
Atlantic Ocean, 86, 95, 96, 242, 246, 279
atmosphere(s), 6, 73, 144, 155, 246
   carbon dioxide in, 169–71
   Earth's, 130–1, 134, 193, 244, 295
   energy movement in, 167–9
   and K-T boundary event, 235–7
   on Mars, 183, 186
   oxygen in, 172, 215–16, 217
   on Titan, 145
   on Venus, 174–5, 177–81
atmosphere-ocean global circulation models (AOGCMs), 287, 290, 292, 293, 295
atomic mass, 40
atomic number, 18, 36
atomic radius, 197
atomic weight, 20
atomism, 17
atoms, 17, 18
ATP. *See* adenosine triphosphate
Aubrey holes, as observatory, 10–12
Australia, 9, 58f, 266, 273
Australian plate, 94–5, 96
Australopithecenes, 257, 258
autocatalysis, vesicles and, 156–8

311

Bacon, Francis, 85
Bacon, Roger, 17
bacteria, 134, 138–9, 149, 151, 152, 190, 204, 215
   and eukaryote development, 220–1
   photosynthesizing, 141, 218
banded iron formations (BIFs), 214–15, 217, 218
Bar-Nun, Akiva, 131
Barron, E., 247
basalt, 86, 95, 132, 134, 172, 199, 200, 201, 206
bases, in DNA, 136–7
batholiths, 202
Becquerel, Henri, 76
Bermuda, 226
beryllium, 37, 40, 42
beta (β) decay, 29, 39, 41
Beta Pictoris, 105
Big Bang, 16, 38–9, 101
bicarbonate ions, 173
BIFs. *See* banded iron formations
biochemistry, 142
bilayer membranes, 156
bimodal distribution, of continents and oceans, 86, 88f
bioenergy, 215
biological processes, and carbon sequestration, 54, 173
birds, 249
blackbody radiation, 30, 32, 39
black holes, 35, 42
blue stars, 15, 38
body forms, 229–30
bonding patterns, 18, 20, 21
boron, 42
boundary clay, 233
Bracewell, Ronald, 113
Brahe, Tycho, 13
British Columbia, 229
Broecker, W. S., 278
brown dwarfs, 114
Burgess Shale, 229–30

Cairns-Smith, A., 142
calcite, 54, 55, 198
calcium, 54
calcium silicates, 173
calendar, Mayan, 13
California, 86, 91, 95, 96, 276, 277f
calories, 139
Callisto, 63, 66, 69, 72
Cameron, A.G.W., 104
Cambrian period, 229, 230–2
Campbell, I. E., 208
Canada, 132, 172, 196, 229
Cann, R., 260
carbohydrates, 136, 159–60
carbon, 20, 42, 54, 109, 142–3, 212, 230

carbonaceous chondrites, 52, 131, 196, 197
carbonates, 149, 150, 190, 246
carbon cycle, 173, 250, 302–3
carbon dioxide, 54, 115, 131, 134, 155, 196, 245, 246, 249
   in Earth's atmosphere, 169–71, 172, 173
   and global warming, 289, 291–2
   and greenhouse effect, 188–9, 244
   in ice cores, 269, 270
   increasing, 281–2
   on Mars, 146, 183, 186, 190–1
   and oceans, 279, 288f, 294–5
   and oxygen, 211–12
   paleosols in, 171–2
   and plate tectonics, 172–4
   on Venus, 174–5, 181, 283–4
carbon fusion, 40
carbon monoxide, 131, 143
carbon-nitrogen-oxygen (CNO) cycle, 38, 40
carbon-silicate cycle, on Earth, 172–4, 175
carbon-12, 40, 49, 149
carbon-13, 40, 49, 54, 149
carbon-14, 49–50, 51f, 149, 277
Carr, Michael, 183
Cassini/Huygens mission, 73, 145, 147–48f
catalysis, 156
catastrophism, 75
Catholic Church, 17
cations, 198
Cech, T., 159
cells, essentials of, 161–3
Celsius, 29–30
Cenozoic era, 82
Centaur objects, 110f
Cepheids, 15, 16
CFCs. *See* chlorofluorocarbons
chalcophiles, 124
chaos, 252, 286
Charon, 109, 129
cheetahs, 227
chemical differentiation, 107
chemical elements, 17, *see also* elements; *by name*
chemical reactions, 37–8
chemisynthesis, 141–2
chert, temperature history in, 56, 58
Chicxulub crater, 237–8
China, 223, 299, 305
chirality, 155–6, 159, 160
chlorine, 296
chlorofluorocarbons (CFCs), 169, 193, 281, 288, 296
chlorophyll, 139, 141
chloroplasts, 139, 220

chondrites, carbonaceous, 52, 109, 131, 196, 197
chondrules, 109
chordates, 229–30
chromosomes, 220, 228
chronology, 47. *See also* absolute chronology; age dating; relative chronology
circulation models, climate, 285–7
citric acid cycle, 139
clay minerals, and origin of life, 160
cleveite, 33
climate, climate change
   circulation models and, 285–7
   during Cretaceous period, 245–8
   Eemian interglacial, 271–2
   factors in, 284–5
   greenhouse effect and, 283–4, 287–91
   Holocene, 269, 277–80
   in ice ages, 243–5
   ice core data on, 269–72
   on Mars, 190–1, 192f
   and mass extinction, 232, 235
   oceans and, 292–5
   packrat midden studies of, 272–3
   and plate tectonics, 96, 240, 242–3
   Pleistocene, 249–52, 255, 257–8, 260–2
   pollen studies of, 272–3
   and stable isotopes, 54–6
   during Tertiary, 248–9
   tree ring data on, 273, 275–7
   and weather, 291–2
cloning, 228
clouds, 246, 284–5
   of Mars, 188–9
   of Venus, 177
Clovis hunters, 279
CNO cycle. *See* carbon-nitrogen-oxygen cycle
coal, as energy, 300–301, 303
coccoliths, 54, 55
Cocos plate, 95
codon, 137
Colorado River, 82
comets, 6, 73, 109–10, 132, 232, 233
   and Earth, 130–1, 223
   and Jupiter, 235, 236f
Comet Shoemaker-Levy 9, 235, 236f
conjugates, nucleotide bases, 137
conservation of energy, 29–30
continental drift, 86, 89, 95f, 97f; *see also* plate tectonics
continental crust, partial melting of, 201–3, 204, 206–7
continental shelf, 86f

continents, 8, 196, 197
   during Archean, 203–4, 205–6
   and climate, 242–3, 247–8
   collisions of, 240, 249
   glaciation and, 243–5
   and oceans, 209, 293
   and plate tectonics, 85–6, 88f, 95–8
   rocks in, 200, 201–3
   and subduction zones, 92–3
continuum radiation, 30
convection, 123, 126f, 134, 284
convection currents, 168–9
Copernican model, 10
Copernicus crater, 64f
copper, 305
core, Earth's, 122, 127
core-mantle boundary, 123
Cosmic Background Explorer, 16
cosmic rays, 42, 49, 50
covalent bonding, 20
craters, cratering
   age dating with, 63–9
   and atmospheres, 73
   distribution and size of, 71–2
   evolution of, 70–1
   on giant planets' moons, 72–3
   impact, 6–7, 60–3, 181
Cretaceous period, 96, 240
   climate during, 245–8
   fossils of, 85, 232–3
Cretaceous-Tertiary boundary extinction, 232–9; *see also* K/T boundary event
Croll, J., 250
Cro-Magnons, 264, 265, 299
crust, Earth's, 122, 125, 127, 197, 199; *see also* continents; oceanic crust; plate tectonics
   Archean, 202f, 203–4, 216
Curie, Marie, 76
Curie point, 89
cyanobacteria, 214–15, 221
cytoplasm, 138–9

Dalton, John, 17
Darwin, Charles, 226
   The Descent of Man, 256
daughter products, 49. *See also* parent-daughter radioactive dating
day length, on Earth, 209–10
Daziel, I.W.D., 96
decay, 212, 213
Deccan Trap, 242
deep water, 293, 295
   North Atlantic, 242–3, 279
delamination, 207
delta, 75
Democritus, 17

Denton, G. H., 278
deoxyribonucleic acid (DNA), 136–8, 151, 155, 158, 159, 220, 225
   in cells, 139, 161–3
   mitochondrial, 260, 263
deposition of sediments, 77–8
Descent of Man, The (Darwin), 256
Des Marais, David, 219
deuterium, 20, 38, 55, 178, 270
   fractionation of, 56, 57f
Devonian period fossils, 81f
diamond anvil cells, 123
diatomic compounds, 18
dinosaurs, 232–3
disks, planet formation in, 104–7
distances
   to galaxies, 14–15
   outside Milky Way, 15–16
   to planets, 13–14
DNA. *See* deoxyriboneucleic acid
domains, in taxonomy, 229
Doppler shift, 15, 16, 112, 113
Douglass, Andrew Ellicott, 275
droughts, 276, 278, 294
dry convection, 284
dust, interstellar, 105–6, 107, 108, 247

Eagle Nebula, 107, 108f
Earth, 5, 116
   atmosphere formation on, 130–1
   basalt formation on, 198–9
   bimodal topography of, 86, 88f
   composition of, 117, 122–3
   day length on, 209–10
   elements on, 196–7
   eukaryotic life on, 220–2
   geologic history of, 69, 75–84
   gravity on, 24–5
   impact craters on, 6, 62f, 63, 66f
   magnetic field of, 26, 89
   motions of, 8, 10–13, 14
   orbital cycles of, 249–50, 251f, 252–3
   oxygen levels of, 213–19
   radioactive heating in, 126–7
   radioisotope dating of, 81–2
   spin of, 249–50
   size of, 3–4
   topography of, 85–6
earthquakes, 86
   distribution of, 88f, 91–2
   measuring, 120–2
East Pacific Rise, 203
eclipses, 10
Eemian interglacial, 270, 271–2, 278
Egyptians, ancient, 3
Einstein, Albert, 25, 29
Einstein mass-energy relation, 29

Eldredge, N., 226, 227
electromagnetic energy, 15–16, 31f
electromagnetic force, 25–6, 27
electromagnetic radiation, 30
electromagnetic spectrum, 30, 32–3
electromagnetism, 23, 25
electrons, 17, 18, 20, 21, 28
element production, 43f, 101
   in Big Bang, 16, 38–9
   in fusion, 36, 39–40
   and life, 42–3
   and stars, 39–42
elements, 17, 18, 26, 109, 199
   artificial, 27–8
   differentiation in planets of, 124–5
   in Sun, 33–4
   in terrestrial rocks, 196–7
El Niño Southern Oscillation (ENSO), 293, 294–5
empirical model, Kepler's, 14
enantiomers, 155
Enceladus, 72
energy, 40, 76
   alternative sources of, 303
   in chemisynthesis, 141–2
   conservation of, 29–30
   in Earth's atmosphere, 167–9
   fossil fuels as, 300–303
   future use of, 303–5
   in giant planets, 119–20
   life and, 139–41
   and nuclei, 35–6
   and star formation, 103–4
   Sun's, 23, 35
   in vesicles, 157–8
England, 299, 301
enstatite, 116
entropy, 153, 154, 157
enzymes, 136, 156
Eocene epoch, 81f, 248, 249
eons, 82
epochs, 82, 223
eras, 82, 223
Eratosthenes, 3, 4
erosion, 77, 78, 174, 196, 299
Ethiopia, 256
eubacteria, 151
eukaryotes, eukaryotic cells, 138f, 139, 149, 151, 152, 163, 193
   genetic complexity and, 230–1
   rise of, 220–2
Eurasia, 96
Europa, 66, 72, 118, 144–5, 146f
Europe, 85, 95, 271, 295, 298, 299
   climate of, 242, 272
   Little Ice Age in, 277–8
   Pleistocene climate of, 260–2
European Space Agency, 110
eutectic solutions, 122

evolution
    Darwinian, 134, 136, 225–6
    of hominids, 255–66
    punctuated equilibrium, 226–9
exponential functions, 48
extinction, 255, 279, 280; see also
        Cretaceous-Tertiary boundary
        extinction
extrusive igneous rocks, 200

faults, fault systems, 86, 91, 94, 95
Feinberg, J., 142, 143
felsic rocks, 132, 200, 201–2
Fenchel, T., 221
fermentation, 139, 140f, 215, 220
ferric iron, 214
ferromagnetic elements, 26
ferrous iron, 172, 214
51 Pegasi, 113
Finlay, B. J., 221
fission, 26, 29
force, 23
    electromagnetic, 25–6
    gravitational, 24–5
    strong nuclear, 26–7
    weak nuclear, 28
Forget, Francois, 189
formaldehyde, 160
forsterite, 116
Forterre, P., 152
fossil fuels, 213, 280, 281, 295, 298
    energy from, 300–303
fossilization, 79–80, 228
fossils, 79–81, 85, 215, 228, 231f
    in Burgess Shale, 229–30
    Cretaceous, 232–3, 245
    and evolution, 225, 226
    hominid, 255–66
Frail, D., 113
France, 272
Fraunhofer, Josef, 33
free energy, 141
free radicals, 306
frequency, 30
Froelich, P., 249
fusion, fusion processes, 29, 35, 165
    and element production, 39–40
    helium, 36–8
    and star formation, 103–4
fusion reactors, 303

gabbro, 200, 23
"Gaia" hypothesis, 174
galaxies, 14, 15, 101, 247
Galileo spacecraft, 120, 145
gamma (γ) decay, 29
Ganymede, 3, 66, 68f, 69, 72, 144
gas chromatograph/mass
        spectrometer, 145
gases, planetary, 5, 6, 106–7

gas hydrates, 303–4
General Circulation Models (GCMs),
        286–7, 293
general relativity, 25
genes, 138
    trigger, 225, 228, 231
genetic code, 137–8, 225
genetic complexity, 230–1
genetic variation, 138
geology, 78f, 80f, 245
    of Chicxulub crater, 237–8
    cycles in, 76–8
    Earth's, 69, 70
    of giant planet moons, 72–3
    of Mars, 183–7
    relative age dating and, 75–84
    succession in, 78–9
    timescale in, 82–4
    of Venus, 73, 181–3, 206–9
geothermal energy, 303
Germany, 272
giant impact, 127–8
giant molecular clouds, 102
giant planets, 106, 144; see also
        Jupiter; Neptune; Saturn; Uranus
    and comets, 131–2
    composition of, 118–20
giardia intestinalis, 220
Gilgamesh crater, 63
glaciation, glaciers, 174, 243, 250,
        269, 283
    and continent position, 244–5
    ice core records of, 270–1
    on Mars, 186, 187f
    Pleistocene, 249–52, 255
Gliese 229, 113–14
global warming, 245, 283, 295, 300
    population growth and, 298–9
    predicted impacts of, 287–91
glucose, 139
gluon, 26
Gondwana, 96
Gould, Stephen Jay, 226–7
Grand Canyon, 82, 84
granites, 86, 126, 134, 196
    formation of, 200, 201–3, 204, 208
granulites, 203
graphite, 233–4
gravitational fields, 16, 118
gravity, 25, 76
    force of, 24, 102–3, 106
Great Lakes, 279
Greeks, classical, 3, 4, 14, 17
greenalite, 172
greenhouse effect, 167, 169, 201, 244,
        246
    and habitability, 181, 193
    on Mars, 188–9
    physics of, 166–9, 283–4
    on Venus, 175, 178–9

greenhouse gases, 169, 170f
Greenland, 56, 270, 271, 281, 305
Green Revolution, 299
greigite, 149–50
Gulf of California, 96
Gulf of Mexico, 234, 237, 285

Hadean era, 115
half-life, 47–9
Halley's comet, 110
halobacteria, 152
halogens, 18
handedness, 155–6; see also chirality
Hansen, V. L., 207
Hawaii, 127, 199, 281
Hawkins, Gerald, 11
heat, heating, 29, 124, 154, 157, 169,
        190, 207, 285
    of Earth, 130, 203
    in Europa, 144–5
    flow of, 204, 240, 241
    in giant planets, 119–20
    and plate tectonics, 96, 98
    radioactive, 125–7
heavy hydrogen. See deuterium
helium, 16, 23, 29, 33, 39, 102
    in giant planets, 119, 120
    in nuclear fusion, 36–8, 40
    and star formation, 103–4
Helmholtz, Herman von, 76
Herodotus, 75
heterochirality, 156, 160
Himalayan plateau, 95
HL Tauri, 107f
Holocene, 249, 269, 271, 275
    climate during, 277–80
homeostasis, 158
Hominidae, 256–7
Homo erectus, 258, 265
Homo heidelbergensis, 263
Homo neanderthalensis, see
        Neanderthals
Homo sapiens, 258, 260, 264, 265
hot spots, 98, 199, 203, 206
Hoyle, Fred, 11, 12–13
Hubbard, W. B., 120
Hubble, Edwin, 16
Hubble Space Telescope, 15, 16, 101,
        107, 108, 120, 235
humans, 264
    and agriculture, 299–300
    and energy resources, 300–305
    evolution of, 256–66
    future of, 307–8
    and global warming, 290–1
    during Holocene, 279–80
    origin of, 255–6
    and pollution, 306–7
    population growth, 298–9
    as travelers, 263–4

Hutton, James, 76
Huygens probe, 73
hydroelectric power, 303
hydrogen, 16, 23, 42, 54, 102, 119, 130, 143, 211, 270
　and fusion, 37–8, 39, 40, 103–4, 165
　on Venus, 178, 181
　in water, 55–6
hydrogen bomb, 29
hydrogen sulfides, 236–7
Hyperion, 72

ice(s), 66, 108, 118; *see also* water ice
ice ages, 55, 174, 232, 240
　causes of, 243–5, 249–52
Iceland, 203
ice cores, 281
　climatic data in, 269–72
ice sheets, 56, 243–4, 260–1, 269
icy bodies, in Kuiper Belt, 109–10
icy material, 5, 6
IDP. *See* interplanetary dust particles
igneous rock(s), 77, 81, 95, 200f; *see also* andesite; basalt; granite; volcanoes, volcanism
impact cratering. *See* cratering, impact
impact events, 234–5, 239
industrialization, 281, 282, 292, 305–6
Industrial Revolution, 213, 299, 305
India, 95, 96, 240, 242, 248, 249
integration, 48
interferometry, imaging, 113
interglacial periods, 270, 271
Intergovernmental Panel on Climate Change (IPCC), 287
interplanetary dust particles (IDP), 60, 108, 111, 112f
interstellar medium, 38
interstellar space, 102
intrusive igneous rocks, 200
inverse exponential curve, 69, 70f
Io, 66, 72, 118, 144
ionic bonding, 20, 197
ionic radius, 197, 198
ions, 32
IPCC. *See* Intergovernmental Panel on Climate Change
iridium, 234, 238
iron, 40, 117, 172, 211
　abundance of, 197, 199
　in banded iron formations, 214, 215
　in Earth's core, 122, 123, 127
iron carbonates, 172
iron meteorites, 117
iron silicates, 122, 172
iron silicide, 123
Ishtar Terra, 207–8
isobars, 40

isolation, in evolution, 226–7
isotopes, 20, 40, *see also* radioactive decay
　stable, 54–6, 245

Japan, 201, 298, 299
Jeanloz, Raymond, 123
Jenkins, G., 172
joules, 37–8
Joyce, Gerald, 134, 160
Jupiter, 5, 14, 72, 106, 117f, 120
　comet impact on, 235, 236f
　density of, 118, 119
　moons of, 6, 63, 66, 144–5
Jupiter family short-period comets, 110
Jurassic, 95–6

Kant, Immanuel, 76
Kasting, James, 166, 191, 217, 219
Kelvin, 29, 30
Kelvin, Lord, 35, 76
Kepler, Johannes, 13, 14
Kepler's laws, 25
kinetic energy, 29
kingdoms, in taxonomy, 229
Knauth, Paul, 56, 58
Knittle, Elise, 123
Krupp, Edward, 13
K/T boundary event, *see also* Cretaceous-Tertiary boundary extinction
　biological effects of, 235–7
　crater for, 237–8
　as impact, 234–5
　sediments of, 233–4
Kuiper belt, 6, 73, 108, 109–10, 129, 132
Kulkarni, S., 113, 114

lake beds, climate record in, 272
Lakshmi Planum, 207, 209
late heavy bombardment, 69
Laurasia, 96
Lavoisier, Antoine, 17, 18
laws of planetary motion, 13–14
Lay, Thorne, 123
lead, 305
lead isotopes, 52, 127
Legèr, Alan, 113
life, 134, 174, 211, 223, 309
　basic structure of, 136–8
　and element production, 42–3
　energetic processes of, 139–41
　eukaryotic, 220–2
　history of, 135f, 193
　kingdoms of, 151f
　on Mars, 186, 189–90, 191
　and planetary systems, 309–10

　potential in solar system for, 144–51, 191, 193
　raw materials of, 155–6
　and thermodynamics, 153–4
light, 168
　brightness of, 14–15
　electromagnetic energy and, 15–16
　visible, 25–6
lipids, 136, 156
lithification, 76
lithophiles, 124
lithosphere, 209
lithium, 37, 39, 42
Little Ice Age, 271, 277–8, 281
Lovelock, James, 174
Lowe, Donald, 56
l-process, 42
Lucretius, 17
lunar highlands, 63–4, 67f
lunar nodes, 10

McKay, Chris, 189
McKay, David, 149, 150
mafic rocks, 132, 200
Magellan spacecraft, 73, 181, 209
magnesium, 40, 116, 117, 122, 132, 197, 199
magnesium oxide, 122
magnesium silicate, 122
magnetic dynamo, 127
magnetic fields, 25–6, 89, 103, 127
magnetite, 89, 149–50
magnetism. *See* paleomagnetism
mammals, 232, 249, 255, 279
Manicouagan, Lake, 66f
Manitoba, 279
Mann, M. E., 291
mantle, 122, 123, 127, 129, 134, 196
　and continental crust, 204, 240, 241
　partial melting of, 198–9, 200–201
Mare Oriental, 63, 64f
mare, 64, 67f, 129
Margulis, Lynn, 220, 221
Marianas trench, 86
Mars, 5, 6, 47, 82, 106, 115, 116, 120, 124, 125, 126, 131, 175, 178f, 194f, 252, 309
　channels on, 66, 68f
　climatic history of, 190–1, 192f, 244
　geology of, 183–7, 199
　greenhouse effect on, 188–9
　impact craters on, 61f, 65f, 71
　meteorites from, 60, 69, 109
　potential for life on, 144, 145–6, 149–51, 189–90
　water on, 101, 177
Mars Surveyor orbiter, 151
Martin, Paul, 279

mass, 17, 24, 38, 104
mass extinction event, Cretaceous-Tertiary, 232–9
massive vector bosons, 28
matter, 17, 38
Maunder Minimum, 277
Mayans, 13
M-dwarfs, 193. *See also* red dwarf stars.
mean lifetime of radioactive atoms, 48
mean radiating level, 283
Medieval Warm Period, 271, 278
Mediterranean Sea, 95, 96
Meert, J., 245
Melosh, H. J., 149
melt spherules, 233, 238
Mendeleev, Dmitri, 18
Mercury, 5, 71, 106, 116, 117, 124, 125, 126
mesolands, 207, 208f
Mesozoic era, 82
metals, 305–6, *see also* alkali metals
metamorphic rocks, 78–9, 81–2, 203
metamorphism, 80, 206
Meteor Crater, Arizona, 6, 62f
meteorites, 52, 149, 196
   age dating, 52, 53f, 60, 76, 101
   amino acids in, 132, 161
   composition of, 116, 117
   origins of, 108–9
methane, 29, 118, 119, 131, 145, 189, 193, 281, 303–4
methanogens, 151–2
Mexico, 272
Mexico City, 306, 307
Meyer, Bradford, 41
microlensing, 112–13
microwaves, 30
mid-Atlantic ridge, 91, 95
Middle East, 261–2, 273
midocean ridges, 90–1, 92, 94, 96–7, 203
   life at, 149, 190
   mantle melt at, 199, 201
Milankovitch, M., 250
Milankovitch cycles, 250, 251, 279
Milky Way galaxy, 14, 15, 104
Miller, Stanley, 155
minerals, 54, 197–8, 201, 213–14, 305–6
mining, 305–6
Miocene, 249
Miranda, 72f
"missing links," 256
Mississippi River, 279
mitochondria, 139, 220, 221
Mohorovičić (Moho) discontinuity, 122
moist convection, 284
molecular clouds
   and planet formation, 104–5
   and star formation, 102–4
molecules, 17, 18, 136; *see also by type*
momentum, 104
   angular, 104
Montevallo (Ala.), 61f
Moon, 6, 47, 85, 94, 102, 109, 117, 120, 124, 134, 252
   age dating, 52, 60, 63–4, 66, 69, 70
   formation of, 107, 127–30
   gravity on, 24, 25
   impact craters on, 64f, 65f, 67f, 71
   motions of, 8, 10, 11, 12
   orbit of, 209–10
moons, 107, 129
   composition of, 116, 118
   cratering on, 66f, 68f, 72
   in solar system, 5, 6
mountain building, 242, 246
multicellular organisms, 223
multiring basins, 63
Murchison meteorite, 161
mutations, 138, 225, 227–8

nannobacteria, 150
natural gas, 301, 303–4
natural selection, 134, 136
Nazca plate, 95
Neanderthals (*Homo neanderthalensis*), 260, 265
   distribution of, 261–2
   lifestyle of, 263–4
   physical adaptations of, 262–3
nebula, protoplanetary, 104–5
negative feedback, 174, 244
neon, isotopes of, 40
Neptune, 5, 72, 106, 118, 119, 120, 121f, 132, 144
neutrinos, 36
neutron capture, 40
neutrons, 17, 27, 28, 39
neutron stars, 39, 41–2, 113
Nevada, nuclear detonation crater in, 62f
Newton, Isaac, 13–14
   Principia, 23
nickel, 117, 122
Nile River flooding, 3, 75
nitrogen, 42, 131, 145, 183
   isotopes of, 40, 49
nitrous oxide, 281, 306
noble gases, 18, 20, 131f
node stones, 12–13
nonchiralic, 156
North America, 85, 96, 242, 271, 279
North American plate, 95
North Atlantic, 295
   deep water, 242–3, 279
nuclear detonation crater, 62f

nuclear energy, 303, 304f
nuclear reactions, 36–8. *See also* fusion
nucleic acids, 136–7, 155
nucleotides, 137
nucleus, nuclei, 35–6, 139
nuclides, 40, 41, 42, 51

ocean crust, 97, 201, 240
   magnetization of, 89–91
   subduction zones in, 92–3, 242
   topography of, 86, 87f
oceans, 54, 131, 209, 217, 219, 246, 293f
   and climate, 55, 278–9, 285, 287, 292–5
   and greenhouse gases, 289, 291
   iron deposition in, 214–15
Oerlemans, J., 283
oil, 301–2
Oklahoma, 285
Oligocene, 249
olivine, 122, 123f
ONA. *See* other nucleic acid
one-dimensional model, 217
Ontario, Lake, 188f
Oort cloud, 6, 73, 110, 111, 128, 132
orbit(s)
   Earth's, 247, 249–50, 251f, 252–3
   measuring, 10–12
   planetary, 6, 13–14
organelles, 139
organic compounds, 132
organic molecules, 145–6, 155
Orion molecular cloud, 102, 103f, 105
oscillations, and climate, 249–51, 294
other nucleic acid (ONA), 161–2
outflow channels, on Mars, 186
outgassing, 107, 130
Owen, Tobias, 131
oxides, 122
oxygen, 40, 42, 131, 139, 141, 172, 177, 198, 213–14, 219, 230
   free, 143–4
   loss and gain of, 215–16
   reservoirs of, 217–18
   and seafloor sediments, 54–5
oxygen cycle, 211–13
oxygen revolution, 211
oxygen-16, 40
oxygen-17, 40, 55
oxygen-18, 55, 56, 271f
ozone, 172, 179, 219, 236, 289, 295–6, 306

Pacific Ocean, 85, 96, 242, 294
Pacific plate, 94, 95
packrat middens, 272–3
PAHs. *See* polycyclic aromatic hydrocarbons

paleomagnetism, 86, 89–91
paleosols, 171–2
Paleozoic era, 82
palimpsests, 66f
Palomar, Mt., 113
Pangaea, 95–6, 97f, 242, 246
parallax, 14
Paranthropus, 257, 258
parent bodies, 108–9
parent-daughter radioactive dating, 49, 51–2
particles, subatomic, 25
Pathfinder lander, 151, 184, 194f
Patterson, Claire, 52
peptide nucleic acid (PNA), 161
peptides, 139
peridotite, 200
period(s)
 geologic, 82, 83f, 223
 orbital, 14
periodic table, 18–19, 20, 80, 123, 142
peritectic solutions, 122
Permian, 232
perovskite, 122
petrification, 80
Phanerozoic eon, 82, 223–5, 232, 244
phases, lunar, 10
Phillips, R. J., 207
phosphate group, 159
photochemistry, 211, 213, 218, 219, 295–6
photons, 23, 25, 30, 32, 35
 and greenhouse effect, 166–8, 170f
 radiation of, 30–2, 125, 283–5
photosphere, 33
photosynthesis, 134, 136, 139, 141, 210
 and banded iron formations, 214–15
 and oxygen, 211–12, 213, 218–19
phylogenetic tree, 151f
phylum, phyla, Cambrian, 229–30
Pierrehumbert, Ray, 189
Piltdown hoax, 256
Pinatubo, Mount, 173, 292
pines, bristlecone, 275
Pioneer Venus probe, 178
Planck function, 30
Planck, Max, 30
planetary differentiation, 120
planetary systems
 identifying, 111–14
 indirect techniques of identification, 112–13
 life and, 309–10
 and star formation, 101–2
planetesimals, 105, 111f, 115, 127–8, 132
planets, 5, 193, 309

accretion of, 123–4, 125f, 127–8
age dating events on, 63, 66–9
bulk densities and compositions of, 115–20
distances to, 13–14
Giant, 118–20
element differentiation in, 124–5
formation of, 69, 104–5
identifying extrasolar, 111–14
motion of, 10, 13–14
radioactive heating in, 125–7
in solar system, 4–6
solid, 116–18
plants, 223, 272, 290
plasma, 17
plastids, 139, 220
plate tectonics, 85–98, 193, 201, 206, 213
 carbon dioxide cycling and, 172–4
 climate and, 242–3, 246, 248
 driving forces of, 96–8
 evidence for, 86–93
 genesis of, 85–6, 204
 model of, 93–5
 and supercontinents, 95–6, 240–2
 and water, 208–9
Plato, 3
Pleistocene, 255
 in Africa, 256–8
 climate of, 260–2
 ice age oscillations of, 249–52
 in U.S. Southwest, 273, 274f
Pliocene, 248, 255
Pluto, 5, 6, 14, 109, 110, 118, 179
PNA. *See* peptide nucleic acid
polar regions, and global warming, 289
pollen, and climate change, 272, 278
pollution, 306–7
polycyclic aromatic hydrocarbons (PAHs), 149
population growth, human, 298–9
positrons, 36
potassium, 124, 125–6, 197, 201
potential energy, 29
power, 29
p-p chain. *See* proton-proton chain
ppI chain, 36–7
ppII chain, 37
p process, 42
precession, 249, 251f
precipitation, and global warming, 289, 290
pressure-release partial melting, 198–9, 200–1
Primary period, geologic, 82
primates, 255, 256
Principia (Newton), 23
Priscoan eon, 82, 115
prochloron, 221

prokaryotic cells, 138–9, 151, 158f
prokaryotes, 151, 163, 190, 220, 221
proteins, 136, 137–8, 156
Proterozoic eon, 82, 196, 204, 215, 244
 day length during, 209, 210
 eukaryote evolution during, 221–2
 glaciation during, 174, 245
 oxygen production during, 211, 219
protium, 20
protocontinents, 203–4, 206, 209
proton-proton (p-p) chain, 36–7
protons, 17, 27, 28, 29, 39
protoplanetary disks, 104–5
protostars, 104
 disks around, 105–7
Proxima Centauri, 4
PSR1257 + 12, pulsar, 113
Ptolemy, 4
pulsars, 113
pulsation, in Cepheid-variable stars, 15
punctuated equilibrium, 226–9
P-waves, 120, 121–2, 200
pyrite, 213, 214
pyruvate, 139

quantum mechanics, 18, 20, 21
quarks, 26–7

radar, 181
radial velocity techniques, 112, 113
radiation, 16, 283, 284. *See also* photons
radicals, 220
radioactive decay, 20, 28, 29, 47–9, 51
radioactivity, 28–9, 35, 76
 heating and, 125–7
radiocarbon dating, 49–50
radioisotope dating, 49–52, 69, 81–2, 91, 101
radio static, and Big Bang, 16, 38–9
Ramsey, William, 33
rare-earth elements, 197, 203
Raymo, M., 249
reactants, 156
reactions, nuclear, 36
recombinant DNA, 225
redbeds, 215
red dwarf stars, 38. *See also* M-dwarfs
red giant, 40
red shift, 16
reduced carbon, 212
reducing compounds, 217
reflection, 32
relative chronology, 47, 60
 of cratering events, 63–6
 of Earth's history, 75–84
 of planetary surfaces, 66–9

respiration, 139, 140f, 212, 213, 215, 218, 219
rhyolite, 95
ribonucleic acid (RNA), 136, 137–8, 139, 142, 151, 152, 153, 155, 156, 220, 225
　action of, 158–61
　in cells, 161–2
ribose, 159–60
ribosomes, 139, 220
ridge push, 96, 97
ridges, oceanic, 90–1, 92, 94, 96
rifting, rift valleys, 95, 241–2
ring systems, planetary, 6
RNA. See ribonucleic acid
rocks, 89, 122, 190, 200
　Archaen eon, 134, 203
　elements in, 196–7
　origins of, 76–7
　radioisotope dating of, 51–2, 81–2
　and weathering cycle, 172–3, 196, 211, 213f
Rodinia, 96, 97f
Roman Empire, 305
Röntgen, Wilhelm, 76
Rosetta mission, 110
r process, 41–2
rubidium–87, 51, 53f
rubidium-strontium dating, 51–2, 53f
Ruddiman, W., 249
Russia, 298, 303
Rutherford, Ernest, 17
Rye, R., 172

Sagan, Carl, 142, 144
St. Helens, Mount, 95, 173
St. Lawrence River, 279
Salpeter, Edwin, 120
samarium-neodymium, 52
San Andreas fault, 86, 91, 92, 94, 95
Sapa Mons, 182f
Saturn, 5, 6, 72, 106, 117f, 118, 119, 120
scanning tunneling microscopy, 21
scattering, of light, 168, 189, 285
"Scopes Monkey Trial," 256
Scott, David R., 24
seafloor
　sediments on, 54–6
　spreading of, 90f, 91f, 241–2, 246
sea level, 289
Secondary period, geological, 82
sedimentary rocks, 76, 78, 82, 214
sediments
　K-T boundary, 233–4
　recycling buried, 212–13
seismic waves, 120–1, 200
seismometers, 91, 120, 121–2, 123
Seno, Nicolaus, 76
Sequoia, Giant, 276

Shapiro, R., 142, 143
shelly marine fauna, 232, 233f
Shergottites-Nakhlites-Chassigny (SNC) meteorites, 60, 69, 149
shocked quartz, 233, 238
Siberia, 73
Siccar Point (Scotland), 76
siderite, 172
siderophiles, 124
Sierra Nevada, 276, 277f
silica, 56, 200, 214
silicates, 66, 118, 142, 173
silicon, 40, 116–17, 122, 142–3, 146, 197, 199
silicon dioxide, 173
slab-pull, 97
snails, 226
SNC. See Shergottites-Nakhlites-Chassigny meteorites
sodium, 40, 124, 197, 201
sodium chloride, 197
soil moisture, 289–90, 300
solar energy, 303
　oscillations in, 250–1
solar wind, 34, 105, 107, 176
solar system, 4, 33f, 114f, 309
　absolute chronology of, 69–73
　age of, 52, 53, 60, 76
　bodies in, 5–7
　motions in, 8, 10–11
　potential for life in, 144–51
solid planets, composition of, 116–18
sonar, 86
Sonett, Charles, 209
South America, 85, 91, 95, 273, 294
South American plate, 95, 96
Southern Oscillation, 294
Spain, 264
speciation, human, 260
species, 225, 229
　and punctuated equilibrium, 226, 227–8
spectra, spectrum, electromagnetic, 30, 32–3
　of extrasolar planets, 113, 114
spectrometers, spectrometry, 16, 30
spinel, 122, 123f
s process, 40
Stardust probe, 110
star formation, 76
　molecular clouds and, 102–4
　planetary systems and, 101–2
starlight, and extrasolar planet identification, 112–13
stars, 3, 35, 106f, 193
　brightness of, 14–15
　element production in, 38–42
　formation of, 107f, 114

　and companion planet identification, 112–13
　and protoplanetary disks, 104–5
　static, from Big Bang, 38–9
　stellar nucleosynthesis, 38, 40
Stevenson, David, 120
Stone Age, 11
Stonehenge, as observatory, 10–13
strandlines, 243
strata, stratigraphic section, 79
stratigraphy, 76
stratosphere, 179, 287–9
stromatolites, 82, 134, 149, 204, 219
strong nuclear force, 26–7
strontium-86, 51, 53f
strontium-87, 51
Subcommittee on Global Change Research, 287
subduction, subduction zones, 92–3, 95, 96–7, 201, 203, 240, 242
sulfate aerosols, 292
sulfothermophiles, 152
sulfur compounds, as greenhouse gases, 189
sulfur dioxide, 141–2
sulfur oxides, 236
Sun, 6, 14, 58, 76, 107
　and Cretaceous climate, 246–7
　elements in, 33–4
　energy in, 23, 35, 38
　luminosity of, 165–6, 167f, 176, 193, 285
　motions of, apparent, 10–11, 12
　solar energy from, 277–8
　temperatures in, 39–40
Sung dynasty, 305
sunspots, 277
supercontinents, 95–6, 98, 219
　cycling of, 240–2, 244
　and oxygen production, 219
supernovas, 40, 42, 51
　Type 1A, 15, 16
S-waves, 120–1, 122
symbiosis, bacterial, 220, 221

taxonomic classification, 229
Taylor, S. R., 208
tectonics, 204, 207, 209; see also continental drift; plate tectonics
temperature, 29–30, 38, 57f
　and global warming, 288f, 289
　recording history of, 55, 56, 58, 282–3
　scales of, 29–30
　solar, 39–40
　of Venus, 283–4
terraforming, 193
terrestrial planets, 5, 116
Tertiary period, 82, 248–9
Tethys sea, 96

tetrahedral coordination, 198
Texas, 285
Tharsis volcano, 184
thermodynamics, second law of, 153–4
Thermoplasm, 221
Thompson, William. *See* Kelvin, Lord
thorium, 125–6, 201
Tibetan Plateau, 240, 249
tidal heating, 144
tidal wave action, 234
tides, 25, 209–10, 250
Titan, 6, 73, 144, 145
Tonga subduction region, 92f
topography, 123
   of Mars, 184–7
   of Venus, 180f, 181, 182f, 196, 206–7
transform faults, 94
tree rings, 50, 273, 275–7
trenches, ocean, 86, 92–3, 94, 96–7
Triassic period, 81f, 232
triatomic compounds, 18
triple junction, 95
tritium, 20
Triton, 5, 72, 118
tropopause, 179
troposphere, on Venus, 179
T-Tauri stars, 102, 105, 107
Tucson, 273, 275f, 283
Tunguska River asteroid, 73

ultraviolet radiation, 181, 183, 211, 219–20, 236
unconformity, 76
uniformitarianism, 75–6
United States, 201, 298
   energy use in, 301, 303
   mining in, 305–6
   southwestern climate data for, 272–3, 274f, 278
universe, 15–16, 38

uraninite, 213, 214, 218
uranium, 201, 204, 125–6, 127, 213–14
uranium-lead dating, 52
uranium oxides, 213
Uranus, 5, 6, 72, 106, 118, 119, 120, 121f, 128, 132, 144
Urey, Harold, 155

Valhalla crater, 63
Valles Marineris, 82
Van der Voo, R., 245
velocity, 23
Vendian-Cambrian revolution, 229, 230–2
Vendian period, 229, 230
Venera orbiters, 181
Venus, 5, 6, 13, 116, 117, 124, 125, 126, 129, 131, 191, 196, 199, 309
   atmosphere of, 171, 174–5, 177–81
   craters on, 71f, 73
   formation of, 106, 115
   geology of, 204, 206–9
   and plate tectonics, 207–8
   surface of, 181–3
   temperature of, 283–4
vesicles, 156–8, 162
Viking spacecraft, 66, 120, 145, 146, 184
volcanoes, volcanism, 127, 182f, 198, 200, 201, 232, 234
   Archaean, 203, 216
   on Mars, 184, 190
   and oxygen, 211, 219
   and plate tectonics, 95, 173, 242
Vostok station, 270–1
Voyager spacecraft, 66, 120, 144, 145

Wahlen, M., 281
Walker, J.C.G., 173

water, 6, 57f, 115, 131, 141, 143, 169, 174, 201, 211, 218, 288f
   atmospheric, 179, 181
   and Earth, 101, 199
   on Europa, 144, 145
   and K-T boundary event, 236–7
   on Mars, 177, 183, 184, 186, 190
   and plate tectonics, 98, 208–9
   stable isotopes in, 54–6
water ice, 66, 106, 144
   on Mars, 183, 184, 186, 191
water vapor, and climate, 169, 246, 284
wavefunction, 21
wavelength, 30, 31f
weak nuclear force, 28
weather, vs. climate, 291–2
weathering, 77, 196, 211, 213f, 217, 218
weathering cycle, carbon-silicate, 172–3
Wegener, Alfred, 86
white dwarf, 40
Whitmire, D., 175–6
Wilson, A., 260
Wilson, J. Tuzo, 241
wind, *see* air movement; solar wind
wind stress, 293
wüstite, 122, 123

xenoliths, 202, 203
xenon isotopes, 127

Younger Dryas, 278–9
Yucatan Peninsula, 237f

zero
   absolute, 29
   date, Mayan, 13
   scientific notation, in, 8
zodiacal light, 6